Pi π

Calculated to 1,000,001 Digits After the Decimal Point•

Published by
Obvious Press, LLC
918 5th Street
Ames, IA, 50010-5906
USA
www.obviouspress.com

First published in 2011. Second Edition 2016
PI CALCULATED TO 1,000,001 DIGITS AFTER THE DECIMAL POINT

For information address Obvious Press, 918 5th Street,
Ames, IA 50010-5906 or write to editor@obviouspress.com

ISBN: 978-1-941892-22-0

To MSZ and r, the other two mysteries

Prologue

The What

A book filled with 1,000,001 digits of pi after the decimal point.

The Why

Why a book of pi calculated 1,000,0001 times after the decimal point? Let me count the reasons. There are at least 3.14...:

1. I recently read the essay *Mountains of Pi* by Richard Preston, originally published in the March 2, 1992 issue of *The New Yorker*, and was inspired by the work of the Chudnovsky brothers — who were the focus of the essay — and their work on pi. To appreciate the possibilities, consider what Preston wrote on page 63 of *The New Yorker* article:

> If you were to assign letters of the alphabet to combinations of digits, and were to do this for all human alphabets, syllabaries, and ideograms, then you could fit any written character in any language to a combination of digits in pi. According to this system, pi could be turned into literature. Then, if you could look far enough into pi, you would probably find the expression "See the U.S.A. in a Chevrolet!" a billion times in a row. Elsewhere, you would find Christ's Sermon on the Mount in His

native Aramaic tongue, and you would find versions of the Sermon on the Mount that are pure blasphemy. Also, you would find a dictionary of Yanomamo curses. A guide to the pawnshops of Lubbock. The book about the sea which James Joyce supposedly declared he would write after he finished "Finnegans Wake." The collected transcripts of "The Tonight Show" rendered into Etruscan. "Knowledge of All Existing Things," by Ahmes the Egyptian scribe. Each occurrence of an apparently ordered string in pi, such as the words "Ruin hath taught me thus to ruminate/That Time will come and take my love away," is followed by unimaginable deserts of babble. No book and none but the shortest poems will ever be seen in pi, since it is infinitesimally unlikely that even as brief a text as an English sonnet will appear in the first 10^{77} digits of pi, which is the longest piece of pi that can be calculated in this universe [because that's about how many atoms exist in the universe].

2. 1,000,001 pi digits is doable in a 6 x 9 book, and it's still usable as a book. Of course there exist already at least two similar volumes,[1] but they were obviously designed by aesthetic agnostics without the necessary sensibility required for something as beautiful as the transcendental number of pi...

3. Pi seems a perfectly good reason to help some dignified donkeys. Every penny, cent, dollar, or Euro of profit that comes down the pike from sales of this book will be transferred by the author or Obvious Press to the Donkey Sanctuary in Sidmouth, Devon, UK.[2]

0.14... This is more philosophical than practical: pi is mysterious and open-ended, even if it has an imaginary boundary close to 3.15, it's still infinite within its limitation, and that's worth at least a book of numbers.

---

[1] Google "Pi Calculated"...

[2] For more information go to thedonkeysanctuary.org or to dignifieddonkeys.com

The How (using Umberto Eco's *How to Travel with a Salmon* as an inspiration)

There is no attempt at structuring the digits as in other publications on pi. The reader won't find groupings or line numbers to facilitate easy location of sequences. Instead it's simply one dense page of digits after another, and I suggest to take it slowly (courtesy of Harvey Keitel in the movie *Smoke*) to find what you're searching for. Here are some possible uses:

– Stare at each page for a few minutes with slightly squinting eyes, and ponder the patterns that emerge, slowly, between the same numbers, enhanced by their repetition or singularity. You will soon be in a mildly psychedelic state, no drugs required.

– Start drawing lines between clusters or circle patterns that appear to be meaningful.

– If you need a new, high-strength password for a file you want to keep from prying eyes, open the book on any page, close your eyes and with a quick stab of a pencil find a spot on the page, then circle as many digits as you need (going vertical, horizontal, diagonal, or following a secret pattern, etc.) and use those numbers for your new password. Anything beyond forty digits should be pretty secure.

– If you are interested in locating specific strings of digits in pi, there is an easier way than using this book: go to the pi search page at www.angio.net/pi/piquery (accessed on December 23, 2015) and you shall find what you're looking for...

– For the pattern seekers I've included a few pages from the beginning of pi at the end of the book (529-536) where I've replaced the conventional font with an ESRI font that makes patterns on each page more visible than numbers alone.

Enjoy!
πolytekton
Gainesville, FL, June 30, 2016

π = 3.141592653589793238462643383279502884197169399375
1058209749445923078164062862089986280348253421170679
8214808651328230664709384460955058223172535940812848
1117450284102701938521105559644622948954930381964428
8109756659334461284756482337867831652712019091456485
6692346034861045432664821339360726024914127372458700
6606315588174881520920962829254091715364367892590360
0113305305488204665213841469519415116094330572703657
5959195309218611738193261179310511854807446237996274
9567351885752724891227938183011949129833673362440656
6430860213949463952247371907021798609437027705392171
7629317675238467481846766940513200056812714526356082
7785771342757789609173637178721468440901224953430146
5495853710507922796892589235420199561121290219608640
3441815981362977477130996051870721134999999837297804
9951059731732816096318595024459455346908302642522308
2533446850352619311881710100031378387528865875332083
8142061717766914730359825349042875546873115956286388
2353787593751957781857780532171226806613001927876611
1959092164201989380952572010654858632788659361533818
2796823030195203530185296899577362259941389124972177
5283479131515574857242454150695950829533116861727855

88907509838175463746493931925506040092770167113900984882401285836160356370766010471018194295559619894676
7837449448255379774726847104047534646208046684259069
4912933136770289891521047521620569660240580381501935
1125338243003558764024749647326391419927260426992279
6782354781636009341721641219924586315030286182974555
7067498385054945885869269956909272107975093029553211
6534498720275596023648066549911988183479775356636980
7426542527862551818417574672890977772793800081647060
0161452491921732172147723501414419735685481613611573
5255213347574184946843852332390739414333454776241686
2518983569485562099219222184272550254256887671790494
6016534668049886272327917860857843838279679766814541
0095388378636095068006422512520511739298489608412848
8626945604241965285022210661186306744278622039194945
0471237137869609563643719172874677646575739624138908
6583264599581339047802759009946576407895126946839835
2595709825822620522489407726719478268482601476990902
6401363944374553050682034962524517493996514314298091
9065925093722169646151570985838741059788595977297549
8930161753928468138268683868942774155991855925245953
9594310499725246808459872736446958486538367362226260
9912460805124388439045124413654976278079771569143599
7700129616089441694868555848406353422072225828488648
1584560285060168427394522674676788952521385225499546
6672782398645659611635488623057745649803559363456817
4324112515076069479451096596094025228879710893145669
1368672287489405601015033086179286809208747609178249
3858900971490967598526136554978189312978482168299894
8722658804857564014270477555132379641451523746234364
5428584447952658678210511413547357395231134271661021
3596953623144295248493718711014576540359027993440374
2007310578539062198387447808478489683321445713868751
9435064302184531910484810053706146806749192781911979
3995206141966342875444064374512371819217999839101591
9561814675142691239748940907186494231961567945208095
1465502252316038819301420937621378559566389377870830

39069792077346722182562599661501421503068038447734549
20260541466592520149744285073251866600213243408819071
04863317346496514539057962685610055081066587969981
63574736384052571459102897064140110971206280439039759
51567715770042033786993600723055876317635942187312514
71205329281918261861258673215791984148488291644706
0957527069572209175671167229109816909152801735067127
4858322287183520935396572512108357915136988209144421
0067510334671103141267111369908658516398315019701651
5116851714376576183515565088490998985998238734552833
1635507647918535893226185489632132933089857064204675
2590709154814165498594616371802709819943099244889575
7128289059232332609729971208443357326548938239119325
9746366730583604142813883032038249037589852437441702
9132765618093773444030707469211201913020330380197621
1011004492932151608424448596376698389522868478312355
2658213144957685726243344189303968642624341077322697
8028073189154411010446823252716201052652272111660396
6655730925471105578537634668206531098965269186205647
6931257058635662018558100729360659876486117910453348
8503461136576867532494416680396265797877185560845529
6541266540853061434443185867697514566140680070023787
7659134401712749470420562230538994561314071127000407
8547332699390814546646458807972708266830634328587856
9830523580893306575740679545716377525420211495576158
1400250126228594130216471550979259230990796547376125
5176567513575178296664547791745011299614890304639947
1329621073404375189573596145890193897131117904297828
5647503203198691514028708085990480109412147221317947
6477726224142548545403321571853061422881375850430633
2175182979866223717215916077166925474873898665494945
0114654062843366393790039769265672146385306736096571
2091807638327166416274888800786925602902284721040317
2118608204190004229661711963779213375751149595015660
4963186294726547364252308177036751590673502350728354
0567040386743513622224771589150495309844489333096340
8780769325993978054193414473774418426312986080998886

8741326047215695162396586457302163159819319516735381
2974167729478672422924654366800980676928238280689964
0048243540370141631496589794092432378969070697794223
6250822168895738379862300159377647165122893578601588
1617557829735233446042815126272037343146531977774160
3199066554187639792933441952154134189948544473456738
3162499341913181480927777103863877343177207545654532
2077709212019051660962804909263601975988281613323166
6365286193266863360627356763035447762803504507772355
4710585954870279081435624014517180624643626794561275
3181340783303362542327839449753824372058353114771199
2606381334677687969597030983391307710987040859133746
4144282277263465947047458784778720192771528073176790
7707157213444730605700733492436931138350493163128404
2512192565179806941135280131470130478164378851852909
2854520116583934196562134914341595625865865570552690
4965209858033850722426482939728584783163057777560688
8764462482468579260395352773480304802900587607582510
4747091643961362676044925627420420832085661190625454
3372131535958450687724602901618766795240616342522577
1954291629919306455377991403734043287526288896399587
9475729174642635745525407909145135711136941091193932
5191076020825202618798531887705842972591677813149699
0090192116971737278476847268608490033770242429165130
0500516832336435038951702989392233451722013812806965
0117844087451960121228599371623130171144484640903890
6449544400619869075485160263275052983491874078668088
1833851022833450850486082503930213321971551843063545
5007668282949304137765527939751754613953984683393638
3047461199665385815384205685338621867252334028308711
2328278921250771262946322956398989893582116745627010
2183564622013496715188190973038119800497340723961036
8540664319395097901906996395524530054505806855019567
3022921913933918568034490398205955100226353536192041
9947455385938102343955449597783779023742161727111723
6434354394782218185286240851400666044332588856986705
4315470696574745855033232334210730154594051655379068

6627333799585115625784322988273723198987571415957811
1963583300594087306812160287649628674460477464915995
0549737425626901049037781986835938146574126804925648
7985561453723478673303904688383436346553794986419270
5638729317487233208376011230299113679386270894387993
6201629515413371424892830722012690147546684765357616
4773794675200490757155527819653621323926406160136358
1559074220202031872776052772190055614842555187925303
4351398442532234157623361064250639049750086562710953
5919465897514131034822769306247435363256916078154781
8115284366795706110861533150445212747392454494542368
2886061340841486377670096120715124914043027253860764
8236341433462351897576645216413767969031495019108575
9844239198629164219399490723623464684411739403265918
4044378051333894525742399508296591228508555821572503
1071257012668302402929525220118726767562204154205161
8416348475651699981161410100299607838690929160302884
0026910414079288621507842451670908700069928212066041
8371806535567252532567532861291042487761825829765157
9598470356222629348600341587229805349896502262917487
8820273420922224533985626476691490556284250391275771
0284027998066365825488926488025456610172967026640765
5904290994568150652653053718294127033693137851786090
4070866711496558343434769338578171138645587367812301
4587687126603489139095620099393610310291616152881384
3790990423174733639480457593149314052976347574811935
6709110137751721008031559024853090669203767192203322
9094334676851422144773793937517034436619910403375111
7354719185504644902636551281622882446257591633303910
7225383742182140883508657391771509682887478265699599
5744906617583441375223970968340800535598491754173818
8399944697486762655165827658483588453142775687900290
9517028352971634456212964043523117600665101241200659
7558512761785838292041974844236080071930457618932349
2292796501987518721272675079812554709589045563579212
2103334669749923563025494780249011419521238281530911
4079073860251522742995818072471625916685451333123948

0494707911915326734302824418604142636395480004480026
7049624820179289647669758318327131425170296923488962
7668440323260927524960357996469256504936818360900323
8092934595889706953653494060340216654437558900456328
8225054525564056448246515187547119621844396582533754
3885690941130315095261793780029741207665147939425902
9896959469955657612186561967337862362561252163208628
6922210327488921865436480229678070576561514463204692
7906821207388377814233562823608963208068222468012248
2611771858963814091839036736722208883215137556003727
9839400415297002878307667094447456013455641725437090
6979396122571429894671543578468788614445812314593571
9849225284716050492212424701412147805734551050080190
8699603302763478708108175450119307141223390866393833
9529425786905076431006383519834389341596131854347546
4955697810382930971646514384070070736041123735998434
5225161050702705623526601276484830840761183013052793
2054274628654036036745328651057065874882256981579367
8976697422057505968344086973502014102067235850200724
5225632651341055924019027421624843914035998953539459
0944070469120914093870012645600162374288021092764579
3106579229552498872758461012648369998922569596881592
0560010165525637567856672279661988578279484885583439
7518744545512965634434803966420557982936804352202770
9842942325330225763418070394769941597915945300697521
4829336655566156787364005366656416547321704390352132
9543529169414599041608753201868379370234888689479151
0716378529023452924407736594956305100742108714261349
7459561513849871375704710178795731042296906667021449
8637464595280824369445789772330048764765241339075920
4340196340391147320233807150952220106825634274716460
2433544005152126693249341967397704159568375355516673
0273900749729736354964533288869844061196496162773449
5182736955882207573551766515898551909866653935494810
6887320685990754079234240230092590070173196036225475
6478940647548346647760411463233905651343306844953979
0709030234604614709616968868850140834704054607429586

9913829668246818571031887906528703665083243197440477
1855678934823089431068287027228097362480939962706074
7264553992539944280811373694338872940630792615959954
6262462970706259484556903471197299640908941805953439
3251236235508134949004364278527138315912568989295196
4272875739469142725343669415323610045373048819855170
6594121735246258954873016760029886592578662856124966
5523533829428785425340483083307016537228563559152534
7844598183134112900199920598135220511733658564078264
8494276441137639386692480311836445369858917544264739
9882284621844900877769776312795722672655562596282542
7653183001340709223343657791601280931794017185985999
3384923549564005709955856113498025249906698423301735
0358044081168552653117099570899427328709258487894436
4600504108922669178352587078595129834417295351953788
5534573742608590290817651557803905946408735061232261
1200937310804854852635722825768203416050484662775045
0031262008007998049254853469414697751649327095049346
3938243222718851597405470214828971117779237612257887
3477188196825462981268685817050740272550263329044976
2778944236216741191862694396506715157795867564823993
9176042601763387045499017614364120469218237076488783
4196896861181558158736062938603810171215855272668300
8238340465647588040513808016336388742163714064354955
6186896411228214075330265510042410489678352858829024
3670904887118190909494533144218287661810310073547705
4981596807720094746961343609286148494178501718077930
6810854690009445899527942439813921350558642219648349
1512639012803832001097738680662877923971801461343244
5726400973742570073592100315415089367930081699805365
2027600727749674584002836240534603726341655425902760
1834840306811381855105979705664007509426087885735796
0373245141467867036880988060971642584975951380693094
4940151542222194329130217391253835591503100333032511
1749156969174502714943315155885403922164097229101129
0355218157628232831823425483261119128009282525619020
5263016391147724733148573910777587442538761174657867

1169414776421441111263583553871361011023267987756410
2468240322648346417663698066378576813492045302240819
7278564719839630878154322116691224641591177673225326
4335686146186545222681268872684459684424161078540167
6814208088502800541436131462308210259417375623899420
7571362751674573189189456283525704413354375857534269
8699472547031656613991999682628247270641336222178923
9031760854289437339356188916512504244040089527198378
7386480584726895462438823437517885201439560057104811
9498842390606136957342315590796703461491434478863604
1031823507365027785908975782727313050488939890099239
1350337325085598265586708924261242947367019390772713
0706869170926462548423240748550366080136046689511840
0936686095463250021458529309500009071510582362672932
6453738210493872499669933942468551648326113414611068
0267446637334375340764294026682973865220935701626384
6485285149036293201991996882851718395366913452224447
0804592396602817156551565666111359823112250628905854
9145097157553900243931535190902107119457300243880176
6150352708626025378817975194780610137150044899172100
2220133501310601639154158957803711779277522597874289
1917915522417189585361680594741234193398420218745649
2564434623925319531351033114763949119950728584306583
6193536932969928983791494193940608572486396883690326
5564364216644257607914710869984315733749648835292769
3282207629472823815374099615455987982598910937171262
1828302584811238901196822142945766758071865380650648
7026133892822994972574530332838963818439447707794022
8435988341003583854238973542439564755568409522484455
4139239410001620769363684677641301781965937997155746
8541946334893748439129742391433659360410035234377706
5888677811394986164787471407932638587386247328896456
4359877466763847946650407411182565837887845485814896
2961273998413442726086061872455452360643153710112746
8097787044640947582803487697589483282412392929605829
4861919667091895808983320121031843034012849511620353
4280144127617285830243559830032042024512072872535581

1958401491809692533950757784000674655260314461670508
2768277222353419110263416315714740612385042584598841
9907611287258059113935689601431668283176323567325417
0734208173322304629879928049085140947903688786878949
3054695570307261900950207643349335910602454508645362
8935456862958531315337183868265617862273637169757741
8302398600659148161640494496501173213138957470620884
7480236537103115089842799275442685327797431139514357
4172219759799359685252285745263796289612691572357986
6205734083757668738842664059909935050008133754324546
3596750484423528487470144354541957625847356421619813
4073468541117668831186544893776979566517279662326714
8103386439137518659467300244345005449953997423723287
1249483470604406347160632583064982979551010954183623
5030309453097335834462839476304775645015008507578949
5489313939448992161255255977014368589435858775263796
2559708167764380012543650237141278346792610199558522
4717220177723700417808419423948725406801556035998390
5489857235467456423905858502167190313952629445543913
1663134530893906204678438778505423939052473136201294
7691874975191011472315289326772533918146607300089027
7689631148109022097245207591672970078505807171863810
5496797310016787085069420709223290807038326345345203
8027860990556900134137182368370991949516489600755049
3412678764367463849020639640197666855923356546391383
6318574569814719621084108096188460545603903845534372
9141446513474940784884423772175154334260306698831768
3310011331086904219390310801437843341513709243530136
7763108491351615642269847507430329716746964066653152
7035325467112667522460551199581831963763707617991919
2035795820075956053023462677579439363074630569010801
1494271410093913691381072581378135789400559950018354
2511841721360557275221035268037357265279224173736057
5112788721819084490061780138897107708229310027976659
3583875890939568814856026322439372656247277603789081
4458837855019702843779362407825052704875816470324581
2908783952324532378960298416692254896497156069811921

8658492677040395648127810217991321741630581055459880
1300484562997651121241536374515005635070127815926714
2413421033015661653560247338078430286552572227530499
9883701534879300806260180962381516136690334111138653
8510919367393835229345888322550887064507539473952043
9680790670868064450969865488016828743437861264538158
3428075306184548590379821799459968115441974253634439
9602902510015888272164745006820704193761584547123183
4600726293395505482395571372568402322682130124767945
2264482091023564775272308208106351889915269288910845
5571126603965034397896278250016110153235160519655904
2118449499077899920073294769058685778787209829013529
5661397888486050978608595701773129815531495168146717
6959760994210036183559138777817698458758104466283998
8060061622984861693533738657877359833616133841338536
8421197893890018529569196780455448285848370117096721
2535338758621582310133103877668272115726949518179589
7546939926421979155233857662316762754757035469941489
2904130186386119439196283887054367774322427680913236
5449485366768000001065262485473055861598999140170769
8385483188750142938908995068545307651168033373222651
7566220752695179144225280816517166776672793035485154
2040238174608923283917032754257508676551178593950027
9338959205766827896776445318404041855401043513483895
3120132637836928358082719378312654961745997056745071
8332065034556644034490453627560011250184335607361222
7659492783937064784264567633881880756561216896050416
1139039063960162022153684941092605387688714837989559
9991120991646464411918568277004574243434021672276445
5893301277815868695250694993646101756850601671453543
1581480105458860564550133203758645485840324029871709
3480910556211671546848477803944756979804263180991756
4228098739987669732376957370158080682290459921236616
8902596273043067931653114940176473769387351409336183
3216142802149763399189835484875625298752423873077559
5559554651963944018218409984124898262367377146722606
1633643296406335728107078875816404381485018841143188

5988276944901193212968271588841338694346828590066640
8063140777577257056307294004929403024204984165654797
3670548558044586572022763784046682337985282710578431
9753541795011347273625774080213476826045022851579795
7976474670228409995616015691089038458245026792659420
5550395879229818526480070683765041836562094555434613
5134152570065974881916341359556719649654032187271602
6485930490397874895890661272507948282769389535217536
2185079629778514618843271922322381015874445052866523
8022532843891375273845892384422535472653098171578447
8342158223270206902872323300538621634798850946954720
0479523112015043293226628272763217790884008786148022
1475376578105819702226309717495072127248479478169572
9614236585957820908307332335603484653187302930266596
4501371837542889755797144992465403868179921389346924
4741985097334626793321072686870768062639919361965044
0995421676278409146698569257150743157407938053239252
3947755744159184582156251819215523370960748332923492
1034514626437449805596103307994145347784574699992128
5999993996122816152193148887693880222810830019860165
4941654261696858678837260958774567618250727599295089
3180521872924610867639958916145855058397274209809097
8172932393010676638682404011130402470073508578287246
2713494636853181546969046696869392547251941399291465
2423857762550047485295476814795467007050347999588867
6950161249722820403039954632788306959762493615101024
3655535223069061294938859901573466102371223547891129
2547696176005047974928060721268039226911027772261025
4414922157650450812067717357120271802429681062037765
7883716690910941807448781404907551782038565390991047
7594141321543284406250301802757169650820964273484146
9572639788425600845312140659358090412711359200419759
8513625479616063228873618136737324450607924411763997
5974619383584574915988097667447093006546342423460634
2374746660804317012600520559284936959414340814685298
1505394717890045183575515412522359059068726487863575
2541911288877371766374860276606349603536794702692322

971868327717393236192007774522126247518698334951510198642698878471719396649769070825217423365662725928440620430214113719922785269984698847702323823840055655517889087661360130477098438611687052310553149162517283732728676007248172987637569816335415074608838663640693470437206688651275688266149730788657015685016918647488541679154596507234287730699853713904300266530783987763850323818215535597323530686043010675760838908627049841888595138091030423595782495143988590113185835840667472370297149785084145853085781339156270760356390763947311455495832266945702494139831634332378975955680856836297253867913275055542524491943589128405045226953812179131914513500993846311774017971512283785460116035955402864405902496466930707769055481028850208085800878115773817191741776017330738554758006056014337743299012728677253043182519757916792969965041460706645712588834697979642931622965520168797300035646304579308840327480771811555330909887025505207680463034608658165394876951960044084820659673794731680864156456505300498816164905788311543454850526600698230931577765003780704661264706021457505793270962047825615247145918965223608396645624105195510522357239739512881816405978591427914816542632892004281609136937773722299983327082082969955737727375667615527113922588055201898876201141680054687365580633471603734291703907986396522961312801782679717289822936070288069087768660593252746378405397691848082041021944719713869256084162451123980620113184541244782050110798760717155683154078865439041210873032402010685341947230476666721749869868547076781205124736792479193150856444775379853799732234456122785843296846647513336573692387201464723679427870042503255589926884349592876124007558756946413705625140011797133166207153715436006876477318675587148783989081074295309410605969443158477539700943988394914432353668539209946879645066533985738887866147629443414010498889931600512076781035886116602029611936396821349607501116498327856353161451684576956871

9002999769841263266502347716728657378579085746646077
2283415403114415294188047825438761770790430001566986
7767957609099669360755949651527363498118964130433116
6277471233881740603731743970540670310967676574869535
8789670031925866259410510533584384656023391796749267
8447637084749783336555790073841914731988627135259546
2518160434225372996286326749682405806029642114638643
6864224724887283434170441573482481833301640566959668
8667695634914163284264149745333499994800026699875888
1593507357815195889900539512085351035726137364034367
5347141048360175464883004078464167452167371904831096
7671134434948192626811107399482506073949507350316901
9731852119552635632584339099822498624067031076831844
6607291248747540316179699411397387765899868554170318
8477886759290260700432126661791922352093822787888098
8633599116081923535557046463491132085918979613279131
9756490976000139962344455350143464268604644958624769
0943470482932941404111465409239883444351591332010773
9441118407410768498106634724104823935827401944935665
1610884631256785297769734684303061462418035852933159
7345830384554103370109167677637427621021370135485445
0926307190114731848574923318167207213727935567952844
3925481560913728128406333039373562420016045664557414
5881660521666087387480472433912129558777639069690370
7882852775389405246075849623157436917113176134783882
7194168606625721036851321566478001476752310393578606
8961112599602818393095487090590738613519145918195102
9732787557104972901148717189718004696169777001791391
9613791417162707018958469214343696762927459109940060
0849835684252019155937037010110497473394938778859894
1743303178534870760322198297057975119144051099423588
3034546353492349826883624043327267415540301619505680
6541809394099820206099941402168909007082133072308966
2119775530665918814119157783627292746156185710372172
4710095214236964830864102592887457999322374955191221
9519034244523075351338068568073544649951272031744871
9540397610730806026990625807602029273145525207807991

4184290638844373499681458273372072663917670201183004
6481900024130835088465841521489912761065137415394356
5721139032857491876909441370209051703148777346165287
9848235338297260136110984514841823808120540996125274
5808810994869722161285248974255555160763716750548961
7301680961380381191436114399210638005083214098760459
9309324851025168294467260666138151745712559754953580
2399831469822036133808284993567055755247129027453977
6214049318201465800802156653606776550878380430413431
0591804606800834591136640834887408005741272586704792
2583191274157390809143831384564241509408491339180968
4025116399193685322555733896695374902662092326131885
5891580832455571948453875628786128859004106006073746
5014026278240273469625282171749415823317492396835301
3617865367376064216677813773995100658952887742766263
6841830680190804609849809469763667335662282915132352
7888061577682781595886691802389403330764419124034120
2231636857786035727694154177882643523813190502808701
8575047046312933353757285386605888904583111450773942
9352019943219711716422350056440429798920815943071670
1985746927384865383343614579463417592257389858800169
8014757420542995801242958105456510831046297282937584
1611625325625165724980784920998979906200359365099347
2158296517413579849104711166079158743698654122234834
1887722929446335178653856731962559852026072947674072
6167671455736498121056777168934849176607717052771876
0119990814411305864557791052568430481144026193840232
2470939249802933550731845890355397133088446174107959
1625117148648744686112476054286734367090466784686702
7409188101424971114965781772427934707021668829561087
7794405048437528443375108828264771978540006509704033
0218625561473321177711744133502816088403517814525419
6432030957601869464908868154528562134698835544456024
9556668436602922195124830910605377201980218310103270
4178386654471812603971906884623708575180800353270471
8565949947612424811099928867915896904956394762460842
4065930948621507690314987020673533848349550836366017

8487710608098042692471324100094640143736032656451845
6679245666955100150229833079849607994988249706172367
4493612262229617908143114146609412341593593095854079
1390872083227335495720807571651718765994498569379562
3875551617575438091780528029464200447215396280746360
2113294255916002570735628126387331060058910652457080
2447493754318414940148211999627645310680066311838237
6163966318093144467129861552759820145141027560068929
7502463040173514891945763607893528555053173314164570
5049964438909363084387448478396168405184527328840323
4520247056851646571647713932377551729479512613239822
9602394548579754586517458787713318138752959809412174
2273003522965080891777050682592488223221549380483714
5478164721397682096332050830564792048208592047549985
7320388876391601995240918938945576768749730856955958
0106595265030362661597506622250840674288982659075106
3756356996821151094966974458054728869363102036782325
0182323708459790111548472087618212477813266330412076
2165873129708112307581598212486398072124078688781145
0165582513617890307086087019897588980745664395515741
5363193191981070575336633738038272152798849350397480
0158905194208797113080512339332219034662499171691509
4854140187106035460379464337900589095772118080446574
3962806186717861017156740967662080295766577051291209
9079443046328929473061595104309022214393718495606340
5618934251305726829146578329334052463502892917547087
2564842600349629611654138230077313327298305001602567
2401418515204189070115428857992081219844931569990591
8201181973350012618772803681248199587707020753240636
1259313438595542547781961142935163561223496661522614
7353996740515849986035529533292457523888101362023476
2466905581643896786309762736550472434864307121849437
3485300606387644566272186661701238127715621379746149
8613287441177145524447089971445228856629424402301847
9120547849857452163469644897389206240194351831008828
3480249249085403077863875165911302873958787098100772
7182718745290139728366148421428717055317965430765045

3432460053636147261818096997693348626407743519992868
6323835088756683595097265574815431940195576850437248
0010204137498318722596773871549583997184449072791419
6584593008394263702087563539821696205532480321226749
8911402678528599673405242031091797899905718821949391
3207534317079800237365909853755202389116434671855829
0685371189795262623449248339249634244971465684659124
8918556629589329909035239233336474352037077010108433
8800329075983421701855422838616172104176030116459187
8053936744747205998502358289183369292233732399948043
7108419659473162654825748099482509991833006976569367
1596893644933488647442135008407006608835972350395323
4017958255703601693699098867113210979889707051728075
5855191269930673099250704070245568507786790694766126
2980822516331363995211709845280926303759224267425755
9989289278370474445218936320348941552104459726188380
0300677617931381399162058062701651024458869247649246
8919246121253102757313908404700071435613623169923716
9484813255420091453041037135453296620639210547982439
2125172540132314902740585892063217589494345489068463
9931375709103463327141531622328055229729795380188016
2859073572955416278867649827418616421878988574107164
9069191851162815285486794173638906653885764229158342
5006736124538491606741373401735727799563410433268835
6950781493137800736235418007061918026732855119194267
6091221035987469241172837493126163395001239599240508
4543756985079570462226646190001035004901830341535458
4283376437811198855631877779253720116671853954183598
4438305203762819440761594106820716970302285152250573
1260930468984234331527321313612165828080752126315477
3060442377475350595228717440266638914881717308643611
1389069420279088143119448799417154042103412190847094
0802540239329429454938786402305129271190975135360009
2197110541209668311151632870542302847007312065803262
6417116165957613272351566662536672718998534199895236
8848309993027574199164638414270779887088742292770538
9122717248632202889842512528721782603050099451082478

3572905691988555467886079462805371227042466543192145
2817607414824038278358297193010178883456741678113989
5475044833931468963076339665722672704339321674542182
4557062524797219978668542798977992339579057581890622
5254735822052364248507834071101449804787266919901864
3882293230538231855973286978092225352959101734140733
4884761005564018242392192695062083183814546983923664
6136398910121021770959767049083050818547041946643713
1229969235889538493013635657618610606222870559942337
1631021278457446463989738188566746260879482018647487
6727272220626764653380998019668836809941590757768526
3986514625333631245053640261056960551318381317426118
4420189088853196356986962795036738424313011331753305
3298020166888174813429886815855778103432317530647849
8321062971842518438553442762012823457071698853051832
6179641178579608888150329602290705614476220915094739
0359466469162353968092013945781758910889319921122600
7392814916948161527384273626429809823406320024402449
5894456129167049508235812487391799648641133480324757
7752197089327722623494860150466526814398770516153170
2669692970492831628550421289814670619533197026950721
4378230476875280287354126166391708245925170010714180
8548006369232594620190022780874098597719218051585321
4739265325155903541020928466592529991435379182531454
5290598415817637058927906909896911164381187809435371
5213322614436253144901274547726957393934815469163116
2492887357471882407150399500944673195431619385548520
7665738825139639163576723151005556037263394867208207
8086537349424401157996675073607111593513319591971209
4896471755302453136477094209463569698222667377520994
5168450643623824211853534887989395673187806606107885
4400055082765703055874485418057788917192078814233511
3866292966717964346876007704799953788338787034871802
1842437342112273940255717690819603092018240188427057
0460926225641783752652633583242406612533115294234579
6556950250681001831090041124537901533296615697052237
9210325706937051090830789479999004999395322153622748

4766036136776979785673865846709366795885837887956259
4646489137665219958828693380183601193236857855855819
5556042156250883650203322024513762158204618106705195
3306530606065010548871672453779428313388716313955969
0583208341689847606560711834713621812324622725884199
0286142087284956879639325464285343075301105285713829
6437099903569488852851904029560473461311382638788975
5178856042499874831638280404684861893818959054203988
9872650697620201995548412650005394428203930127481638
1585303964399254702016727593285743666616441109625663
3730540921951967514832873480895747777527834422109107
3111351828046036347198185655572957144747682552857863
3493428584231187494400032296906977583159038580393535
2135886007960034209754739229673331064939560181223781
2854584317605561733861126734780745850676063048229409
6530411183066710818930311088717281675195796753471885
3722930961614320400638132246584111115775835858113501
8569047815368938137718472814751998350504781297718599
0847076219746058874232569958288925350419379582606162
1184236876851141831606831586799460165205774052942305
3601780313357263267054790338401257305912339601880137
8254219270947673371919872873852480574212489211834708
7662966720727232565056512933312605950577772754247124
1648312832982072361750574673870128209575544305968395
5556868611883971355220844528526400812520276655576774
9596962661260456524568408613923826576858338469849977
8726706555191854468698469478495734622606294219624557
0853712727765230989554501930377321666491825781546772
9200521266714346320963789185232321501897612603437368
4067194193037746880999296877582441047878123266253181
8459604538535438391144967753128642609252115376732588
6672260404252349108702695809964759580579466397341906
4010036361904042033113579336542426303561457009011244
8008900208014780566037101541223288914657223931450760
7167064355682743774396578906797268743847307634645167
7562103098604092717090951280863090297385044527182892
7496892121066700816485833955377359191369501531620189

0888748421079870689911480466927065094076204650277252
8650728905328548561433160812693005693785417861096969
2025388650345771831766868859236814884752764984688219
4973972970773718718840041432312763650481453112285099
0020742409255859252926103021067368154347015252348786
3516439762358604191941296976904052648323470099111542
4260127343802208933109668636789869497799400126016422
7609260823493041180643829138347354679725399262338791
5829984864592717340592256207491053085315371829116816
3721939518870095778818158685046450769934394098743351
4431626330317247747486897918209239480833143970840673
0840795893581089665647758599055637695252326536144247
8023082681183103773588708924061303133647737101162821
4614661679404090518615260360092521947218890918107335
8719641421444786548995285823439470500798303885388608
3103571930600277119455802191194289992272235345870756
6246926177663178855144350218287026685610665003531050
2163182060176092179846849368631612937279518730789726
3735371715025637873357977180818487845886650433582437
7004147710414934927438457587107159731559439426412570
2709651251081155482479394035976811881172824721582501
0949609662539339538092219559191818855267806214992317
2763163218339896938075616855911752998450132067129392
4041445938623988093812404521914848316462101473891825
1010909677386906640415897361047643650006807710565671
8486281496371118832192445663945814491486165500495676
9826903089111856879869294705135248160917432430153836
8470729289898284602223730145265567989862776796809146
9798378268764311598832109043715611299766521539635464
4208691975673700057387649784376862876817924974694384
2746525631632300555130417422734164645512781278457777
2457520386543754282825671412885834544435132562054464
2410110379554641905811686230596447695870540721419852
1210673433241075676757581845699069304604752277016700
5684543969234041711089888993416350585157887353430815
5208117720718803791040469830695786854739376564336319
7978680367187307969392423632144845035477631567025539

0065423117920153464977929066241508328858395290542637
6876689688050333172278001858850697362324038947004718
9761934734430843744375992503417880797223585913424581
3144049847701732361694719765715353197754997162785663
1190469126091825912498903676541769799036237552865263
7573376352696934435440047306719886890196814742876779
0866979688522501636949856730217523132529265375896415
1714795595387842784998664563028788319620998304945198
7439636907068276265748581043911223261879405994155406
3270131989895703761105323606298674803779153767511583
0432084987209202809297526498125691634250005229088726
4692528466610466539217148208013050229805263783642695
9733707053922789153510568883938113249757071331029504
4303467159894487868471164383280506925077662745001220
0352620370946602341464899839025258883014867816219677
5194583167718762757200505439794412459900771152051546
1993050983869825428464072555409274031325716326407929
3418334214709041254253352324802193227707535554679587
1638358750181593387174236061551171013123525633485820
3651461418700492057043720182617331947157008675785393
3607862273955818579758725874410254207710547536129404
7460100094095444959662881486915903899071865980563617
1376922272907641977551777201042764969496110562205925
0242021770426962215495872645398922769766031052498085
5759471631075870133208861463266412591148633881220284
4406941694882615295776253250198703598706743804698219
4205638125583343642194923227593722128905642094308235
2544084110864545369404969271494003319782861318186188
8111184082578659287574263844500599442295685864604810
3301538891149948693543603022181094346676400002236255
0573631294626296096198760564259963946138692330837196
2659547392346241345977957485246478379807956931986508
1597767535055391899115133525229873611277918274854200
8689539658359421963331502869561192012298889887006079
9927954111882690230789131076036176347794894320321027
7335941690865007193280401716384064498787175375678118
5321328408216571107549528294974936214608215583205687

2321855740651610962748743750980922302116099826330339
1546949464449100451528092508974507489676032409076898
3652940657920198315265410658136823791984090645712468
9484702093577611931399802468134052003947819498662026
2400890215016616381353838151503773502296607462795291
0384068685569070157516624192987244482719429331004854
8244545807188976330032325258215812803274679620028147
6243182862217105435289834820827345168018613171959332
4711074662228508710666117703465352839577625997744672
1857158161264111432717943478859908928084866949141390
9771673690027775850268664654056595039486784111079011
6104008572744562938425494167594605487117235946429105
8509099502149587931121961359083158826206823321561530
8683373083817327932819698387508708348388046388478441
8840031847126974543709373298362402875197920802321878
7448828728437273780178270080587824107493575148899789
1173974612932035108143270325140903048746226294234432
7571260086642508333187688650756429271605525289544921
5376517514921963671810494353178583834538652556566406
5725136357506435323650893679043170259787817719031486
7963840828810209461490079715137717099061954969640070
8676671023300486726314755105372317571143223174114116
8062286420638890621019235522354671166213749969326932
1737043105987225039456574924616978260970253359475020
9138366737728944386964000281103440260847128990007468
0776484408871134135250336787731679770937277868216611
7865344231732264637847697875144332095340001650692130
5464768909850502030150448808342618452087305309731894
9291642532293361243151430657826407028389840984160295
0309241897120971601649265613413433422298827909921786
0426798124572853458013382609958771781131021673402565
6274400729683406619848067661580502169183372368039902
7931606420436812079900316264449146190219458229690992
1227885539487835383056468648816555622943156731282743
9082645061162894280350166133669782405177015521962652
2725455850738640585299830379180350432876703809252167
9075712040612375963276856748450791511473134400018325

7034492090971243580944790046249431345502890068064870
4293534037436032625820535790118395649089354345101342
9696175452495739606214902887289327925206965353863964
4322538832752249960598697475988232991626354597332444
5163755334377492928990581175786355555626937426910947
1170021654117182197505198317871371060510637955585889
0556885288798908475091576463907469361988150781468526
2133252473837651192990156109189777922008705793396463
8274906806987691681974923656242260871541761004306089
0437797667851966189140414492527048088197149880154205
7787006521594009289777601330756847966992955433656139
8477380603943688958876460549838714789684828053847017
3087111776115966350503997934386933911978988710915654
1709133082607647406305711411098839388095481437828474
5288383680794188843426662220704387228874139478010177
2139228191199236540551639589347426395382482960903690
0288359327745855060801317988407162446563997948275783
6501955142215513392819782269842786383916797150912624
1054872570092407004548848569295044811073808799654748
1568913935380943474556972128919827177020766613602489
5814681191336141212587838955773571949863172108443989
0142394849665925173138817160266326193106536653504147
3070804414939169363262373767777095850313255990095762
7319573086480424677012123270205337426670531424482081
6813030639737873664248367253983748769098060218278578
6216512738563513290148903509883270617258932575363993
9790557291751600976154590447716922658063151110280384
3601737474215247608515209901615858231257159073342173
6576267142390478279587281505095633092802668458937649
6497702329736413190609827406335310897924642421345837
4090116939196425045912881340349881063540088759682005
4408364386516617880557608956896727531538081942077332
5979172784376256611843198910250074918290864751497940
0316070384554946538594602745244746681231468794344161
0993338908992638411847425257044572517459325738989565
1857165759614812660203107976282541655905060424791140
1695790033835657486925280074302562341949828646791447

63227740055294609039401775363356554719310001754300047
50471914489984104001586794617924161001645471655133370
74073950260442769538553834397550548871099785205401117
51697475813449260794336895437832211724506873442319889
87884412854206474280973562580706698310697993526069333
92135685881391214807354728463227784908087002467776300
36055512323866562951788537196730346347012229395816006
79250915321748903084088651606111901149844341235012466
46928028805996134283511884715449771278473361766285066
21697787177438243625657117794500644777183702219991066
69502165675764404499794076503799995484500271066598788
13603802314126836905783190460792765297277694043613022
30517870805465115424693952651271010529270703066730244
44712597393995051462840476743136373997825918454117644
13327906460636584152927019030276017339474866960348699
49765417524293060407270050590395031485229213925755944
84507886797792525393176515641619716844352436979444733
55964260633391055126826061595726217036698506473281266
67245219890605498802807828814297963366967441248059822
19214633956574572210229867759974673812606936706913400
81559412016115960190237753525556300606247983261249888
12881929373434768626892192397778339107331065882568133
77717232831532908252509273304785072497713944833389255
52081175608452966590553940965568541706001179857293811
39982583192936791003918440992865756059935989100029699
86446097471471847010153128376263114677420914557404188
15908800064943237855839308530828305476076799524357399
16312218860575496738322431956506554608528812019023633
64471270374863442172725787950342848631294491631847533
47531435041392096108796057730987201352484075057637199
92536504709085825139368634638633680428917671076021111
15982887553994012007601394703366179371539630613986366
55492213741597905119083588290097656647300733879314677
89131814651093167615758213514248604422924453041131600
65270097433008849903467540551864067734260358340960866
05533747362760935658853109760994238347382222087292466
44976845605795625167655740884103217313456277358560522

3582363895320385340248422733716391239732159954408284
2166663602329654569470357718487344203422770665383738
7506169212768015766181095420097708363604361110592409
1178895403380214265239489296864398089261146354145715
3519434285072135345301831587562827573389826889852355
7799295727645229391567477566676051087887648453493636
0682780505646228135988858792599409464460417052044700
4631513797543173718775603981596264750141090665886616
2180038266989961965580587208639721176995219466789857
0117983324406018115756580742841829106151939176300591
9431443460515404771057005433900018245311773371895585
7603607182860506356479979004139761808955363669603162
1931132502238517916720551806592635180362512145759262
3836934822266589557699466049193811248660909979812857
1823494006615552196112207203092277646200999315244273
5894887105766238946938894464950939603304543408421024
6240104872332875008174917987554387938738143989423801
1762700837196053094383940063756116458560943129517597
7139353960743227924892212670458081833137641658182695
6210587289244774003594700926866265965142205063007859
2002488291860839743732353849083964326147000532423540
6470420894992102504047267810590836440074663800208701
2666420945718170294675227854007450855237772089058168
3918446592829417018288233014971554235235911774818628
5929676050482038643431087795628929254056389466219482
6871104282816389397571175778691543016505860296521745
9581988878680408110328432739867198621306205559855266
0364050462821523061545944744899088390819997387474529
6981077620148713400012253552224669540931521311533791
5798026979555710508507473874750758068765376445782524
4326380461430428892359348529610582693821034980004052
4840708440356116781717051281337880570564345061611933
0424440798260377951198548694559152051960093041271007
2778493015550388953603382619293437970818743209499141
5959339636811062755729527800425486306005452383915106
8998913578820019411786535682149118528207852130125518
5184937115034221595422445119002073935396274002081104

6553020793286725474054365271759589350071633607632161

0345334500256244101006359530039598864466169595626351
8780606885137234627079973272331346939714562855426154
6765063246567662027924520858134771760852169134094652
0307673391841147504140168924121319826881568664561485
3802875393311602322925556189410429953356400957864953
4093511526645402441877594931693056044868642086275720
1172319526405023099774567647838488973464317215980626
7876718380052476968840849891850861490034324034767426
8624595239589035858213500645099817824463608731775437
8859677672919526111213859194725451400301180503437875
2776644027626189410175768726804281766238606804778852
4288743025914524707395054652513533945959878961977891
1041890292943818567205070964606263541732944649576612
6519534957018600154126239622864138977967333290705673
7696215649818450684226369036784955597002607986799626
1019039331263768556968767029295371162528005543100786
4087289392257145124811357786276649024251619902774710
9033593330930494838059785662884478744146984149906712
3764789582263294904679812089984857163571087831191848
6302545016209298058292083348136384054217200561219893
5366937133673339246441612522319694347120641737549121
6357008573694397305979709719726666642267431117762176
4030686813103518991122713397240368870009968629225464
6500638528862039380050477827691283560337254825579391
2985251506829969107754257647488325341412132800626717
0940090982235296579579978030182824284902214707481111
2401860761341515038756983091865278065889668236252393
7845272634530420418802508442363190383318384550522367
9923577529291069250432614469501098610888999146585518
8187358252816430252093928525807796973762084563748211
4433988162710031703151334402309526351929588680690821
3558536801610002137408511544849126858412686958991741
4913382057849280069825519574020181810564129725083607
0356851055331787840829000041552511865779453963317538
5320921497205266078312602819611648580986845875251299
9740409279768317663991465538610893758795221497173172
8131517932904431121815871023518740757222100123768721

9447472093493123241070650806185623725267325407333248
7575448296757345001932190219911996079798937338367324
2576103938985349278777473980508080015544764061053522
2023254094435677187945654304067358964910176107759483
6454082348613025471847648518957583667439979150851285
8020607820554462991723202028222914886959399729974297
4711553718589242384938558585954074381048826246487880
5330427146301194158989632879267832732245610385219701
1130466587100500083285177311776489735230926661234588
8731028835156264460236719966445547276083101187883891
5114934093934475007302585581475619088139875235781233
1342279866503522725367171230756861045004548970360079
5698276263923441071465848957802414081584052295369374
9971066559489445924628661996355635065262340533943914
2111271810691052290024657423604130093691889255865784
6684612156795542566054160050712766417660568742742003
2957716064344860620123982169827172319782681662824993
8714995449137302051843669076723577400053932662622760
3236597517189259018011042903842741855078948874388327
0306328327996300720069801224436511639408692222074532
0244624121155804354542064215121585056896157356414313
0688834431852808539759277344336553841883403035178229
4625370201578215737326552318576355409895403323638231
9219892171177449469403678296185920803403867575834111
5188241774391450773663840718804893582568685420116450
3135763335550944031923672034865101056104987272647213
1986543435450409131859513145181276437310438972507004
9819870521762724940652146199592321423144397765467083
5171474936798618655279171582408065106379950018429593
8799158350171580759883784962257398512129810326379376
2183224565942366853767991131401080431397323354490908
2491049914332584329882103398469814171575601082970658
3065211347076803680695322971990599904451209087275776
2253510409023928887794246304832803191327104954785991
8019696783532146444118926063152661816744319355081708
1875477050802654025294109218264858213857526688155584
1131985600221351588872103656960875150631875330029421

1868222189377554602722729129050429225978771066787384
0000616772154638441292371193521828499824350920891801
6855727981564218581911974909857305703326676464607287
5743056537260276898237325974508447964954564803077159
8153955827779139373601717422996027353102768719449444
9179397851446315973144353518504914139415573293820485
4212350817391254974981930871439661513294204591938010
6231421774199184060180347949887691051557905554806953
8785400664533759818628464199052204528033062636956264
9091082762711590385699505124652999606285544383833032
7638599800792922846659503551211245284087516229060262
0118577753137479493620554964010730013488531507354873
5390560290893352640071327473262196031177343394367338
5759124508149335736911664541281788171454023054750667
1365182582848980995121391939956332413365567770980030
8191027204099714868741813466700609405102146269028044
9159646545330107754695413088714165312544813061192407
8211886900560277818242350226961893443525476335735364
8561936325441775661398170393063287216690572225974520
9192917262199844409646158269456380239502837121686446
5617852355651641277128269186886155727162014749340522
7694659571219831494338162211400693630743044417328478
6101777438379770372317952554341072234455125555899998
6461838767649039724611679590181000350989286412041951
6355110876320426761297982652942588295114127584126273
2790798807559751851576841264742209479721843309352972
6652100156625145529947451276315509176367302594621329
3019040283795424632325855030109670692272022707486341
9005438302650681214142135057154175057508639907673946
3351462090828889349383764393992569006040673114220933
1219593620298297235116325938677224147791162957278075
2395056251581603133359382311500518626890530658368129
9881086632632719806112715488587980934879129137074982
3057592909186293919501472119758606727009254771802575
0337730799397134539532646195269996596385654917590458
3335857991020127132045839032008538788816336376851820
8372788513117522776960978796214237216254521459128183

1798216044111311671406914827170981015457781939202311
5638719508050246797257924976057726259133285597263712
1120190572077140914864507409492671803581515757151405
0397610963846755569298970383547314100223802583468767
3501297754132795320609711545064842121859364909979177
6687477448188287063231551586503289816422828823274686
6106592732197907162384642153489852476216789050260998
0452664839295423572873439776804957740914495383915755
6548545905897649519851380100795801078375994577529919
6700547602252552034453988712538780171960718164078124
8478472579124078245443616823452395706895142722697504
3187363326301110305342333582160933319121880660826834
1428910415173247216053355849993224548730778822905252
3242348615315209769384610425828497149634753418375620
0301491570327968530186863157248840152663983568956363
4657435321783493199825542117308467745297085839507616
4582296303244243282377374505170285606980678895217681
9815671078163340526675953942492628075696832610749532
3390536223090807081455919837355377748742029039018142
9373115293346444681512129450975965343062842153194457
2711861490001765055817709530246887526325011970520947
6159416768727784472000192789137251841622857783792284
4390843011811214963664246590336341945406571835447719
1244662125939265662030688852005559912123536371822692
2531781458792593750441448933981608657900876165024635
1970458288954817937566810464746141051424988702521399
3687050937230544773411264135489280684105910771667782
1238332810262185587751312721179344448201440425745083
0639447383637939062830089733062413806145894142276947
4793166571762318247216835067807648757342049155762821
7583972975134478990696589532548940335615613167403276
4724692125057591162515296545685446334981143176702572
9566184477548746937846423373723898192066204851189437
8868224807279352022501796545343757274163910791972952
9508129429222053477173041844779156739917384183117103
6252439571615271466900581470000263301045264354786590
3290733205468338872078735444762647925297690170912007

87418373673508771337697768349634425241994995138831507
4877537433849458259765560996555954318040920178497184
6854973706962120885243770138537576814166327224126344
2398215294164537800049250726276515078908507126599703
6708726692764308377229685985169122305037462744310852
9343052730788652839773352460174635277032059381791253
9691562106363762588293757137384075440646896478310070
4580613446731271591194608435935825987782835266531151
0650416232953290477721740835593497237585521380483050
9000964667608830154061282430874064559443185341375522
0166305812111033453120745086824339432159043594430312
4312274713858420303901060709403152355561727679941600
2039397509989762933532585557562480899669182986422267
7502360193257974726742578211119734709402357457222271
2125268523842958742735015636600931880454933389897415
7149054418255973808087156528143010267046028431681923
0392535297795765862414392701549740879273131051636119
1375770089295648233236482982630246079758757677453771
6010249080462430185652416175665560016085912153455626
7602192689982855377872583145144082654583484409478463
1787773747946535801699607794055687011923286080411309
0462935087182712593466871276669487389982459852778649
9569165464029458935064964335809824765965165142090986
7552038083092032304873427034682887516040715466538346
1961122301375945157925269674364253192739003603860823
6450762698827497618723575476762889950752114804852527
9508450339585708381304769378813211236742813194879502
2806632017002246033198967197064916374117585485187848
4012054844672588851401562725019821719066960812627785
4859648183696214107217142149863619187747545096503089
5709947093433785698167446582826791194061195603784539
7855839240761276344105766751024307559814552786167815
9496570625597550743065210853015979080733437360794328
6675789053348366955548680391343372015649883422089339
9971641479746938696905480089193067138057171505857307
1488156499207140867582596028760564597824237702424698
0532805663278704192676846711626687946348695046450746

2021937394525926266861355294062478136120620263649819
9999498405143868285258956342264328707663299304891723
4007254717641886853513723326678779217383475414800228
0339299735793615241275582956927683723123479898944627
4330454566790062032420516396282588443085438307201495
6721064605332385372031432421126074244858450945804940
8182092763914000854042202355626021856434899414543995
0410980591817948882628052066441086319001688568155169
2294862030107388971810077092905904807490924271410189
3354281842999598816966099383696164438152887721408526
8088757488293258735809905670755817017949161906114001
9085537448827262009366856044755965574764856740081773
8170330738030547697360978654385938218722058390234444
3508867499866506040645874346005331827436296177862518
0818931443632512051070946908135864405192295129324500
7883339878842933934243512634336520438581291283434529
7308652909783300671261798130316794385535726296998740
3595704584522308563900989131794759487521263970783759
4486113945196028675121056163897600888009274611586080
0207803341591451797073036835196977766076373785333012
0241201120469886092093390853657732223924124490515327
8095095586645947763448226998607481329730263097502881
2103517723124465095349653693090018637764094094349837
3132513218620802148099226855029484546618147155574447
0966953017769043427203189277060471778452793916047228
1534379803539679861424370956683221491465438014593829
2773933960327540480095522318166673803571839327570771
4204672383862461780397629237713120958078936384144792
9802588065522129262093623930637313496640186619510811
5834711733120258058667276399927635790780638188130691
5636627412543125958993611964762610140556350339952314
0323113819656236327198961837254845333702062563464223
9527669435683767613687119629218187545760816170530315
9072882870071231366630872275491866139577373054606599
7437810987649802414011242142773668082751390959313404
1558262667895108467761186659576601659981780894149857
5497628438785610026379654317831363402513581416115190

2096499133548733131115022700681930135929595971640197
1960536250335584799809634887180391116128135959685654
7886832585643789617315976200241962155289629790481982
2199462269487137462444729093456470028537694958859591
6067892824910544125159963007813683674902093749157328
9627002865682934443134234735123929825916673950342599
5868970697267332582735903121288746660451461487850346
1428277659916080903986525757172630818334944418201935
3338507129234577437557934406217871133006310600332405
3991693682603746176638565758877580201229366353270267
1006812618251729146082025418928859352444910701382062
1155382779356529691457650204864328286555793470720963
4807372692141186895467322767751335690190153723669036
8653891612916888787640752549349424973342718117889275
5993159671935475898809792452526236365903632007085444
0784544797348291802082044926670634420437555325050527
5228337788870408040335319234076856301093477721256390
8864041310107381785333831603813528082811904083256440
1842053746792992622037698718018061122624490909242641
9858208617511771137890516091403815750033664241560952
1632819712233502316742260056794128140621721964184270
5784328959802882335059828208196666249035857789940333
1522748177769528436816300885317696947836905806710648
2808359804669884109813515865490693331952239436328792
3990534810987830274500172065433699066117784554364687
7236318444647680691428280045510746866453928053994091
0875493916609573161971503316696830992946634914279878
0842257220697148875580637480308862995118473187124777
2919100702275888934869394562895158029653721504096031
0776128983126358996489341024703603664505868728758905
1406841238124247386385427908282733827973326885504935
8743031602747490631295723497426112215174171531336186
2241091386950068883589896234927631731647834007746088
6655598733382113829928776911495492184192087771606068
4728746736818861675072210172611038306717878566948129
4878504894306308616994879870316051588410828235127415
3538513365895332948629494495061868514779105804696039

0693726626703865129052011378108586161888869479576074
1358553458515176805197333443349523012039577073962377
1316030242887200537320998253008977618973129817881944
6717311606472314762484575519287327828251271824468078
2421521646956781929409823892628494376024885227900362
0219386696482215628093605373178040863727268426696421
9299468192149087017075333610947913818040632873875938
4826953558307739576144799727000347288018278528138950
3217986345216111066608839314053226944905455527867894
4175792024400214507801920998044613825478058580484424
1640477503153605490659143007815837243012313751156228
4015838644270890718284816757527123846782459534334449
6220100960710513706084618011875431207254913349942476
1711563332140893460915656155060031738421870157022610
3101916603887064661438897736318780940711527528174689
5764015810470169652475577408916445686777171585005832
6994340167720215676772406812836656526412298243946513
3197359199709403275938502669557470231813203243716420
5861410336065245369391600506449530601612678226489424
3739716671766123104897503188573216555498834212180284
6912529086101485527815277625623750456375769497734336
8460156077270355096290493924870884062810679436224187
0474700836884267102255830240359984164595112248527263
3632645114017395248086194635840783753556885622317115
5209472230654370926067973510005655493812245754837285
4571179739361575616764169289580525729752233855861138
8322171107362265816218842443178857488798109026653793
4266642169909140565364322493013348679881548866286650
5234699723557473842483059042367714327879231642240387
7764330192600192284778313837632536121025336935812624
0868666997382759773656822279072158324788886423693463
9616436330873013981421143030600873066616480367898409
1335926293402304324974926887831643602681011309570716
1419128306865773235326396536773903176613613159655535
8499939860056515592193675997771793301974468814837110
3206503693192894521402650915465184309936553493337183
4252984336799159394174662239003895276738133306177476

2957494386871697845376721949350659087571191772087547
7107189937960894774512654757501871194870738736785890
2006173733210756933022163206284320656711920969505857
6117396163232621770894542621460985841023781321581772
7602222738133495410481003073275107799948991977963883
5307344434575329759142637684054422647842160631227696
4696715647399904371590332390656072664411643860540483
8847161912109008701019130726071044114143241976796828
5478855247794764818029597360494397004795960402927462
9920357209976195014034831538094771460105633344699882
0822120587281510729182971211917876424880354672316916
5418522567292344291871281632325969654135485895771332
0833991128877591722611527337901034136208561457799239
8778325083550730199818459025958355989260553299673770
4917224549353296833000022301815172265757875240588322
4908582128008974790932610076257877042865600699617621
2176845478996440705066241710213327486796237430229155
3582007801411653480656474882306150033920689837947662
5503654982280532966286211793062843017049240230198571
9978948836897183043805182174419147660429752437251683
4354112170386313794114220952958857980601529387527537
9903093887168357209576071522190027937929278630363726
8765822681241993384808166021603722154710143007377537
7926990695871212892880190520316012858618254944133538
2078488346531163265040764242839087012101519423196165
2268422003711230464300673442064747718021353070124098
8603533991526679238711017062218658835737812109351797
7560442563469499978725112544085452227481091487430725
9869602040275941178942581281882159952359658979181144
0776533543217575952555361581280011638467203193465072
9680799079396371496177431211940202129757312516525376
8017359101557338153772001952444543620071848475663415
4074423286210609976132434875488474345396659813387174
6609302053507027195298394327142537115576660002578442
3031073429551533945060486222764966687624079324353192
9926392537310768921353525723210808898193391686682789
4828117047262450194840970097576092098372409007471797

3340788141825195842598096241747610138252643955135259
3118850456362641883003385396524359974169313228947198
7830842760040136807470390409723847394583489618653979
0594118599310356168436869219485382055780395773881360
6795499000851232594425297244866667668346414021899159
4456530942344065066785194841776677947047204195882204
3295380326310537494883122180391279678446100139726753
8921951191178365876625280836900532490045974109470687
7291232821430463533728351995364827432583311914445901
7809607782883583730111857543659958982724531925310588
1150263075425714939430244539318701799236081666113054
2625399583389794297160207033876781503301028012009599
7252222280801423571094760351925544434929986767817891
0455590630159538097618759203589373419789623589311259
8390259831026719330418921510968915622506965911982832
3455503059081730735195503721665870288053992138576037
0353771051780212801295668419841403628727256232144287
5430221090947272107347413497551419073704331827662617
7275996888826027225247133683353452816692779591328861
3817663498577289369009657495622871030243625907724122
1909430087175569262575806570991201665962243608024287
0024547362036394841255954881727272473653467783647201
9183039987176270375157246499222894679323226936191776
4161461879561395669956778306829031658969943076733350
8234990790624100202506134057344300695745474682175690
4416515406365846804636926212742110753990421887161276
1778701425886482577522388918459952337629237791558574
4549477361295525952226578636462118377598473700347971
4082069941455807190802135907322692331008317595106590
1912129479540860364075735875020589020870457967000705
5262505811420663907459215273309406823649441590891009
2202966805233252661989113118420162916310768940847235
6436680818216865721968826835840278550078280404345371
0183651096951782335743030504852653738073531074185917
7056103973950626403554422751561011072617793706347238
0499066692216197119425912044508464174638358993823994
6517395509000859479990136026674261494290066467115067

17542217703877450767356374215478290591101261915755558
70238957001405117822646989944917908301795475876760l6
80941001358376135785913569244556477644641786671153 91
9513576961048649224900834467154863830544779143300976
8048687834818467273375843689272431044740680768527862
5585165092088263813233623148733336714764520450876627
6149503899495048095604609896043291233583488599902945
2640028499428087862403981181488476730121675416110662
9995553668193123287425702063738352020086863691311733
4697317412191536332467453256308713473027921749562270
1468732586789173455837996435135880095935087755635624
8810493852999007675135513527792412429277488565888566
5132473025147102105753525165118148509027504768455182
5209633189906852761443513821366215236889057878669943
2288816028377482035506016029894009119713850179871683
6337441392759736440170070147637066557035043381211135
7641501845182141361982349515960106475271257593518530
4332875537783057509567425442684712219618709178560783
9361445113833356491032564057338986671781239722375193
1643061701385953947436784339267098671245221118969084
0236327411496601243483098929941738030588417166613073
0400675883804321115553794406054977217059428215148861
6567277124090338772774562909711013488518437411869565
5449745736845218066982911045058004299887953899027804
3835962824094218605562877884288021275538848037286400
1944161425749990427200959520465417059810498996750451
1936471172772220436102614079750809686975176600237187
7483480161203102346805671126447661237476278521902412
0256994353471622666089367521983311181351114650385489
5025120655772636145473604426859498074396932331297127
3771573470997139522911826534851555871373366291202427
1430250376326950135091161295299378586468130722648600
8270881333538193703682598867893321238327053297625857
3827900978264605455985551318366888446282651337984916
6783940976135376625179825824966345877195012438404035
9140849209733754642474488176184070023569580177410177
6969250778148933866725578985645898510568919609243988

4156928069698335224022563457049731224526935419383700
4843183357196516626721575524193401933099018319309196
5829209696562476676836596470195957547393455143374137
0876151732367720422738567427917069820454995309591887
2434939524094441678998846319845504852393662972079777
4528143994182567894577957125524268260899408633173715
3889626288962940211210888442737656862452761213037101
7300785135715404533041507959447776143597437803742436
6469732471384104921243141389035790924160364063140381
4983148190525172093710396402680899483257229795456404
2701757722904173234796073618787889913318305843069394
8259613187138164234672187308451338772190869751049428
4376932502498165667381626061594176825250999374167288
3951744066932549653403101452225316189009235376486378
4828813442098700480962271712264074895719390029185733
0746010436072919094576799461492929042798168772942648
7729952858434647775386906950148984133924540394144680
2636254021186143170312511175776428299146445334089209
7696169909837265236176874560589470496817013697490952
3072082682887890730190018253425805343421705928713931
7379931424108526473909482845964180936141384758311361
3057610846236683723769591349261582451622155213487924
4145041756848064120636520170386330129532777699023118
6480200675569056822950163549319923059142463962170253
2974757311409422018019936803502649563695586642590676
2685687372110339156793839895765565193177883000241613
5395624377778408017488193730950206999008908993280883
9743036773659552489130015663329407790713961546453408
8791510300651321934486673248275907946807879819425019
5826223203951312520141099605312606965554042486705499
8678692302174698900954785072567297879476988883109348
7464426400718183160331655511534276155622405474473378
0492462149521332585276988473362691826491743389878247
8927846891882805466998230368993978341374758702580571
6349413568433929396068192061773331791738208562436433
6353598634944968907810640196740744365836670715869245
2118299789380407713750129085864657890577142683358276

8978554717687184427726120509266486102051535642840632
36848180728794071712796682006072755955559040402331787
4944734645476062818954151213916291844429765106694796
9354016866010055196077687335396511614930937570968554
5593815137895690392510149532656281470119983269922000
6639287537471313523642158926512620407288771657835840
5219646054105435443642166562244565042999010256586927
2791427529311720827939377513261060528812353734510683
7293989358087124386938593438917571337630072031976081
6604464683937725806909237297523486702916910426369262
0901996052041210240776481903160140858635584276095370
8655816427399534934654631450404019952853725200495780
5254656251154109252437991326262713609099402902262062
8367521323050651839340574501120993414649184333236465
6937172591448932415900624202061288573292613359680872
6500045628284557574596592120530341310111827501306961
5098355156320043107846019065654938065425252291619918
1995960275232770224985573882489988270746593635576858
2560518068964285376850772012220347920993936179268206
5901421656159253067379445689490708532635681968318617
7226824991147261573203580764629811624401331673789278
8689229032593349861797021994981925739617673075834417
0985592221701718257127775344915082052784309046194608
3521740200583867284970941102326695392144546106621500
6410674740207009189911951376466904481267253691537162
2907913854039375600778351533741677479421003840023089
5185099454877903934612222086506016050035177626483161
1153325587705073541279249909859373473787081194253055
1214369797499149518605359204038302357163527276308746
9321962219006426088618367610334600225547747781364101
2691906569686495012688376296907233961276287223041141
8136100602640440300359969889199458273976241146137448
0405969706257676472376606554161857469052722923822827
5186799156983390747671146103022776606020061246876477
7288190967916133540198814027579921741676787992316039
6356949285151363364721954061117176738737255572852294
0054361785176502307544693869307873499110352182532929

7260445532107978877114498988709115112372506042387537
3484125708606406905205845212275453384800820530245045
6517669518576913200042816758054924811780519832646032
4457928297301291053183856368212062155312886685649565
1261389226136706409395333457052698695969235035309422
4543865278677673027540402702246384483553239914751363
4410440500923303612714960813554905315390210022995957
5658370538126196568314428605795669662215472169562087
0013727768536960840704833325132793112232507148630206
9512453950037357233468070946564830892098015348787056
3349109236605755405086411152144148143463043727327104
5027768661953107858323334857840297160925215326092558
9326556006721243594642550659967717703884453961816328
7961446081778927217183690888012677820743010642252463
4807454300476492885553409062185153654355474125476152
7697726677697727770583158014121856880117050283652755
4321480348800444297999806215790456416195721278450892
8489806426497427090579129069217807298769477975112447
3059914060506299468942809310342164166299356148281309
9887074529271604843363081840412646963792584309418544
2216359084576146078558562473814931427078266215185541
6038702068769804617474008083243436653823545551094494
9843109349475994467267366535251766270677219418319197
7196378015702169933675083760057163454643671776723387
5886434056448715669643210412825956453498413884128904
2068204700761559691684303899934836679354254921032811
3363184722592305554383058206941675629992013373175489
1220372303490726810685344540359935618235763128377676
4063101312533521214199461186935083317658785204711236
4331226765129964171325217513553261867681942338790365
4689080018271352835848884441117612341011799187092365
0718485785622102110400977699445312179502247957806950
6532965940383987369907240797679040826794007618729547
8359634927939045769736616434053597922192858705749574
8169669406233427261973351813662606373598257555249650
9807260123668283605928341855848026958413772558970883
7899429105498003311138846034019391661221866960584915

71485733568286149500019097591125218800396419762163555
93757437180114805594422987304181968080856472657135547
61283162920044988031540210553059707666636274932830899
16880932359290081787411985738317192616728834918402422
97212904349655269427264025596414635259143484006758677
69035038232057293413298159353304444649682944136732344
42158380761694831219333119819061096142952201536170299
85751055943264614685054526849757648078080092213358111
37819774927176854507553832876887447459159373116247066
01091244609829424841287520224462594477638749491997844
04468292573609685345498432665368628444893657041118177
79380644161653122360021491876876946739840751717630755
16849856359201486892943105940202457969622924566644888
19675762943495353263821716133957577907663707645695700
25973880043841580589433613710655185998760075492418722
11714889295221737721146081154344982665479872580056677
47240511220073834592715757277152185899469481179406444
46639943237004429114074721818022482583773601734668533
00744985564715420036123593397312914458591522887408711
95087086322188372882628228846318437172619033057771477
65156414382230679184738603914768310814135827575585366
43597721650028277803713422869688787349795096031108899
91961433866640684506974207877002805093672033872326299
63785603865321643234881555755701846908907464787912244
36375556668678067610544955017260791142930831285761255
44819444494732448190937953690082063846316782250648099
53181040657025432760438570350592281891987806586541211
84299217273720955103242251079718077833042609086794277
34289557355592527238055114404380012390416877164451800
22649168164192740110645162243110170005669112173318944
23400547959684669804298017362570406733282129962153688
48814041021944634246462207455756439604529853130714099
08460849965376780379320189914086581466217531933766599
70114330608625009829566917638846056762972931464911499
37046244693519840395344491351411936679333019366176633
65255514917498230798707228086085962611266050428929699
66535652516688885572112276802772743708917389639772255

7564890533401038855931125679991516589025016486961427

9616294134180012951306836386092941000831366733721530
0835269623573717533073865333820484219030818644918409
3723944033405244909554558016406460761581010301767488
4750176619086929460987692016912021816882910408707095
6095147041692114702741339005225334083481287035303102
3919699978597413908593605433599697075604460134242453
6824960987725813110247327985620721265724990034682938
8687230489556225320446360263985422525841646432427161
1419817802482595563544907219226583863662663750835944
3148776351561457107455280161596770484427141944351832
7569840755267792641126176525061596523545718795667317
0913319358761628255920783080185206890151504713340386
1003100559148178521103847545429333891884441205179439
6997019411269511952656491959418997541839323464742429
0702718875223534393673633663200307232747037407123982
5620246626519740901997624520561985576257600087081730
8328834438183107005451449354588542267857855191537229
2379555494333410174420169600090696415612732297770221
2179518683763590822551288164700219923488640439591530
1846400471432118636062252701154112228380277853891109
8490201342741014121559769965438877197485376431158229
8385331230717511329619045590079380642766958190148426
2799122179294798734890186847167650382732855205908298
4529806259250352128451925927986593506132961946796252
3739725655841578537445675589980324054921869628884903
3256085145534439166022625777551291620077279685262938
7937530454181080729285891989715381797343496187232927
6147478501926114504132748732429705834084711123337462
7461727462658241532427105932250625530231473875925172
4787322881491455915605036334575424233779160374952502
4930223514819613811625639114156103268449580725082734
3176594405409826976526934457986347970974312449827193
3113863873159636361218623497261409556079920628316999
4200720548115253533939460768500199098865538614334957
8165008996164907967814290114838764568217491407562376
7618453775144031475411206760160726460556859257799322
0703373333989163695043466906948284366299800374145276

2771654762382554617088318981086880684785370553648046
9350958818025360529740793538676511195079373282083146
2689600710751755206144337841145499501364324463281933
4638905093654571450690086448344018042836339051357815
7273973334537284263372174065775771079830517555721036
7959769018899584941301959995730179012401939086813565
8553966194137179448763207986880037160730322054742357
2266896801882123424391885984168972277652194032493227
3147936692340048489760590379580946960417542796137825
5378122394764614783292697654516229028170110043784603
8756544151739433960048915318817576650500951697402415
6447712936566142539493688842305174001299205568542898
5389794266995677702708914651373689220610441548166215
6804219838476730871787590279209175900695273456682026
5133731115180001814341209626016586298210766635233617
7400783778342370915264406305407180784335806107296110
5550020415131696373046849213356837265400307509829089
3646120478911147530370498939528334578240828173864413
2271000296831194020332345642082647327623383029463937
8998375836554559919340866235090967961134004867027123
1765266637107787251118603540375544874186935197336566
2177235922939677646325156202348757011379571209623772
3431370212031004965152111976013176419408203437348512
8526029133349151250831198028501778557107253731491392
1570910513096505988599993156086365547740355189816673
3535880048214665099741433761182777723351910741217572
8415925808725913150746060256349037772633739144613770
3802131834744730111303267029691733504770163210661622
7830027269283365584011791419447808748253360714403296
2522857750098085996090409363126356213281620714534061
0422411208301000858726425211226248014264751942618432
5853386753874054743491072710049754281159466017136122
5904401589916002298278017960351940800465135347526987
7760952783998436808690898919783969353217998013913544
2552717910225397010810632143048511378291498511381969
1430434975001899806816444121232733283071928243624067
3319655469267785119315277511344646890550424811336143

4984604849051258345683266441528489713972376040328212
6602535166939140820499473204860216277597917712347510
9750240307893575993771509502175169355582707253391189
2334070223832077585802137174778378778391015234132098
4894234596136923404979982793041444631627072147961174
5697571968123929191374098292580556195520743424329598
2898980529233366415419256367380689494201471241340525
0722040617943552525552250087487900865683145428351677
5054229480327478304405643858159195266675828292970522
6127628711040134801787224801789684052407924360582742
4674430767216452703134513541676496689012747868010102
9513386269864974821211862904033769156857624069929637
2493097201628707200189835423690364149270236961938547
3724803298550451120891928798298744678641291594175316
7560253343531062674525450711418148323988060729714023
4725520713490798398982355268723950909365667878992383
7125789762487559904432288953883773173489411227570714
1095979004791930104674075041143538178246463079598955
5638991884773781341347070246747362112048986226991888
5174562517325193413520381158633501239130544419100736
2844756751416105041097350585276204448919097890198431
5485280533985777844313933883994310444465669244550885
9463140817512203313906815965925105468580131338381521
7641821043342978882611963044311138879625874609022613
0900849975430395771243230616906262919403921439740270
8947776637024881554993224588259790206312574369109463
9325280624164247686849545532493801763937161563684785
9823715902385421265840615367228607131702674740131145
2610637653833903159219434698176053583803106128878520
5154693363924108846763200956708971836749057816308515
8138161966882222047570437590614338040725853862083565
1769984267745231958241826836982701602374149383634966
2935157685406139734274647089968561817016055110488097
1554859118617189668025973541705423985135560018720335
0790609464212711439931960465274240508822253597734815
1913543857125325854049394601086579379805862014336607
8825219717809025817370870916460452727977153509910340

7364250203863867182205228796944583876529479510486607
1739022932745542678566977686593992341683412227466301
5062155320502655341460995249356050854921756549134830
9589065361756938176374736441833789742297007035452066
6317092960759198962773242309025239744386101426309868
7733913882518684316501027964911497737582888913450341
1488659486702154921010843280807834280894172980089832
9753694064496990312539986391958160146899522088066228
5408414864274786281975546629278814621607171381880180
8405720847158689068369193933818642784545379567192723
9797236465166759201105799566396259853551276355876814
0213409829016296873429850792471846056874828331381259
1619624761569028759010727331032991406238646083333786
3825792630239159000355760903247728133888733917809696
6601469615031754226751125993315529674213336300222964
9064809345820081810618021002276645804002782133367585
7301901137175467276305904435313131903609248909724642
7928455549913490005180295707082919052556781889913899
6251386623193800536113462242946102489540724048571232
5662888893172211643294781619055486805494344103409068
0716088028227959686950133643814268252170472870863010
1373011552368614169083756757476372397631857570381094
4339056456446852418302814810799837691851212720193504
4041804604721626939445788377090105974693219720558114
0787759897720720096893822493032368305158626572811146
3799698313751793762321511125234973430524062210524423
4353732905655163406669506165892878218707756794176080
7129737813351871179316500331555238224877306534441794
5341539520242444970341012087407218810938826816751204
2299404948179449472732894770111574139441228455521828
4249222406587526891722727806071167540469730080370396
1878779669488255561467438439257011582954666135867867
1897661297311267200072971553613027503556167817765442
2874421147298816148027052438068176535732755786025058
4708401320883793281600876908130049249147368251703538
2219619039014999523495387105997351143478292339499187
9366086923013755963685323738067035911442432685615121

0940425958263930167801712866923928323105765885171402
0211196957064799814031505633045141564414623163763809
9044028162569175764891425697141635984393174332702378
1233693804301289262637538266779503416933432360750024
8175741808750388475094939454896209740485442635637164
9959499209808842947903636662975260032438563529458447
2894454716620929749549661687741412088213047702281611
6456044007236351581149729739218966737382647204722642
2212420165601502849713063327958143025160136948255670
1478093579088965713492615816134690180696508955631012
1218491805847922720691871696316330044858020102860657
8585912699746376617414639341595695395542033146280265
1895116793807457331575984608617370268786760294367778
0500244673391332431669880354073232388281847501051641
3311895370364884226902704780527424906034920829547550
5400345716018407257453693814553117535421072655783561
5499874447480427323457880061873149341566046352979779
4550753593047956872093167245365472083816858556060438
0197703076424608348987610134570939487700294617579206
1952549255757109038525171488525265671045349813419803
3906415298763436954202560802776144219143189213939088
3454313176968510184010384447234894886952098194353190
6506555354617335814045544837884752526253949665869992
0584176527801253410338964698186424300341467913806190
2805960785488801078970551694621522877309010446746249
7979992627120951684779568482583341402266477210843362
4375937416105367340419547389641978954253350363018614
0095153476696147625565187382329246854735693580289601
1536791787303553159378363082248615177770541577576561
7593585120166929431111388635821596676188303261041646
5171484697938542262168716140012237821377977413126897
7266712992025922017408770076956283473932201088159356
2862819285635718933849588506038531581797606794798408
7836097596014973342057270460352179060564760328556927
6273495182203236144112584182426247712012035776388895
9743182328278713146080535335744942976217967890345681
6988955351850447832561638070947695169908624710001974

8809205009521943632378719764870339223811540363475488
6268459561597551937654101150140670012269274743938885
8994385973024541480106123590803627458528849356325158
5384383242493252666087588908318700709100237377106576
9850564339288543376583425967506537150053335144899082
9388773735205145933304962653141514138612443793588507
0944688045486975358170212908490787347806814366323322
8194158273456713564431715379678180581958524648400840
3290998194378171817730231700398973305049538735611626
1023999433259780126893432605584710278764901070923443
8846340117355568659035852449193701810416262085042992
5869743581709813389404593447193749387762423240985283
2762266604942385129709453245586252103600829286649724
1749191419889661295580767709795947953060131191590117
7394310420904907942444886851308684449370590902600612
0649425744710353547657859242708130410618546219881830
0906345881870387558562749115873754210646679513464875
8677154383801852134828191581246259933516019893559516
7968932852205824799421034512715877163345222995418839
6804488355297533612868372259353900792016669413390911
6875880398882886921600237325736158820716351627133281
0518187602104852180675526648673908900907195138058626
7351243122156916379022773287054108420378415256832887
1804698795251307326634027851905941733892035854039567
7035611329354482585628287610610698229721420961993509
3313121711878910787668720445488760894101747986471378
824621539559333327556200943958043453791978228059039
5959927436913793778664940964048777841748336432684026
2829324062600819080818043909145563519368560630450891
4228964521998779884934747772913279726602765840166789
0136490508741142126861969862044126965282981087045479
8615595453380212011556469799767857389201862435993267
7768945406050821883822790983362716712449002676117849
8264377033002081844590009717235204331994708242098771
5144497510170556430295428218196700092025156158441742
0593365814813490269311151709387226002645863056132560
5792560927332265579346280805683443921373688405650434

30739657406101777937014142461549307074136080544210029
5600095663588977899267630517718781943706761498217564
1865901161608654086353915130392013168057690341725964
5369235080641744656235152392905040947995318407486215
1210561833854566176652606393713658802521666223576132
2019417013726649660732520107719479312652827633024138
0516490717456596485374835466919452358031530196916048
0994606814904037819829732360930087135760798621425422
0964190043679054790499300783724215819545354183711293
6865843055384271762803527912882112930835157565659994
4741788438381565148434229858704245592434693295232821
8035083337262837918302165918361815542171574484657784
2013432998259456688455826617197901218084948033244878
7258183774805522268151011371745368417870280274452442
9054745182346749195641885512444213377835214238659799
2598820328708510933838682990657199461490629025742768
6038850511032638544540419184958866538545040571323629
6810691468148478696591668618427567984600418687622980
5556296304595322792305161672159196867584952363529893
5788507746081537321454642984792310511676357749494622
9525694976603594739624309953433104049942096778838270
0271447849406903707324910644415169605325656058677875
7417472110827435774315194060757983563629143326397812
2189462874477981198072256467146640548501310096567863
1488009030374933887536418316513498254669467331611812
3364854397649325026179549357204305402182974871251107
4040116114058999110930624923128131163405492625713567
2181862893278613883371802853505650359195274140086951
0926167541476792668032109237467087213606278332922386
4136195941213392780361182763241060047409711110481400
0362334271451448333464167546635469973149475664342365
9493496845884551524150756376605086632827424794136062
8760412906449138285194564026431532258586240431418386
6959063324506300039221319264762596269151090445769530
1444054618037857503036686212462278639752746667870121
0033929848733750144756003221006223580293437749550320
37012738468163061026570300872275462966796880890587

2767636106622572235222973920644309352432722810085997

7680533336983284156138692363443358553868471111430498
2483989918031654586382893537991305352228334301379533
7295401625762322808113849949187614414132293376710656
3492528814528239506209022357876684650116660097382753
6604054469416534222390521083145858470355293522199282
7276057482126606529138553034554974455147034493948686
3429459658431024190785923680224560763936784166270518
5551787029040735573046206396924533077957822459497104
2018804300018388142900817303945050734278701312446686
0092778581811040911511729374873627887874907465285565
4347488868310641100510230208751077689187815256227352
5155037953244485778727761700196485370355516765520911
9339343762866284619844026295252183678522367475108809
7815070989784130862458815226609635514018744958369269
1779904712072649490573726428600521140358123107600669
9518536124862746756375896225299116496066876508261734
1784847893372950567390078786179253514406210453662506
4046372881569823231750059626108092195521115085930295
5654967538862612972339914628358476048627627027309739
2020014322487075823373549152460856082103288829741839
0647886992327369136004883743661522351705843770554521
0815513361262142911815615301758882573594892507108879
2621286413924433093837973338678061317952373152667738
2085802470143352700924380326695174211950767088432634
6442749127558907746863582162166042741315170212458586
0562336314931646469139465624974717419583542186077487
1105733845843368993964591374060338215935224359475162
6239188685307822821763983237306180204246560477527943
1047961897242995330297924974816840528937910449470045
9086499187272734541350810198388186467360939257193051
1968645601855782450218231065889437986522432050677379
9661969554724405859224179530068204517953700434724517
6289356677050849021310773662575169733552746230294303
1203596260953423574397249659211010657817826108745318
8748031874308235736991951563409571627009924449297491
0548985151965866474014822510633536794973714251022934
1882585117371994499115097583746130105505064197721531

9293548753711916302620303285886585284801935092258757
7559742527658401172134232364808402714335636754204637
5182555252494432965704386138786590196573880286840189 4
0876728167141370336617326501205786539157807030887142
6151907500149257611292767519309672845397116021360630
3090542243966320674323582797889332324405779199278484
6333397777376559018705748068286783479656241461028995
0848739969297075043275302997287229732793444298864641
2725348160603779707298299173029296308695801996312413
3049393504933254123550710544611825911411164545347103
2988104784406778013807713146540009938630648126661433
0858206811395838319169545558259426895769841428893743
4670841079463189325391069639557807060212459748982935
6461356078898347241997947856436204209461341238761319
8865352358312996862268948608408456655606876954501274
4866314050547353517468730098063227804689122468214608
0672762770840240226615548502400895289165711761743902
0337584877842911289623247059191874691042005848326140
6773337510271956539946971625172483122306339193287079
8380074848572651612343493327335666447335855643023528
0883924348278760886164943289399166399210488307847777
0480457284914563033532650700295889062659154985094079
7276756712979501009822947622896189159144152003228387
8773485130979081019129267227103778898053964156362364
1691549857684083984688616843754070651210390625061281
0766379904790887967477806973847317047525344215639038
7201238806323688037017949308954900776331523063548374
2568166533616066419800301882871237674818983302468363
7148830925928337590227894258806008728603885916884973
0693948020511221766359138251524278670094406942355120
2015683777788518246700256517085092496237477268136942
8435006293881442998790530105621737545918267997321773
5029368928065210025396268807498092643458011655715886
7004435039765053234782873273688408635400027406767838
2196352222653929093980736739136408289872201777674716
8118195856133721583119054682936083236976113450281757
8302029348459829250008956826302712632958662921476531

4223335179309338795135709534637718368409244442209631
9331295620305575517340067973740614162107923633423805
6468500920371671526425563718538895714164197723874226
1059666739699717316816941543509528319355641770566862
2215217991151355639707143312893657553844648326201206
4243380169558626985610224606460693307938478588143674
0700059976970364901927332882613532936311240365069865
2160638987250267238087403396744397830258296894256896
7418643361349794752455262914265228424192430833881035
8005378702399954217211368655027534136221169314069466
9513186928102574795985605145005021715913317751609957
8655519818861932112821107094422872404424811534060558
9595835581523201218460582056359269930347885113206862
6627588771446035996656108430725696500563064489187599
4665967728471715395736121081808415472731426617489331
3417463266235422207260014601270120693463952056444554
3291662986660783089068118790090815295063626782075614
3888157813511346953663038784120923469428687308393204
3233387277549680521030282154432472338884521534372725
0128589747691460808314404125868181540049187772287869
8018534545370065266556491709154295227567092222174741
1206272065662298980603289167206874365494824610869736
7225547404812889242471854323605753411672850757552057
1311566979545848873987422281358879858407831350605482
9055148278529489112190538319562422871948475940785939
8047901094194070671764439032730712135887385049993638
8382055016834027774960702768448802819122206368886368
1104356952930065219552826152699127163727738841899328
7130563464688227398288763198645709836308917786487086
6761854856800476725526754147428510281458074031529921
9781455775684368111018531749816701642664788409026268
2824448258027532094549915104518517716546311804904567
9857132575281179136562781581112888165622858760308759
7496384943527567661216895926148503078536204527450775
2950631012480341804584059432926079854435620093708091
8215239203717906781219922804960697382387433126267303
0679594396095495718957721791559730058869364684557667

6092450906088202212235719254536715191834872587423919
4108904441159599327600445065562064611646556654875942
4736925233695599303035509581762617623184956190649483
9673002037763874369343999829430209147073618947932692
7624451865602395590537051289781634554233201149759948
9627842432748378803270141867695262118097500640514975
5889650293004867605208010491537885413909424531691719
9876289412772211294645682948602814931815602496778879
4981377721622935943781100444806079767242927624951078
4153446429150842764520002042769470698041775832209097
0202916573472515829046309103590378429775726517208772
4474095226716630600546971638794317119687348468873818
6656751279298575016363411314627530499019135646823804
3299706957701507893377286580357127909137674208056554
9362464641260024379684543777339026472512819416320076
8487362517640659675406936217588793078559164787772747
3927200291034294956244766130820072925073452917076422
6621047673037863169954237455117456522022783324096803
5246676631908610112067458562873174135111622920788651
3294124481547162818207987716834634132236223411778823
1027659825109358892359162055108763298087993165172528
9380012378174348968321515905624933473702068322321001
1863739577056747386710217321237522432524162635803437
6253606808669163571594551527817803921774322823436633
7728111863905118930759016666507429527583840085446354
1931719053136365972490515840910658220181473479902235
9067138146905116051922301269482316113417439944714833
0408624842691395023367134124251238640266572581309439
6762193965540738652422989787978219863791829970955792
4747320303239116410445906907977862315518349593035305
9237898175158914576504080251094791234217584828418819
5013854616568030175503558005494489488487135160537559
3402345748979516602442338321406030095937105588457052
5157042662846003544028236787685509826781617655203757
9565548167789603892749835560879154117774942357340076
4161093294003899982199267257086957326068774974224802
0233075251876502559684207606932299885875798988964607

4438178817008154889522651672283404527721910699141576
4639485231126794730865803195076455197675628957428881
7968120900263871452578583152776151090886317402436956
8056787301523542780479341426649522383370711751126537
5503942372098784668049139473446530714079622597287130
5030772587148755705025825734668666138023514260561161
9740554343654869800544487929597028759035225840978268
3598666446586045694241390729095266249932902973440568
1606838057266260572770884070734714960600645614540707
3443278251408747427550672230484535700609221439000299
2981608211717047917614505191008132670375214930740567
8533111060583529127810073917499491978451129159136811
0739405517520801963053935074024850955377250036705466
5162330430425087442324262404632115078997336929985407
0416562610419767002024150948924118560924096376044296
1200236459070644977062720791901923596480704892363697
9860198283087284228564752353162882791324295524814447
5055219096720460806895451817122049303218537406272474
2151974030576904360268636078079200477623242955182947
3522027244376339027721392087767065716241639751785859
2544269234285352743288563368507896519620725194165560
6187037055021846284543425785038300009537451829295844
0464918838685793483961151297160581665745096703677495
8366666931218817636796449436171304160372430506584851
3174926405585519401800518090847521186822461697614924
3238319486434415908558011073070311201502243416073157
9295287529368358203970033891121141706852193665897894
5950315438958901530382714300192958907414994359289408
3097077078362875914484037045038618966975811201852319
2318686599680385838123703291562075788359487809416882
0553160512819015264759280757495815456422134145937816
7056992868299895611982353837157880480478704584175394
6654976901732203108900703033629117673084484503721456
6964440146954517385743415781015861878383927855260939
9130570255575559060947051498093487773320072797573038
2459894668096808222213484858738229992817940908256652
0958165547247524456674369759447468637633242890426977

6106791933910983300422310293728298798903209391092682
8363061736101738781236798986451493117024371282858826
3048629888449220741564060714705913740552466575697187
0217355287245439427714809179364437650637861861324348
6357974112585208634599278036887924983543632984576876
5016506511534500869572123950754478568317363155715352
7046524235259737513408825461609661440746675514226836
0319598010721524635510691718713357316854856312808578
3443562367095965094994696882066118511808603420282133
1801249410991502601435450017432730793625113070298250
4994179942844511464793291545995559095878076216366685
9179106543596606525352532027365072598912125568684280
2077246487722010996631829559552903393312284364864475
9735608598407609472983895424339326231532399189818522
6418083129633354635687482886346561850481063228880559
6737844562000941465603499280879405115310057587129552
5719641115068503407737106043803712595755969859493620
5847751202635494734753474818926225419035267161442928
4899857536740692165271630086060654373736823556588626
4863436891532180955722044567771373683104580755845296
1283283260631962972852796667436297480082131862792186
9044284342630735760703999669430789508147269730253817
3756949227517953543261569120405948328609499923664122
8788122641914850485632807206641855705952037503032291
6894489427578306090910852410601400683274205583969773
8231507349961087587637042555649640868550719422563449
6673243065625925047458176273328181601701969816654242
6378763601453035946538450325476674999737340835665138
1860251565202836373891710165454148826744480091057041
8616262683797112088614135727961109908829297022969212
8180978798951391504270936786444983196420134566833908
7759430064424856230121246145116979219396344095080832
2928129427043659914648274998437594211302041829730841
7178813090379558545603247170819195302771465794555475
5447542844344081393889086097760178573893075186619065
0501807716500184074432585402418436050111824299070232
3417243674525365349594799063334540754371812699399833

7192184854187359798453489345922685150681826624900780
2933501265882497422624188535252663670282766249934982
9488748331061764208429016923052899608978604130065109
0281798050405871076711790411302174827966823530019602
2025318557678984331758680637835996879160153892222023
6575765581586611409199394861599209159917553341783033
3476431316350127053906970793265678124159064342847213
6023521823674121473312449994433415591527431593168747
7882533155092770336202901222597794809855392200064527
1622808553982789065842334475528212765176505726632676
9114107503484587189699643487577513847914818363510062
1466818585096348887081456976722020167991199462417776
6889079171368659459607264685388107787830021613682766
9702622345941873747673353799888440342704680304255169
4127158739320398444374604547816113056625176412759821
1819396611018505628805559425660600323121161809946221
2930100247091334715068226843045868030090424286168202
5562140946087900065191099495570815816505828983340739
4660844575657806366902728434620185873282529247965052
8668140850353851983752363745192562279549029055790703
0283950104854835929834542814487304358047053315081510
5030015214281171753936491331661726212354055278633080
0208317705563029496359420165433309409417719632623411
9387105161570101798053551679370860291366756986097124
1203685838129576953077981413657001747613569669861460
6849143969957383763169582460251334210807262171360194
3018087209888551415024163818325975259593165531865833
1171268579415272066122184226614118251546574848783126
1034783454674925830872998544742120644509523324505087
7431496166555251797168020991720026409374921907569936
8963302813916472089635817717355558485927065245048625
1641954055080134351032338981337830249770182275490638
1499964723334079613041469739476372650869273347108415
6856084309213162404346298639208416600559045985064912
4350526476606760034444161818640367008377411410109432
0588955598658670077863671896944089622321374034113597
1991331359465536854466923676525890121084137774324821

9181274784789228726489297003237187345615798159983483
9100412601050746964599430331978810634913923812490503
0614334079183280040639070986725961970983112659601474
7372533052685371774214655400587392462372761736490519
8713368067723952570781360686683261395014329509474851
5947246675272016843165866088075127685847555411843811
6901162200555211348448896066825922743131900796301158
7084670117654935393046563356225311244727796669005831
1906161019726630739705425314398184573794494867801346
1821787593907699960202908396567728784690573640156401
5047696448993947541474608339918696889271156942345492
6512466455077925540281050376220359675305586018564920
5606287909076945333920880884947782889485112215474323
0191383245562993881020614490266876010207753210915684
9778307408596498579671526170100394754945399176987913
2354655010640735581699940975624814996744327842920276
2644189793918158394562708173301582160225519659898769
3761640198612074667550488611108557267645070526224461
3022233585207227362048505728923881588493875453522918
6399714380884061757286220950122506515863104258884134
3554319737298562177530720226294755524830444453404348
8887858117034134534252235431940787797284676018158322
7097745180929342193189815812482832658950040704855206
0998937839003419141630446391638805496587865013750463
4169565515661829887863070584230696766025405302481147
1007899784211830489010464056896539702885595530925558
6360521589573751140895649058441567749371058596480143
1587461449125054925319116465382158519737009328019453
0320572628452658046046337816631429933076646646530760
5905489628887241897160602258826175775399220551315093
7720062486308556282049357575272499556708922163423398
3602565328731029194007041176919220850015116735670101
9589710017970195781208929109694177543699043682025630
2405482262540190569650771058157424072149633956036527
0283334407305750073674562260584649886115101689612181
1190584717144610687197610174565873737967406971374232
3875383903031720020020720592848878512391174647167374

3737923283881966201687622191346233893762599527025672
1386221124589802121305014072889043003225355040958668
1872413936993819306914874471718664618311194260316166
4070377316487001864799600243044003242241809402278533
3090115098808706782688353172007675225531380088187804
3169019007280483179928741412547612308960683309582837
7667688287578688683092976001011974533898331952588619
6301329170943858166153741717944963191771543125069598
5348128568461937766989427745917091880252001274990555
9407289696594793331672243621567896776966708035229039
0184857308062756708676586271047694092035655930253527
4341896592700222704923318682999156093641375700498853
7304596396152734629396974951748062696451793018719986
7885375814159757993148066085572325683743052827641756
7005028804048942989958094810353483393414492788592526
2192415547231997143385086637320926632728243514933640
7045896838523456247443611752567669877675972234392063
5750747155291810276261401299248042288399029787992541
8517499129630283990729635588579890593317795908769073
9056460256235335672215522594688382984528829229662751
3716242217295467867071584092418408414755758253938524
0963302051349704740695399567897981727860920462286839
7357798151118681526598846069497589654813146511503926
2637774951376155724819511611987725034456471073851343
5927355538712462375598193813214238441581929070046389
7716838872079163617414324970791096581627464297170728
7172514274589835689709553462682016908535610894489840
7100581920302176945120771774588795519510473384184739
9807963067678858451675757299043069715426423834980098
7086993367091210839445350624592243231234827854966037
4657188014892937945147870540607924575900601219622123
9287200172155886663457349714095337211516559857579417
2441988902616701610161155783431502546032878119842402
7484608510722406676778760855247617773833089502610064
3883505502054563243461678594519417956698749685152448
8384751361818066710831616556420936927052061189851729
2617141714434655508706306063551012949400309759167799

1584260491971209543227026784326542965724032720887143
2199964531320258710967716512854966996255269860731176
3718207498827399770601991362093083230736838206455732
5637659829125781314922242204279712414416299512659456
3979275938038380478262316042432539913285112303224703
7561942321733047854078576244013291717992979240783390
7157579814268168646553829468473992058886316559349198
6789696284044734496802407709283137640810335225524271
7404107673565424441004483347440101726441052954787296
3458986405012036080244511903509949744939736171815752
7709378020923666813584163626831926340671418279742134
2546220705415600050959674045616840451771747952790353
2549325891204833857465900967817304160005210889346107
6875400424197780308288518120017336955912713771419501
1361304409753279190504891583246399143483531648681548
5791786329351239255525102111827885736960602769313014
6966143344964230211438248370563353279385889526767207
6688971274435815632088106650149568143558796576909857
7659027687074536592763649755534496173080781609871032
4801379513617036776345759497568620801399637455176242
5147780628722265971455482906769295713643572152674468
9878894188207512922257565091435528288746141950978624
2752788157156640076372103780319404309584427254926998
7169234331890022141503113998765260688761566740210197
2017196023908610829749276395695411530322754601738707
9562599357978530244347671639959146231793123998998692
8437975702492369551587297683854005227651495614447105
9719628898881571094151717015181147435136438540051162
4620213117480079198374970010047136343252328157891135
5450453371905275068229156185003328469567926226208190
4424733403625038927920715859600393631533688427243753
6679969864793474113319832861944146065392278409990314
3840354565047056789552024827176011874335643690243503
0856313095590552503904927316133117349225846446090245
3507919018441129932169977045183285358648042855682220
8737213616490586303256368913084103760215679927020005
3223554398046531193397754590440450785680213984650096

9342954731026924994758646605809166998416068464608729
3943808274308285817479694172872990311013192675573897
9840913642534796949434803777033646349584768629825901
0347072786121862300198660798778268424593383563891957
0206853521603211635230064988744600200170413056985365
1546687520238593751832803728511432748116996836928492
2044738057063349661871124094783591586962685864358914
1359854253577688774932743634514754488640868818030369
6524317556883002058607732569597160864854158344684324
8996307701137134467515693024488548207712413355773230
6949458067267845235943631507872728157901573070033178
7968544362795257190236232746142628687327380094977411
2285623766321490465329407202619753907174042225953924
2888164559796570030957141389106936845036268231053986
7437532400527015347458933256795149418545378088270634
5729596216908538353537038141811557381637820903256151
9869745357646412125498076005156141707298046994813593
4831505681166427932193352798227147157673401860887215
1879966935025270075755609971988286306428544812827513
9280694702750148163289727314347348528529504604883271
6739789815636788047804436021090073207273697493446304
9973144257156043313369038761810094887312071348271081
5889857483265854207510077953118326861708037070935927
6149367825308583404823510036321663789574262025503501
1686154340737950451648289675569835893552202017367954
8075781909502697981271148703431190363112246128295303
8205128704309294719745946908210256347889954317715243
7969621128122450342606639926885213307919637027778044
8857920573046990800923440186638113252097123096476059
9899479257598510081730396068222199753273016065826285
2758257669507854726034938298133582528178670608512656
0022688717811253597829337347791412736284188656175920
8328794474109697038798547369840254580632948350223593
9354358748022398976091629625011047393116944910066690
7230634693130169711820632535269244043840093724284428
2097093648569094689200873717532525570305435398287278
1230113980809386701547488580344563187131960267854879

3893316205007675264112044390237583342724298699654786
3685341028488573702547255023656634186809190383886707
8790720840361940216467012153483797815183282647257862
8815207101081499558980338118961569441756761340717046
5385121709021237778843336496518721199054075818773943
9752836414395304424591390317881300418879188711455314
8267469987055587931040240388884083850687341625071657
2741851349520849636709555424504394839480459791562282
8248378793415272036226336956180555637107681488889361
9275742659935823559431530887933052767558747512365065
8439694756042971920023198680243517199378681003611023
1256836425607959741057415362829718004649774857371837
8639037039015397374911654685499716453941611216417610
7171454017651905650525206622778831290457196932059902
4137539598386198260320549583950167555250964413711822
2561496014003023035407899209698677507867200038074267
9705303071679322960156486228085184033523501706085895
1291222324611783025316362894394607365277133651163164
6446199099021224922412315168992767855863736315526002
5034884878132330019101893996167027314169996265119457
4263676196500243473717272902846220979839487106598227
0009954918877696188505432653211802219444282228425152
5561411874340180419461413945147128725275923912559644
3735683397289633126767823491035633296129471910151571
4311579549093390326141191865475237624721531102079369
1158487422058227473432017355850771224379698579654915
8062795027409771688611480761631516185530685669245717
1769220443668433127398933794111629722451699985468562
2157024175947117699529165502116855001089857619346394
5590882627077531146577522388463435193765397349848024
5497607602440308084489010683878697261237097835782451
6680117148598367940552904619826216566917202742628548
2393396001825459940925430816969103297841123402288560
0190549342750223185294712829609693976813734197704278
1213001473286776057194059699792755124617184349569856
4171287248118346542064231871455182415286763056751311
6267717735061751124546338799426529127010578995671805

7214365579183506917779307040757329043974949958224106
2381051491765023850418273009662017175094059080540895
7283755406355152219965820757351315707592361539863945
9211155864000988097552610538382568992721584785041746
0651615113378833609760121148487005560165812492470682
5684427204547289630942030665044529864622359422600855
4991589149953606498428034579492757009497959450602378
7750194706246323949549578230822830668408188025210766
3907423097372091628533717680621644693543231791785530
5833171420847988630340846572642693955700268576057539
3478885870946005827232305191081175142349126873365859
6079989173292891589600181509181633740080603547520005
1511751029012299248709615459280262060761698272181029
1673155489294237408519674330791660784990557821019357
1366243599088361385980851615641747694605478554008195
3530670803089697630452946868233210532878237438944115
6851762717116363094014799096494563545929501307390036
2682100732637008235615069126964318335171625439030469
8989314261544263595113634660573786549512445747526216
7895470362890483048499680403772251343193737344123661
8586944588064018584073147633792940386340435919419872
3552630156546080518686760680431608451284591604244132
6987912538560299159967278766195195053176488313469325
7366894644382558139108486209663742674579831301222343
8725831244220330945714575414704792938758582389977385
1521352372389559664312235643262628601147489086817159
2810668727084008203377186921535235269263472268090825
9898898400262081521782826112293131182086600709968603
6540981832680755824776706950410997586143624355216194
5353029200254667367996485043373133495208210751199258
9266389956475698587079018561237915788643744690378715
0950011255021003884531192365296559946190047484662064
2347942329670060529003709175578188708193522146871427
2352776325598980869487211138459800141238421638278244
1273654244674883338167971620112886191415401936712909
47899026466664431560983729615019686242282506723061667
2094354657142514930864248877859868275958874906507726

0250951829536765181182368616944724360783764294762469
2263194989219646440683169287661615060508138463194151
1620257790786307180123115945860389656252655422334623
4454507394788690268159497513116885143694521021688319
0446168629763325229863851818850049286935727647668238
5556463655449640063176482855757858666102285515648599
0882095868944436254698679523822686115969910056366082
9267915337538160661122478695313261585318717638859893
7792918890299879387981000369730784895927062541048485
9315854323395683104239029907026344379787569185543408
9764407601308444819786265079476440830134942435834281
8859152592934714363175337495897010728735012707889804
8163504567666769320755305184043244610074032167647183
6083708475065126930707660849825299000317850305853682
1395127350386382460564251033777558098646433980171862
0814266307417259222600051109134268107467012901430165
4101064933212283790827515001003530015654597508323772
9654396973820477416265710657408216499606262274961879
5334790706598897487177956433406484174564574790692517
0149499810095353413548908754836327579522407206986291
0246717035792514417667038866099069857262605812408253
3622521899200041897574576531512300006444571593170177
1688635483333051921582055946117357716321132233931965
3203861990051161781713340010705766526899197081692022
1946470432379535641186606392055860903445706415179778
2145054722278852987210197858846070047420028468873795
8442289499743336562718779917211379161644925413297156
5287952953263975953853592095013863338050756136953089
9547584883024261962758985941513780515805025767540401
7857958524488311721050892770892272734319738238846873
0716823024878868858551010807352278140537140652075810
7270848167263977098731455162646911423286103036932984
3303003236761627142640675878067318839715150027981633
7477907877503830798675940459107392103458740421961703
4925808189907205961291586420202885734009114955238865
1079113714953346397639881839488045300750747403722809
3682053543049495194833283347007516197900868728543996

2981575605891637624723069162871111376760864803237524
5966493041175394613646433780467116505550467067183622
1285795048067165630427626711429999113487698447050370
6379001810968886297217579517324338027806174704963020
4249291661917188624335559928209324391944571188632155
6320161654247055375938696624656334121541014032286990
9301591328858088312412428828763738727428380385907102
9274863335150309044532805259779565892055456243429798
2794134891756382400771612173324736428540160610044337
6414572207859217155914010378320201321338330963807789
0409572381055882939279637438166068683519505927701951
5361601722158904287856784820682919441698718192862730
8270444163039625471305328438833791337476873582612211
6258360272896162455904189677024745382758396652299371
2351630489833012421417455788591594256059792427721819
9085562798486056174536844789237969079755945551546468
5316302446232567403489584546225674485820204245739199
4253094264224504202689038150152683602412559807597523
6481628093048912746151196231546114008220563967806585
3540766868822754265038122599916207601708955674744652
4234452017661650325945665912966786324621379919222961
4586714224824928806476803210864779941004100600339067
9275237362546027742960073478803835668752200348245769
4908456862696057715701919174892260635208129738797443
8354832861369395624503929768057832234021716765559177
6684037572348440946176293128849268993687138983882227
1060279037990019045583360079739277410926655739233147
0259092338906543884223513241153880185592349561399302
2391964505045036935292701156630515335191864186482344
2499919272027295345959906304872360804159576002966812
1116831723660381105428035914457202482564561057140554
6242082134352094810841715828957244507206354681600230
5120140848054358742526171017681853883557558717415424
7754497722214192613155252691091755633319323222432185
2542218272914915981058368970250352281300214119248601
4248068079753699647771939490680468355280834732761030
6049409733091690316783097934636611832784531868716460

2680738833656704566010423768505801395074436479639222
8411269794513477300492498786496563679490992913271252
8977651918175427962806084932375520815361113240339713
1655043918879601983821385850007732424617788491875814
5964264233788979333081948816004011312652563569324465
9398400636890315254722923991414474377069633893576192
6039189247936317800831026114195485436051577871600495
5788656579706658855104288246636305720777890226677704
2512681571979533225107638903681976284402861025880539
2339329474672024088541276492386447602161162620824212
9916603622991849237822363009834781195229138218473263
4228575912097980547828525059183798336801787411242644
7460022562414980691400740979721023278539575615128345
8061654111179267104279905793944971349463289504565128
6884784187175802050458328387485313736911351025506201
0277534580943910500102183397324565047288947687929892
5945019875076712236379187586472012149660611512804870
9648863056228440839369443872169212084920085155838125
1070741955187208093746942459731172811721051928903896
3703942357768621276682109318276366498404212493814409
7959863114225436483965499983479084307021764385554351
2574368282281530322223808347679511135570148063182004
5322072379489186357214910624252699399467101536684623
4105153338142684770627585203524099207972086991453730
1095516415033176282001969164115460268207236692552751
4184299699205398534330730680573723805041671972211273
7405078927266340638850686734458560773266648384578027
7189114758013231055198784133652185190714606813898688
6710314759826461129379543952667286727599483359025974
4587868768496462683484434414135917714587766088077845
3571839329371937393236408356337576688468211117993505
5410208556188490102016005056395416874510822060355541
0817666460524124966224422804545243216032036019464135
6097920019590240497929236732989245539901019801121402
9086869992057589177718807414612220502472858571536753
0747814389730571787268366360157613610077228631963885
2646235125538077319459563567965382362499926551804330

7963596211067455285214290262949826567553352731004687
8865731047246649332656792733134512295505918623293739
3326086077451350775309015744438294873397796053228493
5830136183795862648032129736847481751647691366211036
0369509106666505171711508278200932788358722598394046
3068376318118089044236262199881236826807857952621972
1668720174551747262781803268305854880397097704793483
1035439855907843552776676033139884605271503138856332
4676889271045958519328951391678238577357726581004798
2563935519352005520408002870596782497393747886052835
6493591497838037796496000521244583477900175604246586
6651998077028839438516380955043049219603244360903400
8517466042962743097683871519459826447359402342482110
4475729111777958773134155360952759570898612586771456
2523994500759380206093550248920084767332293085742222
5502064556902391265436635785242724290560532057540308
2101451238209021746697579765347517250146583747884808
0537735150422224042957603613754324861996558919392205
0469998210629316096756517907513229607778575533102658
5842576086686764535520927748275567545177169950878941
1805936305249944967012375980065534998739666395394417
0170596981015127193331184076792327185395398097640485
2784674387231643291002906549530861283330266400758012
9618499207022002555972156957588376168784364346792755
8635739722535648841330601192895746428093578580811323
3143311528748217976603971257952890036407198923328131
6116404169377366280132597382222374268189176489596422
7033803905929596496964821331144731667650419767811084
9096646942571706945700787126401448652242846948897617
2567465352205061621073001019262483146821203551699501
5220073163840041320303332423121670826854689317584366
3043078435078592810447849266395265239871864417338008
5681692321347429754583269402161253332837900960648627
7854941266795136740458774169455961407626566250299006
9226726787603658713793279604184883939339346926354341
5480951836233233175229370352102914641331275203711716
6754872063473892329378510729029514462927415467619479

4274716691603049782928896147458702649979707920638724
0825023006425544995904011974108535167844409018806462
9374835443961440035352331030404117845722890295818058
1032123743825898702747370401068377771592512645357065
0830092147925834989247512745362200610585457599736931
3529707814374284134055195444672148941505745283917160
3715453082525558343202512542416624457524562964457910
7697171521470951850550035505439063168825810578507463
5656204791466768055698438455202770996971988980723371
4869563567031776877637897432734928293439051455670607
4460797047693164627812141713818274378561462197088087
0210642110573778514713588373773882407652804519142713
7488110559744718310093937519765980210024101251123081
3682603384744910877161322857660263938849284959898236
5657272042635720263748256494949126291419171306462805
9566982549360326132019252804346170439028926027993140
4361370265820121312851488158573111782104131033572888
7181729526271120008147506402683046418988769747879173
1737038139991888242416994212152776045185956711909418
0737347933109970928315546816563952710104611376254066
4495861838546389822089967783295501114314995936803982
2230371363295742321735744647342109741491743641994731
9588400526387269592318364232549184559550453437784670
9470450959420120211422086419127904935994521373924871
1074323149511380429379365543637217263481907571135312
7093079527295221124795314989699080894665747695565124
3605611420086639905609900038030250612423607750329341
3472890501316772809713162683495963409292243031195084
8788671035335200237127302029165929752526570392104214
9634952385708560572343462157695698513406830454833154
5907536471146996824209102321431171769227738534770417
7940764410013010485960927072113205231853822274448702
4332710398781147912754608083611568779215131131045008
3663631007517511025900280864277150209627136623974010
7528844546833161821150278926430729763557610551124620
3324800531059951115054314848295534329598305742724517
3788652719300073232173623758732731489091094553740270

4811855571990516839387453520679708592118964078548950
4109405699659887159886336207795504521932156336124685
3031747054439402941829263552401554523160986825531389
7018801539704596250169179664812501555932311482673005
6338357972603286017784741496004569725783495620587328
7301245145557634523029864814954410090788352980120701
2654109525184606662017674204525736799469077190845378
7482060802904825167017661982073061833123921935356900
4070521549893903446593880904750772416954365185807506
6490459443188862978723571603022481352204601090635214
5082806397492755128476943549962033991644887919743790
2095718886320024750207910237907307296374632633667459
4275563784535691367345524014897125909480368566282321
0050039400731066320752572831471151926332892852069672
3934717509829526021254947643301953574383509258283111
3391153906337661737307723630279889869985799450165923
7690675488379889294006051628261400481504694828140330
8391643424865093963545890913280595111633455036563482
4519150583179498083182728134795050772717335949663371
8821491928378711646390356692577994245739435547304493
5559396848032790208614196815082606481092468854338332
9866390745478052636291615627988031878282707451630327
8639075665336219750632242486457694597535966732006038
9826293000076125149479800895671245256955982758548576
9012463686594942242277271771518496417510715984163572
0724122437196806720392706478942789421712842641334271
1831847944133460647243141150155098551171241466824331
2352062840657226926069047479196447297528322749569819
6327787281625954012020538073295825004974459308097824
0952991296542331849879880077168163198608651208831586
7256506594414061844683749631892913745993421603484822
8831582897309421614736892558516992715531155888876007
2170341024458744020844342827300467309795555666811501
3003388895838023146431382900260076322850347583078087
8895180313981020762788985174353478225120846759497430
0244378958428956807526632036276962994601808349419949
1270655913084000586265639963911040685104128200715324

6256426371456355757694528492711263557719632506589654
5536482125459263355257292595281499341587877651569223
1191510233734407169916564763982000896984629843997759
3853981121332181032819896994579261764935829748373387
7523528594640351382382306269453634581003193672502069
8280738433341175283157314342639896416347127053034775
6991558003118159180911378802688385475769729233988828
6032302997704306662886955301210272705763395989768941
0249968479498168420119925613480756440406559462383708
7236888125489491487948734808614168105521140018455170
0844448429484755073273664282722206336582401745498808
2913018839140156809050000849546573730003274779720991
7507461785951579953202237285235920400742515225638616
6756203188398117618611960221628474319079702503674592
8280467817853664739356003540382782818457669478233745
7113822121932616729501042706940952026502805228985909
3500239449087456262053452217311940957783019536051850
3854961406218253061820365182733706211198939024488975
3863581809944918157848783365288654365422483020278924
1704968965110417275947501781226785814391748649424357
3009091712648771605959209744581146295542231002200851
2052258976477811482703942677666427827462593951174380
7198618722265586504030028469146927864680031836034638
1726405702707422620342971875558099386871240465622333
8914646583055430131550952851097263005080518826527268
5335372937338569182693717167730316118647494810424215
1279159101460656979533313377409593674932644146370242
7524539335030130992833648540706984034399121245249275
5802997988240920664464042585966200888741916498773027
54037292042158109378147131362226288666694547421244955
2849091492193371936234029433712557556998865296623645
0353519202677763794248208286056893623152152317885014
5213132149146986854835944706865850109813142058926764
1611516210940535678073681008973424587293270521085357
2676380564228840929665884477795279546710735193295474
7130150792208403282322044289446782183965471109021173
4072513972475735700855531274321999675125958256806323

5880883884366203262266191414934740436498000247398332
0924118386674296092694607014183881781107142824396577
9638843986478231371542498947258304114514952687242361
8996763058816820846327437441210390552765218710735564
5257133601145580455856845586504328599176765196193271
1434986654077745145004730727117147957122275720181288
6446440777517460328242317338533765298981044232240467
7246320479517980971576025800885768975134059480548268
7728847762938464549604027037050853941909276993706680
4551719416040376351180185513657545109524703460226002
0741742823849481782254906365992084749037583205744677
9591067556606407750093471298170058187694080279926904
6059498721176341519148822518670439557310017937100046
6572921803728487979715692278888397041982545657064289
0898582795862565990137596875007856985342094439959715
2366767355991155709006141301885395600693305082611578
8315979018829128777653969640675392080848582290475561
9051863754905941764720809084852392996636537774687098
5680142361370763704674236180292186795924769777652926
2929041798392750534329433844765333985012282836279851
5026374542796671771484197573390657287154305432157523
5449320534653754238204844850884634590853386677292538
5204449844131368637518941176848626136036819373635133
9325408068522692147430732913446762529322640845330844
9386471515618139413634350364817794755097633925598827
8690369632386330342579445292292377520328744890200405
3266813935475285501746453171721459950814556136469252
6650227115337381817597855795041988075485811336289154
9009039080607754157573613737559880187573075362487370
0129122382611343810392343723135368988915337494937863
2498494176428141704528408296939917243232867725641504
8376577311449335215538523001781108276163630370902052
5950377909253411047057004656525197792567933141088663
2640592623178893126031528575871642421190333798725775
8742901290375936269727234314893572572418837941862768
6456677586869202760143980501638714352047767388090057
8928363381779738845734410014996643323582225792535171

1059485607891824015219982852269465095876314924712795
2016446764740270468954543510306982617999140223407285
4891546806842095743207506621154487626644675798636443
8802325863608869187594422715214296506641613849638150
2797217307126592057826600278471814003420926569307030
9044570245964675764901852781393148131509203641049845
9690602253144748229457070252704363040611144551422276
6936650125425237207439401827752508941432915215170599
7454593125946821214351062276330331850433948895127672
0637291512493681935703191046935729052762887687825004
8505480059732307532652277925524199131596179115220694
1968547918734156699781096702562993993208164507174173
4905643398652199866390557093521198524390679861502144
8623928438739820187602285471230394945966157258750965
0320071247665759381372124801134153550616754720369579
1055974610671125417117453695430147191419937319722797
1690211613572625243116472289366644142621243854981362
3694963571282116036854416071082317751078012983042538
1419089224920859536461082139564811320531607370777207
6055993498150342406407751233151215899924629749784547
4385785595227089267102479199199645043040166005621762
9623401492821816115205046438140512010176327979026932
7122270125927081630457940869593885030885857777676988
0577120277461858372818585997017721116037109827393241
4719793766386484316000841579272530611640850151500165
2030020014274337639041878862263527470225898484946907
7694747613276391052599405660382382371636943555470658
1748273071824741827263627240462399440284444736424586
4447510469029976526749734435698570853905781915995859
9609675061283091019474886565075126139713632927641583
4913042083009508511004140745574437849278985760726105
7697418196336967907551883832201734437643980536829626
8732851893953081597213840998753657746635493253113936
2559789543000911914267407538592549690157973419183710
4016999179009456783596285732244714790732045696471978
6315490862841233325174812784828809848761022100974278
3475164627905539385196688956965108760628729574590889

2017023867207401060245389415195473932814246622312689
2362650272056402643021776903189555520611271146314671
7038915773390065452869232720808111578757374991035324
4466936165351752212468866080593973805468948675560258
8706871030811898922024217495293458219535300991561355
3607315909567346990699248742680019538217524621053498
6270106132159075726024080430082786835629319838427105
2198354727511764233027995892687273053118355805687527
6124091974244476335680956874844410454670283523651415
2765627008043630974774537678098208734980384982599248
8106702977549495352282995165465598506874283176285208
5719613937978285057790149962321392204623415241682380
3889446624267373001896543376476503634125182850951208
8864856294714398779566559280749164896256218592671541
4692176768396054500821642162605610642314443579823069
1965780470574714846007296818237228797756049608915817
8686729363237902415792047283646970210313975180097841
5985500070553649387532125749616748758725832599259576
1507433918622843798830134604454088081780968549119454
1193470268965059919860410997653211196581062966550051
1618365170620292880877609149846167316442686419708923
0648463056754573887202476016525776085293772109335844
5387107402729259191524626762353817978693064215340131
6337011357356351110981418211296622107367262696156726
7483077524887444841676657370240048508393702558385910
1226694835806839154547916601645691486305239359779324
4672558867174160485503871149031760755373219447283058
2219155807880752453696932744601747360524205864696869
7577061218677619720587491045165142715495423853920232
5269751234954654630906132946005665072830987280338737
3515537522356318357025370064940926380803173746348540
3611466000484687624231089472379165007451797052486284
6727663375517303687368385644037049806617909200831710
7882104981833155261485053735407503510822393924744563
0109692042278844737169688950911185736926890336659718
5225377703296220167081065518126758009408525150684775
7921913893213809286961195312209050380181076587488368

3178827814252786261879667606821977039093260067296151
2755712527864370698983544440961391737903545485180403
9733313748052358791095558304048153480453918785403824
3236907304310274062641777762657301034703384021129669
0848180461624964873947345844121553025815222149945822
2499419419547256410317502114422808652302802213424093
1939327276781959906081125986239673394589896190716797
7778025951163147757626402858826251481582164399441350
6196081175890461951158539082613354960388032371352224
5169681180597512189590028591797390866524495280407827
1302700453774372678555325048503974637573946460984085
6589301848223416149865831503466082186223605801948114
5549035154742662660612950268784097547798140726823956
9314724876098280345081189383404096153431486301124867
6465315478758454946522227531877356089083504383708112
0882441759938586466309397048117253004020305813409044
7450511563770541035014166861912485252694933482978510
1811147232987404539612754022221909584405087230662326
8888497042234567000119497518597964940991489713853622
7945887407609904328542281277305818304024945108706336
9869468674008948109753971009084947683041071152955063
8887652490545659994260773886347394552511448972036104
7937572544723966023547748127494160698351013147640236
4194914610598055637570446515566712365256828270157445
2847602207817539723371640969862649205576687615644577
4464466492547734672972555705388285907892317597067686
3982496629455560193873152710362720124293120176425224
6448031819544683337639946131383614457041608883422253
7155878358070161156027177541424723331527813566940098
9800444582389984200640748958923892389275228914732945
5312404247755208380523795101239384358587754549990012
7206828665999857909842930384600732962384262907972182
3337274766946401526920488143042273943883838698807236
5034008809524512726001361525704157749789546427459286
6962164154275190720789657656762047087629102592988877
1283405806131718206887950962735523080228036658853093
0270461940061446449186278566424494208162102038327611

1696224421386397311571301189918531699151581650258342
8128487414927536050735501492751649655689498688144578
2807241540090116176936589862811374592790322578489093
3976881608670857002995345721579420980997220532145751
4271541122093988698745628011653320792545519698519103
8428157268351201092367995242906867995456830838859301
3667218521135364172442283704920603648154449717799886
1873906197012650668437064042512445995190900622608217
9845415139874086156189246593084402747014710167254716
0166860173976919976620111199893015535406281778132823
8679873988318548093651417526904050273992326953229393
1036045698425205947108776022321016774679279356253076
8337722069298099521332754934107640682936962565380979
8299221502007619065671332333071917531109537696743144
5827047452191856565617305618532166042594645538561688
3759934532767382788781222315372811134173554517073553
2082760440774525442307854537481125966546355745960432
7036854215738696224447960925936750083098914000685383
6358817787486427106882578787407992834182519771408422
3048949791551798767827468475408492899386476349839175
3924459329312913808073876500505220066666272734384454
0498968011834325534999762501192176787558098067233241
6782617825708911630179808819558379107540118050962160
1093080422570180549297646784115387691430708824753121
7231379403723659287710434554469626659999262339332986
4113710012680408116027696940228713650729810644525201
6551733860468650406212924578927147227426763861426823
6764085164119476626514371013938556806427007782965968
0486077517949221215629173867163546498898538357515324
9743158354139913221365051551384109030902755433236441
2022530077042821114714191814757096183313782294342072
5434103155582818669328386683660726916383696779320102
1420290468133704915343805924654711497083540122724100
6503949742164188669227447368995062528945027771898946
9132963467585879264235211633546474686426054856131577
8403611431490269544275056480384788879432956556048443
3918406020270451468278242315140650702210485195920723

1200493371767383523709308856526434484194677345382413
2968854306302477825543502819595717543326873583172827
9337741010263471725258000551089980879204274477838536
4274972065430922479605721400330661597939815697061366
0983964055202876699917225472402063960609642994542705
9154600073536731549880773908300158133516035730111114
1092801541228066667058785550927033385009831156762851
6164924255092928303908770988934946072349028658560205
4220670371568046350038260527637108239865979318483093
6764165636079070660523343411137793121612020588095146
1437739476835388395047212945283498654808648378850194
6767694562326701998713318455453483736084512767180056
7875423588719510589565279780453783448465046814695167
7538136951845103083239037496571621433079638601544816
1449552393511121218944302382695405786011646737366479
5652065872508159275305713134383569920048999618043254
9502052195550206179277993056424583665872167535192817
5033449923918332562361626502081490355786124405183440
4038159913582717384337340452974499964059918656664153
5612424308001626179337509214296580882832219570578431
7169794628455133096838246000369899618059298795066037
6071243272559753650882038636095880904003800176047507
8669744332587723215438325998399864395011449541507700
9728226536958394380850912841104162909663701274249881
7616344101667423400506836167648232710388942239482025
3086967222925243407506026512988576358781375008510056
8868743282747187323242898477335425815041625895502385
4489068496767648928297072811584351167607761726048913
5585109814789508429849836055936593710532020599790443
6973534016628764532063718869382189780157321907629981
0361256838764838726985360129448160731761865806680596
8373389411982650087326242669600240908832076226117839
9915744021058427898450630360141993392836245540276835
0998972042185962090201621015651922358421194882020912
3783927557185605541656205455347196978661235058348962
8212860820840349731199881077259045458633766108505095
8238503075128425964285974947159675425924034955860979

6434019664667217572372370707851846466383706717029954
1698329886912472818768027381254962938987607223408465
7095098943201654876047933946794685134373263039223093
3179068730316994180074048000687251365978579585994780
1994965234272868898871781351617155057783915871386404
0578956591823213708140058713808836523047167127182200
6018608811257260339862403542067521276908921081552260
3293004441018906372365919571195303028824858684782564
8830052518126081035421351812247158400462751059244487
0583709540835318975215236103420408450764137674234730
0588220343231604746330435062814232108294872409025947
6441189103223374049794740857827762204826182195142821
7981124372676625846895195106998673740227323002602615
0597064215274602326999497006158235928282229783286840
1997290365378168160028841173067332449662838403243536
5041397536205509105219749095799860595726941384024267
5559674863774293085831406648031844531532908153215494
3458288044293735568005276670180009478873358860913649
4945838526892791365594342881741864555941029617929958
1260809706454746509023426184034501081240335390006107
3469412097838671627721613708361451511050077201170421
4057510295511491370255453350206814116524476917845869
4354034118791350719472868333896624761011830170049726
1895611839898160539092008911727724528273299586808380
1073781314001876067250126926454645097673374700236767
8201352356732426247888048234362900999633010976573057
1074508621321877968280743439896483552427144875730583
0321802494521092319912041786298321106456189823450495
0543971618030395685126538014922516948784795547241863
8278627582327821299397820742867554710924982182446861
4795808140835500466875596261579061717590219271869723
7845472411298557573179374795351829558429913369281405
8848042157153807468531130233549462721418440056323974
4587537727518071466016570650353750000780005476100367
8636991113239858621322182246246434350103632239859670
1728992842523411315434326293039073595342914413933874
2821872148418613127907162685826684720595466403565113

3279272928367042153333781564897878724347231657710811
8905881159220534134477675212977463550655110980181145
4708921701244106349239492424226738349439407865465836
3868597002601991541683855861557896701272200232200316
8619541970289247574216667668015248082402211115619098
2909528829342278406490395339672008649956965447075211
8461343409778577773642631658691698762749541886831332
4751453159002335440951714914081359273191146192006775
7921585633107612547070933961164415088007272939456368
4925327185891551688147209601141540566400389210281186
4854595041190055800792839471619967600301877000729916
613487810389918979927793308260333383340579193386012
5992663543506471009126063462523857434635268474929790
6578001728766596825621946854107798742184455047104825
1138993654279944593202443898985134425672669327861329
5048517020426704168104239887877662828350193125454951
0108703766963812060312761799621889318777830520450194
8120474270520457321254873390393028668085392898551453
9518307016773725339156792769039073362485903433514761
1787051779766471010750245076816165572539548200948091
1058631732989175311841603640219503463573219594755860
0832082926751237884955166725064922072060974120312931
3574353745218554549830258041565179862278016468937481
7239713381123695363735811057393910536917973929343197
7518803252435258608082755374099972101540080046979927
9434223454476897075803131490654997645727199699628033
2692090891558381760321398926448802376910082742090668
0800437399250454122368497194097746704673167378878520
4941656447370713254372831395409623181337647384889412
1827756876058275472115348406411192866091980614228229
5524907588525871140721341401635238119989127477891313
9757468280934247282311021898430070244399964290644450
8447880276686539463578359786330143574307385522480118
0578551630030594803517023052917619376680448974551900
6229814174022546879385980914228583744941429466840567
8447862996873037366863397510139100798455883197189398
4042058517831262556099075164256666091448576606836793

7448065297240370993339629283434833266104136871344725
9629441715366168325692987460751934900436754871245012
5173882289594264322061718377059516656649038896234159
0342836592467623892154316210947396500986925708950750
4114157819718945799485168292399767685260590940847692
5555603209473017988926182294738346886884787742147478
2112462900504876162420975722951786073395988696418605
3995691274261105379964864827288214729865447937270511
4310364153995043024924890389871904738048121737057256
6371346514715413122205631956995297107448454232578540
9319607037480624328873057403741431323821583556267142
7568755755136182019176330108628379725855115674172305
0471906087361627708326296442958048279756363082376436
1615455540616980045819644670667810243347845988069248
4772748952982620451694370037112019129535311291971380
1759557797453217970689981078697996711614064725835573
1385280378144794618645821634745203985589751231713640
7974683851455920414500521772122914466992786476520100
3653978899709419567795422900041438454871434885285565
1763080299251676444247682186490621512191723425686851
6006058597808966236688320128396531227030746548182119
9948225388143004016811445036211672024446204828296777
6160165637897576349795548725510809105781339420347277
4484748769898419218280856304164926029917623036263225
0441829629652154385628760703742186814004738630945015
9109132542103032561351107575582873478656260809325645
0743463372334224085585816338537153069458782692020523
9506727247536900139801149643165945829716486863220484
1795219642449832794880631346462010891393287053134556
1503788769211459272685051467713559958906322386507647
7828269016803601306170856982886336353398216641166133
5548040370382100344583808150558303401797120822493909
5038566095855713953746347628324042175193426566863925
5917743378325548207038610563301262376287698173472822
4250946153189070215082050421810397748940765721499083
2478528545951002467959739308411062725225415696493892
3682735814346077275980334626431259827888944181849173

8026870449603886707186477083156478758911780354308201
3186565820343540734229283474557696514986839150397614
1261336078948099755916482490625516855367948247405098
4649608568188917203699873757964398001165295270277237
2260193575557202326310147686928476263628518930484926
9092640985472493648181412831689383283125795662135988
3554452066740895840923148625755911051962200050308020
4257370028996601241363556488028033999569465609588576
3219926030004685397559802876555831710706399750666047
6148677763563226116127152242671096736184025291082552
4461538857766602779608089830283706877813984923812545
1717898757790676916513246031087551814796001216762016
855436138875351114464644596594898628685003842938167
7597961912729990459134396042836227821457438491080662
6737203981596833114583132775573719396476213947036948
7134483796533672088650760949443106748938628101668608
0935487620406295314268367901622324344216250096191988
6528250184780750093092989616878935144048527844852101
9497293149122933664283836109583591179266973210503286
5863719619130649857332086615243198917751756133072533
6906062894401403624673579168612419076797307215389609
9260914778003921829096605678051574245394812705158278
6560861766280887675485282643534579297510910374324314
8049050997201340093871209967992266732745697219975739
7498352955663444532434557032626027829313689388962967
6914900511179164157396415162234596241438799849972397
2106259104524266556282960145967901286176415352478643
3047855814962571113956032515036318374506194258790732
9747990654033781293234354964770959941597021691810368
1473383333064151387713221517339840938174656833323752
1245212042635149480179573706485748255881296241114146
4692661774781738601561556967768080635428081339262222
6805735860439573916273877143508484770186626531697488
8647386824309419601892875891202138727709615384880950
6565320734420589849785682144810993443271437941292340
7297547932647618296204036144364112746524043691754283
5856614059594332610091323144864164204976494795520171

7108651706981224160848217072171016494824798077491801
6666318076045716395251838609582718327208657052982558
9266492312740506731234877203497799829560941063603051
6581681903848011147030423901820457583727316520859225
3994751093890012112219426665445908677926913711549507
8966657667654609628827777519957055450729792366620852
3507816894340032047543740400762179909188135109499396
6943134279859921580629270421382675621435340592467202
3502064258541096859551282959888016794748534882762322
6089882142602796694948833997353809115310261572752606
1516646757472311267311304563021016442756282782191487
9246698975320978326529216825843304790854783365426975
8433077955719520001012078724019881349498443843676382
7041174210036951169011180168326999466120100860532094
1579019288976139784035165111599346420444148276820545
5063418483061619799460270489648952438970258434171773
1903153309321479805420208961951250759293649016278147
4077322477257322019135045680559997856927754305465787
9842859468408586784134114538241240720656755982648262
5761903033834174251848538540384703710069087650808535
0864021762101015672829143567367711035116439783634404
2830234780735456691438177047450894587211787839154166
5309247269795195268639282330037168506787620787754817
8391081973218290478799329139607887417683308186531819
9940659792678221322713459632471409529463076197396749
9846349363609758067253661551807859814534953582160148
0260233176252015063663993913514287751153532124112251
5057065723115208537650284322101584061898257004704391
7186490724120891714561202491730043799349994206586637
9857873460604806192281194643315629256867108796971234
9623640619373881121802073791598180109759080113272257
8430025011137880349579204391899288300516242921760033
7641079337196813319206758299182607848524757117752420
1683493481941400539164639352182737104891500365804792
5976158343651355349438431915092146293081995018359167
0942530265403298032496761584396347114353247143703922
1486178438282611386688552159846134450580330263691439

41743559917537871666881400452968934351987652723008 45
84655015656598952113011048528816939415686706351783 19
22185595530500029864832544477477719955016508265889 67
13964088988056795806691606580609404851392801022276 97
61561382608319076033245484652866146494294839667733 00
80707320067510426251414296244714536875097068785066 00
59394026518778610327654702806325729906196897591887 38
66723051101237949329259764957482625519592739447176 40
09255618521185772443088894589313045709752725867071 45
56514236034181989031519545721886211491710345305965 78
45082618680743649773583175770086475879964322744548 95
00780966711961621513676950853089233612386662834811 02
93980460743553427272442810490328075676700337727112 09
49128434487450813568822156033050438835175410814830 37
53443420841220816836058132623457677542793161986045 43
05044485105558004116794337671320558147058727208825 36
04731064967931847963735278844788520587318286600656 33
49325602359088898353777250797020050541440210559461 07
20764924409136337227897399466397512341178836631250 90
06141623227657028541048506797449812718146764308414 10
30023752565373049527672754845459997871633253310506 19
02402151814681001465126285103975983941288236986211 31
83152477649679577744191332394798552871653023199869 80
23983984731981788171333103443398908379580000513196 53
45233833901090970444714347942650262857403151815203 54
65072823118385198658029362135224379754319380198343 29
14312502757766754316869888602865677013500372589696 44
58686834176473878390665444218192358577310787002319 17
44542871416003026828372404946436034787690357332618 81
14310108132188552798589730345053440330372276915140 45
31823618783217199889890550829089662419765855980578 34
14287373064809852907821459411264949921965113612567 77
30769946058020640723918086690020175695641759552721 13
59337589791160475982315587253564456825714374658566 88
98203737054970452907158469737635558706092801201769 78
05329357967838079502279220010520167689887325410838 93
19250907174288810810708623207551018480041769696826 29

0392399839381162366384787130819320185559267865898070
9850229537394942175424696253547043954732413392476485
2103761177731123138500161871304710647783932487585006
3619911967787753268071392468984403882659360510854652
3692226192724034991210383162262972411440835568680499
8074604837135925207539017014469373916409286486391905
3757393294556536775435632948953085479197356181168943
4694434436430308714442549106098294828815811595635629
9337794739220978511040672166448032053106709130375948
8434578734398473707653747404793080903438244339705830
5326958562998479383048081779750890193239788196447472
8134854864856399736790769039302521285919509594533031
3797518529818662620117612609532139263391827182563275
8305911893721069157764383887227842285290091226125140
8052315081202726247737066716153729796236517171183091
8171522805265375933737558128234864296932266784713386
9598876915809508115049936337356905900842892007054825
2546176895416471077801175860714328662404483055236425
9377579855244869608072673059076502488514081476189179
9989629290795406069165098627507033091008866119931836
5347811068950055323212323104099431566975712843211058
9272907562665298306834612688174350276344573487313081
2787853966825948045024450899453850626222815657206656
2590807106009071947415806434289617313151570605581139
9896076568427723954812062465492792246644108673930170
5267840652247504105360432350868815254382188405781522
9519878956064995606982745328922732703853758452092709
2429466734689593377789658067695128590449057399130794
8762539798998946853448670842763284764409804653488551
2094360642889373837105351559587950751036819995860092
4794052205154880777749983061313790264128273715757106
1281736249783647450207227756195212674327358168549611
9698882583112616695052224021881146693062574953847086
9958657459988789278684738719864383790480463746222816
1268712763451130947831661759970759508533257460284937
4001043645034556580449442950345318338129078508883338
5837869771084982066510206279570766983344517793452718

0376911410207557477431542932903262953211497882620351
5987412546422884395277795499289564754347105898585159
0055084900569690369399463805412744078272079588120610
9501826667505282910042864401159690915602602458721174
5604551094076846979736827481459790404552190484180115
4566347833534380881534140372398178819077576306472338
3684807661718788752544407318658305011864756320301713
9833900789875424411026277749259455787263151608748702
5048062038162606284156754299711008457236079436838831
7756971160717747601977362998608470922561241903344303
8680607516077836502789166628360931767596955301493681
2797935466652393898654922082126132776378982029467995
8162439870593623917051175070504939244293712287520721
0047900369520353054174702688100313142753117445624640
7354520013033515441611612845306363822062031827141203
4710573330570609561041999774412943789723336195293680
7116294649741746746061619442841957715064212449115406
7073122138420641412696715456643891594717779694935195
8346843367832214130374310734291743734376354415907377
3807768335545452204160755832450014127128997410149470
5488647243415899129609228298624074551605891496310210
0059587134719209797239836831280110175264318686111835
1701735867540649265791513740582162972420188375102977
2027692280780173235365852486610373552463605197417587
1823490873819774519960413516046880860827255906104828
2225757674635819166290343907054759703480904400430433
3317413461423454127456779872589232409091510873059202
4279001496737015134772151425714802387818972789099319
2321188518040030497628938731198868763397705690319074
1451762975055829507905515712897726034354672225187519
4722777503478062988815802764088305858873211408993562
5654452632562629304285439933282550329502893699077705
4902947079622008390293221444112657382089568543447852
2535584373126933754793765994306991005699082156031450
8198864943894886795977365202377638052694955587145427
0658517474445964682352694105685193373700491448623760
6597957437429493137628495423747696298423620404069903

2232862548282233542016522829128844342157512706020215
3831784521856484115066939436436446339032946286921500
1200331737223159459937024404665464401070954637786273
6676904564259977586034142337627592585363126437089730
7579552699685031320690918306791326542030640031482455
9862392657597573177591286253089465412516622840716337
0149790267384325301619010137297886469540342569455726
3052203876294232648064996238163085500312651680544788
5568199731089679575544268392204851309190268824033771
2017786398604639800256037206069295346015367351300935
1664904759969041534844228406494643578396273959796970
1199959968970550071398026714315391239146116135818340
6808760534667255305042239792809656622109111184778965
0335190031281930814047064787403671555521140340703039
8907223233915942351265297171121449159128746969645445
5709228043473384101385887428050725149320183676549865
4426190687675030397956939024213437475259202844493707
0321982409508528743929412781595864754303669533654646
5043812295538569601870814630360006810223193535677588
4221706627177875228953937497394498460688195899260579
0426632428181883297682570878308901643540546417536779
7521401491698161349930449104204274172990731837969851
3124559586063991996596689961080050494007296397098959
5175746349501131523954054363842477157673057968997809
3511231012700060683156013470561688420818621059065843
8546853522653099408095506864518196431045500569852864
0369722726449640722091072805065651759005363319425718
8261901685209110944462304938727622601300966509801815
0216116189314991755448664845101939640892424535185862
9668535880723702520862903963751354424084167679610625
4077453543971872022038982925881504881746263214401932
4591263846776453878214900321873605288401615814676934
0972434249669596797455129521524754130038382417596775
5422515486890349846758461066315941988121179713345250
9275307014085614263503015271473708797902296635683001
7879902880841938939224918844891176700803803875888780
1697701113453349115348021065850875700255173632568820

0975955274871225357182551697875315095569086898546483
7948430351870614923135734029631365279127615262306104
3140923956535229749326101802357414494002010757529248
8958929324580351889348336232266211070472227951789643
1135332221513311128130269965705654236666071242736067
5833767835191012510994437030462907634661496449559967
3032125852284006812886320601384391535239320911579060
4734139362973234927591808942336565260939483133648102
9064358631183082596587859784715023449078747678799566
7424852051040103999757103940220630691734742020896991
7560052888987367593966293674017209821954183371282339
3286247731719643862566531455126992223677667770819864
3496799841526045194640459016395789604927913929104349
0275683817268404705229814089067131491526250441754527
1011235786801298936828493391396383366078142291794554
3449168097066491318845378120257962155232128839888294
8309602541545158301556456231328450317418576979979910
7895565679960825291655375861223383807006921957963941
9837426117676769100507357501471041273917783596347944
1159241607409649189238641623144315098433799999574123
8608455687906501796604659040119009310649145976455087
4416917093615978054671746589930170413753904682544419
8493069773963033614330040322637044138842456485319600
0910240359148604341956718891985856156554657750890144
3177712861645686219001284594607421607429571045831400
4620124639011021019323688746231874639063905184609082
4746612222586831716989063606402540489350875060019353
8323584777977777184156692711224004455167705419310730
3883694387978890462421759004666609104016362270064506
7167256329813561916957758561333390398277599652250990
4038709327898622159549437997063064807096494177080058
1227055933091621184400463589378563243586419106154006
8204787901621404457877173981029526071730009912179711
3754243334882266618671805934535003597940626101694558
9487985287382394619259273830058657862369012719296382
6592639378195968777634491927813839152734685103171283
5011677541289696340176336880334761324250065479448355

1600242312566460801078670258603760993908004517562600
9065555413098404273574300500668774331352820617072990
3389370532225467004205889640465239361428307918540416
9666783240709559587709423200941561095553433443491385
4388408610824862428985961974125657170424006787671236
8593537227109156704060621943472602040994139547201317
5524491583459442749191929135023558344041872069743458
8605383370185897657206225466863899147406171384409111
4054244489241812528058737844437599033702714432320785
2046419314755947583142919416971906297697904498821308
0192587590485875701028049890092764667431817419312738
7987919086670564601741410451836547363921120183127264
2135299075075316741860411390850791740441726580092889
6640035085618299372472136843413149569295704019813001
6060875412795746641903179733259392402107416767024235
3517422118285715161832976814222607309029636948713308
8077855566632397283342252706565073072518903090395022
9751455450814134444281654143644104921750622706436286
1017571711204836658149705824635780075504562645374462
8052593284156788579850690105804527975626285722083047
8354366813133031723323813526470752577952330152891663
9528654318999573174578016782672814602226403818995669
3799484242109824897420088233111400134104409516093083
1309054655031595551547397748022146240676110527161375
7998628354396966578355245670079360497518767958500434
7859444483487473455259996323925882010445287895767233
3911085208137994842347153189526128187510890512135465
4969246066557673451857177405113980900507493228007094
0569206554428799289768091385328823923127426496379071
9799785249090304609585020328130118819189798753861277
0509831126796867211781006094288603341607408020448532
4414144579454721054698929166499819415997508117083997
5855253125347930707237719482373833676065541850211333
7357535711604984088630696264989015115562982779223043
3349844936783915198562685043252008447985546296912629
9788313029363064633704503315526375204047392241570728
5777998802963532890069840076781896975635402176619429

4424753725649845722655067873590934057238979378191460
8031982711392480497941100492298143175949919931032808
9795747253376814606174543313264489248037013462642669
2631734244357427051774756506755634133360059178313763
7375920389020426517169186542244841360946598636267754
3332217290972806772121812294501776642327167331091927
5233772424505908085892756556434411548443888953213270
2856054060064352403401177439426383149269420366767731
2492334460461527922287117473732068207379504077538070
2487513973697487220807941836272457926715860587568437
3825696601528845015936360579976487955466689796317276
4144582671409839302605844437311832195273578242992380
5304609792532175952943764669671785995655479752601048
1116090019255987703160369289053546421969361790098547
2078551257259753257687803418428239419301167647127802
0013244905417068719406087486425099249004124377799025
6723623875213448754686801805700267716590457174167509
3585374663278466147170222912775381648935814037542041
3163179662046260168300583584027808425084706102856321
4676349216441559656539951244111527225209885576318078
8086283744439537336386628913943990459186935629799316
9132074310849123878269670817379838027600732835237130
5703835388120192178074557053921312504821976693406394
2802345335096980545219196416496966420519223222933252
2498099068094382986082393386867739544526736321944401
5986590406652867020651004940978671302450896050636311
4749789754318632737637932022795300108771768440261821
8003859090607702934597864093296951123385326149456558
5967711754424619462086741789881434778027023789183865
4750736725070123164595421042303682530154992729230497
0650068874455290889366465514819384805637455447031971
7186422720965870630049834366571830309458525285629892
5461326079017237462336013820894764047217671021078363
5306328391852894263097083524184268806663972454351949
8745949527236882351106475938370531361494523326299006
0613544202079008007844359185142381095406246359288007
4917472326014282915066473564694914907531304041194874

6127584251725422993376561913228344136159688451230509
7922908093477806100012024307933675460695671886784758
9029167162815104791098481996879505375747610343839289
2981834755913533728342888492285393965950164594229684
9021635769846036566677068849790614937895866238978513
9503019555252071159479162430380571339104412351279771 7
4258949971813208993972409457630504381765420237749372
9292366664085826356304701889428471366217962807794758
1410647203968690057335883783238393851564367692910953
2126309530237234188776377595132558571988684156351143
4654449213461836258200177391119635659736209174802895
1071191193121616150493566140019891540677191474060450
2008489007852104489840715587249131814241237453147390
9585928549192619551275281540455554894860530439518305
5163865296351143585542678957884332247030322398462969
4037003638667059755189622821668494721551679940102372
6052761920617504560496637170762638459530053344438789
9443285436630146414207515026765299871484148385924204
6843515052858926483541295999641906383622255006162029
8517908079995517161428927433221628069635121629029650
5034545598002429203806611312249987574877781454333495
7813655800830045879054556552375964308994728294156584
6898062943112725975546930218879127310353002168642276
3366103189051108633596398607097473741955293417850780
1365337868776791151473833252513002371910235875886798
0393855297049983218303899853337353315103458044340257
3042275868260972398342231501764080332731762263196756
5989772971839422971652277619673408573444137475914779
3179333989243599405813960322813465924785587565505751
5942861161317673955283481508085185207154795143926672
8751057441386769718902087776119592459359290863839696
2005768650996293038181454912807341809720403203633667
6649944391993626414522714507350370590760938575244000
0947482297136237721592630836022139158855909461407406
9763012970865696906676244218618363552047279035331753
0936077977879847080239021859587448789607452374905628
3747418931026806280481743381300198221277706092184700

8152523271459867234378543104978390436493058607645357
5602598382816254098496399831811297188143863954265440
8280086193057291799568887988182572440923086077708626
3513136094630976774099702847276682966685429084540520
2291900303432247189820499382480607516387279456689408
4973666516228133694882858339531350502170536111817502
1010169609373728821746838926180687188721171220320673
3649831281254572692763105648258790106857520808063392
8751364817583758109695596993384841991062692241136561
1711295479677781238623934934140085470537845837828515
1232798730884093735721100458603654544945301868107329
3761108679782828034366133339780538614863594371635008
7711931195598480218289926777976940913930849268923971
6108751625981364642795191163033917961291900176095322
1655734911037471209457900418602899221551175915683036
2494716086505186322797429561365198303518971428760223
7507160599904685227028482420257009629938279147818015
4066416917925969650230616749677247841947414200924297
7297138883251166552227303389644473052704902414772756
4715409237680664165282261122166055190351049532169538
2999570174114651029984898251643905699093945986820498
5525831287644568389843421093658943105114075558384927
8936997409501389198821247991620988839243558736555452
3753623890641072663136573386887150143766765743928298
3207346131403949526170953084489658005655846309523296
3187124518366166304277498527362023490678591557693622
0534724261112532638914302528937667373926286064609916
6425839998746474341416395267773224693331340898922821
3523671609562825134923485926804073551815360719656739
5027069135357634343767443117249084553776703266698841
4491148088848013249302584381770118785666357293539878
1140646588369417283873657084337575104479912359736597
2434455742718384733620516409860393102195921211225720
3436510013963890649452967142056089086176982883163882
8382522970765818961189545729825810733945401727749783
4540687764110778004407429296639807958026689312946890
8915006186184189218530416433221694927821339211827719

0216752020805967262749463002805388777945962185683074
3842989256463024089063366760647389704968736267734714
3193464378269527837602861465838927893362323616036869
6588859944071709014385765008562370357074728812300427
7647474703779463200055437274736584724026183902508185
0203994131095039708124812210776128324595639956407303
7842282949417904179913536533706092953580412844901956
7717436326558733430284401481499075465103281813878210
9072143398374541509572180872332163931411848854048824
7613315649935403031131319714388566683380217666836082
9503236040595136775927155165679680295859733803613440
6930781375730116130026579702426559178634319436264662
3018687258796305755636607828996953634981452238866007
3014771879198641606614390805777255192448707082910976
7355499112006123175361847813176543955729503854529236
5366941334856217878926154740456150452308853118438978
3350748069944802081583047802929139502742186788691980
1765554681814454430741911022927219318644407497903172
5993977136218099711176146890001377240700923484996330
8320304297321967582789400084665235071155111834810320
1448352947448851886324133039606763958576623927274353
8647655332592611391601058972069491216041943843627692
2502040836018348118271585534392555374582362825523726
2533143596996463666782559338210091759874444027185225
1290642515745359479613052718959948884782435317225627
5413109599850442774753526388871189264977071705502201
0568232530711543895647583031225511628763968835143262
7286298152407558879959620980439465968893295190449010
5741914199814985879000053348961206916311175468253485
8290076839537662641452052739360786513805722415068971
8558277653895215135856589764073014881111337386924588
8909822760229733951244135075100387258948206647854453
9290551160492546556830179223635263775546268409049478
0037268471610264950882620693574846316439689789627008
3376303743017561945738907884810424309285523631029835
5174517454466065297670819947432059951652915960087156
5211546129673541395732775167844348486459833913758485

6250554601750792209883587719338601394896751483376380
2139034152904283626453763745808782815379047085445759
6761421037723612329619640229202289514696881144705828
9309803570015041036948417054728692275197044694697292
0349519697662176641436203210987497172990814400037157
3928189152840831974229437336774778258709218717225298
4069305556935225836786253877699037108329877979505169
1696299576070266248772561115321024522871797030937380
0563354059293108901874004957513305645755468645881925
3443722717048411076045834505257242917684423561001394
5668245660428800047407292619165754409533650005445833
3133334866581972749276001242323822517184689306940857
2324615542392813887422027661693797935663441045037115
5982367577143124912796231411501645288804409423900517
3464563780362939532778780163604410427591864202516771
1823591081249478448954880725855125745496079956918012
9713172054038424909608736362135131097528620986224262
4187424357837677291264179188013767520032056332618924
1018616513034810244006856808606110481129427409009375
2748578585868409229731577936688608619964429073615749
0533034510746792139213935734146937826784053313199235
4433034607195155205700661001169301061146455648916805
0091450055613784940454369313485910618419330189545486
3852198600808200402863322268577928659474990069360751
0578023474201503531773730083649498916728935090935421
2097520747272504366618800613349590930611407110464599
2447597175424019965407306540841697352392504156355003
9026440029227515519108629413672360002694120712781130
8763653161522443351622669190123101713629501331698077
0097670312259233409954235276489844208910924390275648
1617004072173927256602429583715571067168541418713003
5713102305447259377064542064373028381478419102186882
1846656713832612826378069753098860829066330396698468
3962467476885114506491315186155246294792481115987311
0907977115298058809209285816227706527567771953931120
3573194334593434733729518994157214722761903678004308
7959047999266424281962016300988837148845043980122446

2455602660408616133199723284978761593171268140040450
5618966869098470823941370855181326199637688970212415
2313781873313001560112199565703541410653535638452439
6556426727217434505317089708620347654758674128146407
1979228057446954068492795994590458209018731765866162
5178733129693572624876301827480530656002946241810151
4313186902408749384112421582150837307412833731003222
6683957690735069881768275484817730499539131031846532
7838386562671747600180627887580492540088784039279856
4649645155278927345020015410031319050962710809628895
2376898287265139140819451167195916663181885337312798
2279478096384186330812324934368273270884716848408230
6510680498401989966141848682192923712432262932844248
3023610179839100426990407744799190837610211112396067
2507192997931365177067316450479862322551979702992566
5315101596604596690150887068882982527286405989514250
4765564643861397139093020257194585582527139271981132
7758886955446292060520268767522136796682746877587452
8760778634491382699563480082544144131825347204948014
2126543298296784668605437790613389102060765389746783
7990904198226428291713569800434724666969930157511495
3715204374031918107954868943240622909458623262452209
6675744952856601646578736884246540265604576973290012
9583787420171151005742659749253328682586625702458378
1152182201277576078722787536254417678516818491979949
4976491000369349095508194502055638112249647967662496
5078580232712368944622866979631971539024990109911767
2053296592010243274158366462851035194005414571907148
6388246944690382245688388500782441039250163597153748
4905694524560531254037917603310165347750019986495805
8355366142071699741173310552540420552110773185810458
9461046273558070947146683528783522244243951951095964
8019339972822544123729119753352333978820050032094830
7780662833646063246671001800870666288977157613180394
4530851778599796791617562364245799131874799529518736
7560206724336078627831644655047133342557745622032970
5837065208461481461803279556572311289137915061078782

3672417063157427908602758268048328204825305959448653
5530533557360894366837877887790883577331658156656404
6333631178965577553867451359654743792882443277617766
5299775378844321226267587896126638330684384900580057
7613730946043245733141597876165553722630161642335345
1002374635368298942478242558064807664336180523774156
3140378933712699900811546084081424058692844640874238
9124577519366466994637359158441193177950085848065280
5204513861789723299109646117709762971698805474148640
4035888392795004056809668826825267833258753583516005
0579458531484837770296761832636064913660564711850804
9163591118168057356862567675748362796259542314440842
6869444178084654590010983008324701273276732518629652
8101198756674251237185471917419644610996381436922527
6487686524296433284880267104880448880155910644769829
1833644325638379834789224992424734734749255855729315
1861103453413733567227462578276718755128522961571935
0186325172175999942277944125127694916659641176453311
3076783943587557015112683397880778230893276729219673
9065650167909884959899971836201837724669791646815888
4004015083264133901702440286390700883106649068349767
6288008809713157726433416470525153647177306613927224
0563257100139729989909559374773055963634856006159849
6125351831074504282805991011356152764613718732307405
4864438709510376239129317441392679964474732361821363
3118585804069936583777606558414953328326602877854696
8943002292685310193430198737058717358218098006693891
2507662570847465950628991846834694991196205056288100
6235243400502407512125659762183568345522576684049165
2515707584146144132895209700930687290227163705638590
6105921696945735131229699292583567531883445210953757
0173563261816644245918307191732592805373518481830987
2294562621725404441898640397503845136061110062107180
8868929053885565380321231977664500797880892291390719
7183215533766071468815888614665937080218118486409491
2441578015869647372390959585803117354939639342323981
2188385832226906227304369154796477329036203102315846

2282118660828589608169094090006189644213461734468252
1433863060864107649130309630386061561226947756727056
6164198322661282955940541852670099389441814526699815
1219653967190513843135365503213138068242451734889475
9250312419248417525574038182351139026163553709368646
8847101525986682006296660433267158847028467252827367
5136369158934985721514957696957393793129333387868587
0155864384721219088131194713370873382327500056239923
7447717210347921689995870015046980589562365188542682
9398566671272305833174739467989387917984475726396699
7256515093350494496239329894118380951152202738593619
9162089315593735213193801270298481882968245692466401
5891024522408334073529472376766018719083566257343946
8354704836224454619937129219945521607705226537983475
1066676946325551156649491168070523052817308690882682
3801294125418146730584593435812734334074634710981016
9733784511370003614666147779737566767622118782539423
6037065492372256647519270025082488886040622340981154
5113334223901773684113599153372373184676634050715689
6681938103585480799073996134538888268575672435045918
9974006910447041116287865267920106161324719998448237
1523349978363752301431351328269553952901086494205818
6043961590530582597540015734752997498272309538705772
1000539641886970487452897359156879270799441658104944
2687939022278200261738842463895922113926387495411419
5943300270846714237068128137782298487438922580196067
3229557664622256072650083204373463689206974257310148
8777832814597005506211252970943955134820697067809204
5789009005563599319303007467104257017918474679952016
4509853815101539696173545527780430626775794877109799
1362593662234937064837059816841914409008692838417581
3696077026265263739842179275186558553400180249473884
2479507635936962516581598005490110797072695324488613
7434993884408366146859209020138746292972909384539568
9309155247032545649484834255843539275002683980891951
2438571472788922881800472797910594164937166417567650
9443374654097289014406328130189143863392806334434942

4002602288104716999725533939570764107067895059052416
3290221291761570720281337963064985988232242671029826
4264548227932715480458764992124132126818276723090347
5595793031158482489483014173171934310466356199829342
2660845241977274300089475751906443642507040115738131
7794809599533926915579884005478289536523956367416572
9648804634630573673711721580999020989445373325540664
9244555657047977207914581230450618880669347731611549
2135285980811109640356420103206503138783298144430856
3872065793894070562327958687446085284069806283901283
1994040317536981729101193027421648744601861963215944
6845380755709872212964758426105804371014414489107488
1337667213835454142478713666653871820712848476170700
2802307713986200152328528467498051716009417700848306
0781630740674129158570458579809143541609290613494597
0968892571056791005567529007475043799463382111921199
9009122153965563172633298735935838666500189702103768
1056553912581127425650365892142910191935677400796661
2713823071408188284186493254567005047890235799834629
6652053903452672297367971122296475763842795337070307
9415632893117466348996286910518604722726888778758797
9536548113309718525774883625499507808962383116823946
5051168547086261364021782044527622621850946877145846
6676588999479371028457027858288649455781921024708840
9805488404942892027586325135120327683691655093337575
6877423110361610668383215808025643334645427172202495
6218060593586057783683982546182236449833541991908181
7549239621687105280495142246375891201136159799843803
4553898743686379416430030513037889583127924845498683
9906586006407899335281278519409840167197297270699322
1339071842095517824752068026846361653977165123457434
0304432466147817711996108553728243091711263519501191
5381033226170096078197922946035526018787669236212486
3624885129035442839737923251389555064014239130766546
7538114524402470683765280641424872089134513796385999
4493516086771074601432747723851028474946663633461941
7230160773629762889772512830025808468772653015168202

9250873001346219923156538719904106055074193036339018
4442397874423384998260696760570205353684564642727270
3489439236648459002459794947394860416671133571702812
0922680527815688335313264331759029946538574852184710
9720477182480567215619231319966276378282067062797786
4382255808727403553887557637225829990506735915414714
9474372649839787057663311505334211612174534089654152
1554977462478886291183035260403687328220250708935308
4352345808150719569588924126052875718396496305507662
8600911167261753007281738884588123735985372692992626
4266600217297690409322916645780080286157310501383405
9960521518020233746749329410957691399996766385217537
4648850721464227683648609183197332363921592490390006
9678881210112974635837340525868785445702221462087368
5872796641453017626335415588794059073212225394670737
8265467560810746496041804339579538721133064646799286
1229485713933856329761617850891155827661197902337999
8663577047496379682239935095795450820550511189303477
9357024430352830442834702410590461224680811375399707
4287434351207241798271000829331913714192887714098986
3705462711361421706031603887715873410756266034626034
6932057574636326530612059614741096786436632812848924
6217276990604400356483137270171843261107628690706296
2876782483372521816784952087018738888352668188068856
1553821029179384688125975922387175756873776636521727
9182935988891248129048499965476445965554595153192301
9867342143962690524533746344986037179272054279681688
9295558794575534131465881283310245574792805020086669
5716939577801534143906770746884443719972294731420962
4309846450531853965219060267110060566217145056523961
6762915821410039307333892918625670333714472417092407
9448220819579349698115524925573254088088316481951994
8492518859797181791650718864975353694319576366026242
6172292425480056059572174815355934092538283243334477
7942345089465946829548015616400884023550373234965498
7866217107668010625102744723405477738722823370632442
2346571309983353563617904512966453592077279387939270

0954660146105091802732697555135713654909405170986914
3338343735386223956625316705081322123467368781442761
8547883058500581078515556788076939732421220873066182
6200908305041506079878672078008638748314710467962218
0439475755630990862442443828090707516360392136097396
7193494081982005189308463418418513775869421385970025
1923572103523597814756562837006498935806194277478376
7367165686044012425353942546083747346222496082982724
7240218753734151044388427140893032903966317059852727
4357572249198043964068936908330704606903363403761135
6692720080172060165258701662092465653183217835903483
1846684963362317735446303933793489237958382338014835
2466207076888417756468257271713619148355289440361157
9624682534709995778541481648466735735611338031920658
2213549678296294583794899259090657150858589924036877
7247095602252060304105945472235734307619920200387034
2440222349094967180951194798118123176621613281265741
8887926717804023857800559852923256156882467651635904
8340588004483845823024199841762420397502821442033237
8136469561291816090880705226927447850235794371561428
5496103309997013947721460617450078824754170067917818
8133730735538786796010124219243417398732897632280986
7622937453437289981172593008222324624375985400183726
6087383264712072554491306433644995100194782544525542
5611985444689633861923341088611902366362520061671773
4072684448767087078633992885187857488689069559520575
6080655359723625548665768065997300269614499791386394
9137643343951178186561697245750119555271398766633102
4199364961593673423336765935189951082105085455865902
4045243950149586570975168801772980081992225972528916
1528326432871330191207202622505599302105520059364272
0672067436580819591983894686241507538027516566228260
4255844872369634231527370496473601247293747058235189
4637772876085862713952359990692232587035991070927536
0771787312754815094035127013817079487040027946364336
8842771692401282640444753830021680605559739911153275
6743042507916896649365346106649030339264547982624507

5275297035511702938954939260502611673280508063619113
5041503872255354805249503072592208321299167699393857
8960521919040226593296893201528053855848832676736575
6858379942868554314884845987804319937107848408933741
9779080033863696659632700048007534107331302869582860
1359287661350885694130726895270622119444657090135000
2850781700817329693606994478080116508997746983832753
3544622311789004142445612565923619067137782218830990
1262050387138637461107547138243333426066111911249960
4311974873003557846753855809319405365641438724087159
3070280022336202403420926692484103654139246032528151
3910602580690086792469478464151377425304908113337485
9256590325210843787058369018030593385329700109696000
8742504481418458925985696534556980827237127625540048
3792707641017020870006758524443257752645790361826803
6052623878996687562686845758711349482617027872642077
4053277917839660593024680537612528783622421631814764
2047643334565869242915619461741479293032672745331987
9627590510582556439064127960599605162941055835770035
3656324285671397243309359986178485440971818172554477
7914093209591840501649984386128073788718816754788756
5056631963197673047058648946240459492697664532851091
9874433735121566448881451325097829979985682830183029
2718126587579974915259421426066384493476182366943613
0100077834745045443839405946382531641746962167963754
0394915211600835534045873007034167447688538635372417
5911919125730296287576998669830602845055125425577813
0491965735370810975388980514498281958517209632887924
9759668785855762687283638577142823352346656795892694
8548919544874241952228540275810132725725884846046549
5185162225272148589697272632895152661007419195971783
2883659455976857057726284785595448837240757916288363
1484906477914553487265755850111942264869962439109009
5948421950501182854570218987410345718389817904863646
4908296777315083677699733551507417008122025805388524
5195363983453187617781231829233845161492018946787202
1748024528159190124225586516987472590215507624974912

2673765945633076166021194348403239799144070248817343

7977350301533905977647834289074748778151734815736268

9952102264641917059020637215636487044104269833633331
3821695884831981209369368359190411493162347872763662
7592154568410702417405297925694298191498174063952714
4786905118423433719559261231919375790621178580909320
5884794836305795612156010565182075216489529364750499
7836425968788047609259997018653611113136048448104343
1372673271493251764067795912827041809284099302148095
7457866349377922712137553712549498364613219105479011
9508054816377823175531880540483447456823482955282130
6383035954647975531338603713165764078334088593594573
7671967408625251809781817880369860116613883471299915
3771123834154528740489956469028260300688542763451896
2357355461815822221407196786684310268265573811519493
7131618234925304365477918872773039577291667603598906
8298497927532644579302056250041982158178336797583245
8201670334401637519436130793606687706059615507458187
3007405885541857077712937653954611123552017737452675
3650277123601026263711408502493754519972388118497200
4852954076053757550486334985176039343403658952960860
7344805531229553356882145671180576047588419420583496
3384542165377020226288732032814262719241911346980715
3506236406048801061017613964306550666466714597747927
5127501313346596764639606994405703118605608781228063
2896781657653727506296728357626397482838464729501891
7985604824919850079916092323997667196476783301263846
5080880428311109850255466129686185565035001236106852
9743566446561984920921101266375831195462401126619489
3008382843865999999283333794876598213558839333097596
5394351687477025420380520337338231789387828254304773
6859273772357478886566858736092568691056377446851131
5594786736516484921786038920469205739213739659762934
2661799387598861055713847390158695380014400337739425
9635248692636896090870539526251096127290887376798222
4107476784882990262592141720651354432719916459983330
3338205097023670379189129777113900021796464556817013
8089418258462959876893635924393798037012604370001954
5375320947585656686261869137769323655538553373664018

1142604011712634532053725124468808392504553806425476
6280934506039108615119488314246477393594536113462632
5397903053106155154047570431835806988891168508827835
7540626047008133489427756419881104615033910819669743
6038560732670871560877665885891060896072087471582697
0169056266871992681584833517410241095060497613332210
3046832009316294819666417866410892596335403862924130
1524764041519532761824707235278977276957745431491457
2040541799531857883749810850505715767111058158521670
5522011002403121471715798464585433248907341098761099
2956496761565447188062442849337194222747404498379859
6475584133489410742608333615212077501929801512946567
2084211550763881645988966176464369760328924324510530
2982505122426807037312128083935122022554084151320294
7999805751376864933655676167849947133492695625773907
8483718248283155679298197287786862903631056068580099
0722240215387661471364480196561481124538862716542334
2887561979792048557301929997500417588186220355084352
6937422418347754235756053467255495615418988381785602
9267690850613136595288039173555602456879717723106032
1745760497595023225629419637906309379558104490958762
1359167786582953930306576530923070439867570625760671
4270638526055475959525321304780063261071076808321621
0014579464097769268006913907193727253192285262742895
7389504137685477459296033592272625266668352170703189
4962827245652845824142546063037280407774797988541294
6353979992464746913355243372318304535384890808089315
2518135768485272858917328591746450365612006882947050
3204716904153768678001929305063669577855088550542369
8901222980877912670610523562973580602220182943158073
5552190937586577473626473699288881279782933393499869
7735232413759931554636311929820706537274786072589973
1206930627210401572394384260875603932638706392902219
0308589098777220198559385372688147932288292236982590
4643093397881652299859711143887919168112556374983131
6110931906115632552892612058651598514939761270556240
8767671406059062759367897286320465589407531927159129

5117018443755758535236978206034603081114085616222042
9042890528709348719387536681994211967871603447511656
3217044041605351341390173136689463887385553138636824
3369975985970616457062670413045912643728498914835689
0455609093481011580923180730184599840879909046157493
1098614313315919784060635683188419505707596210326850
8407539511046071367743150631865568117504568429109859
3609486346869593672277580773060728837988142468100342
6858744195332034222592225911315687185512988438399771
8184817757527652868727478679975609559814433269798023
2246925174800848043735402673868444648250945683719869
6619833088985878352579323281004784980000165924072903
1466028150564724110345203157652765771714505108046030
5129759639033690487822708390133104005385149373537497
2951613489722639790211988963444866201881902957692950
4346472305784526520058067990645390049554274873960333
1115133434232393928153928575524189254275336899367076
7360327076953407153977831769329985800290247380912222
7024700301497321483099349332418808211182569586232946
5185756368975416357468959866026651728710637317821154
4073283084095822937176862803685645159152570329027569
0368571298831278118747345960741731009788473156283864
9486193104350166181226630376959372676458853838094304
9453023030268014210975502503890721484246009339875439
9153838421377545972464098687379266027941662047086632
8438766273660878272150035989277651707445477065383961
9602834310285238409133872378563979536825788370583048
9472663481348213171908883396336724123153639729520379
9561405420265235573318226053603015161076727016136677
5347202108995240601901907310716711572131531313991087
3460499485588793055573290748667569249917791477776275
2572153315305919154375764020855624311494453725459568
0970256475764244423090474070144938720093148556612673
8641899425494931363104759618933034909499307284324090
0986604296477641606362128947695172656741692210412679
1976202629175585305961605883598150943813988815546473
9539002210859787185924059647802767889239242804773232

4168011508809942907513006728614972737850416001553809
7278691011653816376029956001998756771052874341796486
3494875902284345081024845232428506194564649282883380
2467453143600766539393253169069347153411102590915595
0980999607771081924043400817401909049952241694593670
8415512633504468374235408291264653803549416953846871
9159478644821690719718827904537417589786565396354364
1749642113833239127266085382956774626422043748613750
8696560381441154467817463182415780125489762580240567
2218165190255256466551041784031399315527349701282746
4078379677343103957500116764350123239218721736939561
5725612096294658612581792259971229360156048325293246
6059000746753828911358876966050230432754644157727204
1355353431069230209904095882802842492545660922550473
6786633535977670114754779378951221639503917488370060
6920832143131056511403216591497160545033152608756244
3039751201627044475665497445082910844914275328651257
8843201433719161950742434585426712768110260079969773
2731091087404071388839859302056854770568128370032410
6099488089120372337515691677129447677010573628517526
9226738673324904110576188363433437399317405736193536
9077705806991870011038755068258651233963419298473309
6678757320329048370056903353621683728691586822484931
6458641309955612807613543158394797965036457984422529
3998032521346097286226953626724707628997177963276334
6141120704154148305304401967545816235986063466572733
0574033424675682539988557003842039565097719954100268
3762829751197071569287780588231902617101475800897373
7834649921004305707615859532250733610872957027150743
1229792031372110315120578694581824201741832056515117
5338128457998173296130009722859113082820909053314760
1196781850383675303470470005787483609975909091296303
4418276550511984294261174212501745310883761527721032
0916233088335710208577216259509925298643641820689439
6569085647751243182903018353395109114675137185342463
0585177074434321613169130454456207295577914988904854
7850294942518699230156420482367299678208543277081713

9937297136472855162369102809439490498095711147987353
2633610861544909213621071957862718265898464545958700
9069249248820523435112868738626912529335695565624401
5333447567162409478118265711559547566993684235624999
7922772333285678478624526949813038295767158836825390
3484616714968014138599194055597917917858281975784812
3724780229627342713273807017121315934540225441686146
4162064185495562201758027171741932960403072428557591
4037487524125583648684782653057902112930150460093009
7911328939110209284222126288743972398792999872217126
8024426957043640826917512394728858097663173521903477
4020783010825008230686748165992916214204378559690700
8396343174915704007049111330970230468766158574831350
8014447599285202072786040624690986245818371056631825
4920666633928689416422316813978537417455898355023981
4134762756866162211863675611345401850612301450506414
6476620025479372737016911509105700588058385528775155
3568346135550888143137449856363777369433473077922369
2023281951260198833485319308413912969210345115664615
5817184516091865304897119538011024852574989315864723
3999267453725219148787799788807562673750638723780564
6976435268613067747611615640308898107229900613620291
3855386468368424583544342072490652694313192636306455
7919103281746224652305086811453922379034699935761819
2283841178311127342660931717160547230274858700010478
6605983536876204234909356314679354437007086760444160
8093430388964169122938462935021661100210761640546614
5328261330250989929553919275962994627826326321165658
7431955173359427872479954828722781079314977711035342
5543816635050218200475598457194707642967827158772684
8362361118065924451595282915230181808971672271763496
5228375068073131741445335093301055862157197336759105
1672048856745415728163217259397927018267765927879072
6975958652444479862784876695394914610177605776036071
1075086603455755571296234540663775844877314065805021
8144414570121613889442942543012726143996039751548809
6841753887787099771053156896057795536359670078069985

6501195536169958191091853337403661999066186774586536
5937828951586192168358385372055171819669900290622524
4297196477607657921208349979814831084253380066460564
6546284410595975870105383783766951341441171157658015
2919723932831823741907241827055621142924812595008621
9348254518565539701258406477745909416107789844866798
7879836035943067050826469850650965071424287984166501
3303364759597132945835690587596970583659840237526455
9514284152743093476002848059737445115482304008577453
8194414235491878380929229783184414022384436112322168
8505624335418588432511544720643284962084563281194108
2705883189354288454365054845356330088426685693563642
8902027669230848663361182991429872638798806829980861
2394976329510463591338269125251879466945089415396493
3273454972994489836299473991754744164719717317798726
8394360240105216610149815265541625403854517795215840
0249587987974104952480047535581645441160796496743747
6718422118358157376737048968165761864668447399574573
8638952849566518957447866597778195075225888298702478
9009640653185204742376952389335501218478599660074089
6503838595147018040723457768783856075809561645339216
8848975425983059917537610132320635432534424048860000
3090822619003730634184868861438763736494178874012048
2609505127598633905097702424725298017588263922938707
9367325221116705792644140908543740148530459025037169
6374774586071914054256943815611701443788844188830915
9229271920358412987162286685053246038943565002307341
6708375186459536802527582405209237446765733512706016
0117034908068222327234121408469596673325161565758066
5902431013032064115375116874077567874060359258788617
1973634936771114265430484708113330323186633985509494
3143974804840787647767832770534880159671410169844356
6978084548780518231995756407397883177027113564392420
4452033300760976436796999004095854955620131358480587
5374947256934033090917283239418369219324915186872354
7739392127561179466401851180013807501027772171306420
4253265553611432390788203509453770750843488923010206

9364851728497612938332579316328040240236622477073584
8850558619602148189507568896146498647108584644537329
4965523337264188383262127117827240693226571570786417
5572896145338291644891865204955272952633002810498231
0985733943081602256698171115056421803074943611078136
1389682204877365185667020919787109427227650347063385
0855008421170940405082569924575628282627813751332708
0529455232216084540576543785400717990812768836695374
9752286406714615345649011269387426711403621513820477
5875494285657227853366584872908691749510102375874976
6072301695185736509057949181869154204951481895063313
6723233600179192443975940164167719835945106934272172
9348371331527082522858781476449540661682660663281738
5906468170848098019563095401910023030383772107483227
8139011682082582389277936139561206216213391578640790
4096277774306239458871168135932412443371094483087422
9948965727049696689190976787295678568374918266228075
9470730876390942917918464672898935038166571603238341
3004822149073557310114756043910764230704997141717927
2249889362511853771844565361124353668033415834710999
9781275045931072949201640040438736891084890000220658
9689495098835545433034480634690683626426926225260480
5038222965665856445463817257872024223930603167450160
5397755165542460307432569145384140667700093348172625
3378578369549688018197142075830479025045449329434408
0654706966709208196687180957451822379033311686660106
5885464616222513680755807281783990499382032540352222
1479127873573379240505817047934361116046575203509649
9203009430633851515570103965436156004250209175408368
0251075696272405400706130739148399782154975269620067
7717461253751774740807704214694980724656692103138036
5590139144631933785249560765128958847039568360052405
6037732266484889767598647222236870457260025131465330
2789490736683175428527930436416844913090148229779444
1453977670005047645453944199744253400902206497079506
5778667625625790416787951719322821604842790422281457
4555552585011050511185320512824817044934085006511105

8596796611348054315799010027116370414625588451469531
5016137653098634679351398306442172125391421048484018
0699555558933864698447097220729204416001744645744857
8988521913325497133025482098021992094686705513088504
1123215989403060607764070886215302252839630610614984
4929747045128120643925095268393316301653540689292805
6518715726578741194021747809172799541874118113737353
4823204924028544437285424144786673531720397284099921
0753385213768521899202754763751550880323820345141044
9033687861055113974555644534413352805893314950724154
5365042536863587651146455776385286184222500373544338
6084194572025780836246705161354412193605212492654785
5797901126581591993322554214733610252203564003582790
8575507305278835431594674179374264974074094794894477
9573166096230217323972884026016215508990745102462967
1836859160378905981635743926672782950299181795702806
8636510124544515441318142965418452451978873052020028
8020433895520952126242506820736251646482968883150509
5970100022643721353487858260253357898428499264259849
3826986555915745522772230447836700451292620325907284
4700707182646394299397105796504924027215130909020163
2257892936466206907911418909170955485858170999693984
5824188862304346386468537094692019086644250014237049
0706054794401636362244842049461414540733407720561367
5377994717434641869614416355642947159197095912457298
8939233815001041229439585288124290316381893911829364
0475674801320054837776422413083227337901680551345611
8786526378739084602983248449677767652671446090984272
4092219442087290507772474227128491998627528840954536
1224426081223673026362416664636769565823405093478650
1143545223017211043182967461181271247726747558418347
3918296468924243908358983041077861222164667413927458
0844109344670914076889081154804269904644766179037069
1318643164487293481162475314270947951218371189543080
1606136867423308652068568392614804784456647494574832
3298371127834849457568184823573812967298602509445631
0021387076804904301108841043560659563291355136365953

7905774508634658418379378550213855073066062032361892
0265343796554240913886678051764866023556868010244438
1998217408186830806326579344501366069588311635276590
1963710912216830217994317817811597562569334811817590
1637045395488002543869195029394842963338788023245402
6868311592077147266096408147297425641352377071326558
6567292609352131356326973863345139232379491272741604
4071653328372766636069920782898851581890074068178835
6003383955024910544219136949438402592897576804164798
7388754419071010073882502600250529371571205988217997
5190525154813512892650703503129538879739519680714631
2979739398855224067710747813296611251424440942546205
8656056386484117697376509322232005813738988859893022
3363080952193426522815067530677311683499200307497844
9533317392356287724988901104982913538099432346738706
4792939183829847365091741599344224180136090702185376
8394823719725514881388163528250823780875617730371859
3310237690155181489566802645106695566763562703316375
5042821846935526079312867717163008152297052501399440
4111099523758782168987072283241554043785949364881659
7106019417011177530819779600610206107580954184382263
7717441589308934402454807763589859838646004481913063
2918212125220072806340890562731361562825142597291169
0969621167408247163145189174736006959669914230808783
3837868659015986702232142869157014142480704589721910
5420047904207261838945659167576624337481652334310131
9777787506264814478962379685449183339325445226328238
9839955214350864723998824618234678333412034969696346
5231029709800703127298113002987487588451556284431013
1560990894615878405840038361454306275028384345168367
9399431155194067233688033261838130190651593168620191
8396364388118286970411649458769422113657698149517318
6043944768192239400670145512792825405653032464235241
9083789115209165207534501147751337617613160303463500
1583043241198303450459731115480235291472675565285396
1549825173221870281189147558219251097518814749962701
8320123866466554470962703221196735206682568834873759

6450725120796914516873963998729508929286150574509391
8352489864171151563371077207043719429897852585410651
2202087219851152011968200668515495090775699216193168
0576122550841079956447357236211513844260591187852361
1115766746246167605894908847321882511881891653729413
0184756365083622904096877270759063075951737344653812
3581672056998615449337441355115808285999797250700054
2569584482904215703296329695418372061125327781850782
4353239187267379753901060421898213335680014917629276
3589739749151033610294485487554126594588308262730872
9741581359987850589708156429324159565205722438860158
4207810475042628112904425526350548296613431983475578
8519322226718693036456672710264959940051166308663731
7274044545694973748748521103317754936462538061133447
4310806832630846622039370773105244279995137450193526
6142352255141868055104005021438767785929901108592518
6749913131450008725837116693698249769940841616062428
4063083328979971618705057651962404924316599951518966
4975475039001147398903189687832645578474537251804522
3597268776687624285075381661679248800082340903203480
7146522890222308061496574270447722125026619237142356
2609291226018250583731811971039075175338577137807762
1317724528794791583171484322731473506837177881579852
0230352800599998697766693700822670880420433042717610
3604436021195740531832397750825376243533599258744806
6952313140950826729742008271959187161696015340654578
1475710124329470340498901172403145627070070858913555
1306594748305010926753310504767668510068727953244323
6896493872434914018868580217669706551588502561741520
7031509272651458735885771669074118956676294168134057
8424067733886652984335828209920927960002560537316119
5748651729717114043583683023331026924475563496301826
7857351110563974947335708175806329870766803421309668
2726128479506043615265442170363554065832901954741126
3216179414368623878244681088510060879820657196947315
3168872765582925484100600262887084707264146369814546
7602306906484800019508915292088347520029483301183570

7147486046003231803664663011378346148102080104082416
2464398628580275352540541481178772578449824401215358
0883263111576793883443994167425526718127068704857905
0017001882766115402598966456382269528408612570000031
2015134146214627435881881137521596235509096186934825
3038196808508496757130802652210017544521504388244696
3539135452229483822752193978161006308157139473475716
4331002885720115617471919226677195436928312826604396
0699254637219602914253777973983167443812080972188188
3123622660338707532678942538559169182977283327312615
5084174849512359891579860193104630204088365812328283
3932828775274859787053647329515614114298532461034302
5553130194964301167037928656376695698547963744374046
9514404752486274767380255896740849630272538858173832
0957777270442659676450234624195887257359338615526808
1204775136402786059671489936812371201186212349054817
1292454815430238041036501487535674543111800604500426
1307876822158851442673029620840482261369497426208176
0999935003344619768841879030415959515392641119654647
7482084960353618894576122048571862646143232749719188
0858417216502492556122848670444079452809182539144469
8761813663319439606463782245081613817787292827839764
8591104634556227172221781769229741153867862146057242
0158898217549455474948636317672274364708980215462007
3250130237057212162666252200530396135167883101300856
8016798771386008087441449608596103041041197485369831
1136710708247974741971708082430169166177077131276 33
3136381545315891337525416839840847864317750667503948
8466367772146792112185361223631672188803806610698593
7023790963186922402591191463458461497417121925501992
5474796004846006334598186460801159374470373166319535
1890879205648107281187772402039744024602129739110134
9926966489897822336465536512949732934154340689469433
7381826637786050347493433270290837561801105493469017
9339428739905663796976347810695528961987646189850722
0863458747577535586844687233572491790476548077510392
3736396185466753334959708917470501031396943809023634

0457990307072485296328514308887866880742498163585636
3393141947625230661525205658963070371420915744678667
3768335155822444226371755529054939532882366689615332
6331493583928128224584932540555941071950713799703563
7423400973161309864621393795308709471653612565080331
5785044573000094141394600147452544140381692099336041
1596583800506303682545663080628250094880200341800214
5584175546348018765356776441151647710438436690085370
6116905032530314683543713358180929240076805095818888
80313192299660498665119235553334427159951307690820852
6629677403102594730225917768201325910777315857844773
1207588645093398775618726625393836235757625158805620
3092312138665780721626116181270037560534462263494983
8625256665242292344365139697208237825995762610809984
9375422735675122410923244793072428280291762353753386
3708763873518155274821112448002459124640511151114996
6446261984339005792546353949622888924362325218640252
4810490595955408365028689357489054200091253386743431
3407342265195998144887626448318552732774941228785613
0622582187812001162857352133808604365252012350790830
1505963245468281892247598913287169435985142267573258
1509249821248990518465907278237639649232119042056438
4917255643187344162296200604471901611612786080691597
0507233831799024001062116474775843902375746789131695
7011822646217702894571191364126858718686358249327174
6562706728075136743159750756577475837640633804494482
0668352178332133327896776383657446746201728839572367
2110981540162132700681687402313661948332501044648564
6460364125317413333237960756729373305212297457933352
5661685589200437596251342030638342943060971584740953
8019741154953001028216505595925945919485334822732715
5444873521365344729423949559645304788053179455862934
1890107779349027602218084991851412571653165137450875
0314014667742519764762046166931133260453878964516572
9084386151944311401615142307022471639399010043790686
4103416236790741850646376825660389550334773489673113
3431362942854314887603124731335419670980008452642740

1420976313695876225859100931112997379360013553352920
7482985367204276126984764006676698661053455207287218
7381806791058162907487010767369652166873448787438277
1997327186492554248066842383302741069609185500711535
4892417444079433704231825456068386702420523393305803
1730647788593322929965546621687057128180663158107596
9880379541902867105158968218399861722645652372721592
1272699856166884308596839602871715385266941479317328
9354584495315021859300866891179713664949241053953017
4013607858891547134085003976803645381111572086129563
9470964557427082387312687498873097059005337318346168
9693417093000008616802780058956741522844366300229652
6507013856265684358886297585892712289731225045019397
5398801959929585946674448852792346410372473341353383
9025948077395517640674147646580145330375512587839152
0600273054598058280083415867508782021829802912417977
3152353857706406771166845213368665010906443991846647
2914384152284355957780524178692213439026209703590303
5025270328397986765487111297164150657689153935090940
4216300292126234234712852108395421664911751887684890
1601635079499087251459442840907695196996180377128279
2923306313946321509657936648852867185365898542823240
4638733828178481530209203088315697267343925583364321
6320660898884580711362776399966495706481333243008044
3070692281796296832861316394983415817887142621966549
9051404499949051322758329020397338902854257513664074
2837719838951375846035685933196763654229787959796756
8283998310181525423666598572785888868064851894597071
6203467370351680456789741083210206877691531050566876
6877329334920023893505744369544516023429794578060306
7189315767951908958081128270486867856517949494253179
8989854558463511016629241506701611762219757292557732
2229957957026951427313412587036021325937476429476772
3385539394960803494329630814590799338159431146102374
3648260905274892609114997817599242523396972869525241
6687315009238204121285426136163532491366251378662874
4172873692777326685338999050914428805931696176825772

8559277785548891224880886696290222200907105319867273
3203501256083276186546860690046121765511410345328312
7120443522951001679479031335053425355678386919223431
2490521332794361256904680330454064259314334859893529
8788225495318574248810376413754148449982952274890279
6950898149864690761644389575234356650649798259415250
3242632552944116596940559895866507612153399297486410
5280830988791971237287616972907302953015863380954319
4018202669104693139303526636283583219629341950220558
2156281151008278370219142231861577528944307401251206
9822362570413511621279344747937375070858534490402518
9467769147420649139024731524047392237570356833125539
7444736369775913101672485564252270498558713299184758
4382118515249153210866087093894774655589097681500909
1552453184371101679704394227200606593472786492376559
4695847171642902578632718343604387060615267993199251
7807196060181997889618914413296815327355365655317827
8789877045484925656831540484336866358934827911537849
9601462943301785359189222687135602115638066888736024
5242861517707711106712851439717394625668407770725858
9195186572002830268782748806462486258045143333445413
3086163786823325729625795380067350910605339652325575
9682415048279519619749459051008217962365670147705645
9027478980181006309518889621379037693653372987268128
2088478870106308255415850421334101495828542771806949
4633813881682451903444805049224355100033141429208942
2576831348019510419539564834283831689946997068936123
9529933647736059673795630161780318422618261992081634
8676196602758664471180876032530070874535085357549089
4833166708013253482497118067652281580236070823339041
4281170229413525360033063302611245516864922753389765
3332750883730873546591411189798341977081211090804713
7442356324199743619581423276740560044467491569494557
8714935547922254176429822307573665159603939567872952
0830762129957290564633327979056087360196683806841521
6005340982287176820543030494829640714377958967789178
5265134420901479656996958603321761028398322325242090

9187497569528250236244494235687350103470187419905300
2938096986090876149456728711268068719599242400646532
7711570046123469550672596301566722909054455688966949
0363819793746846586653406795597194462977563164582434
3862403793489804730057570983951582161392144404188942
2681665534895414328206155392681993338132341431398790
8720655644117610051979103079211594464124822986954039
5866978962963602248076632631118560938170907553225965
8171492545809500486428193072375865331093474102684608
8351017655232979279258864296905772257139082911909071
9641708538459454433599189629618258137957661952533777
0939593093755869597915058546959060081600343557079220
5728418485855996164771561906337685043293655454747429
7930822840340104214779400494818065457292244834261048
0152048933259789368235759477584893907965398613200977
7388783890023066496506731865265056828395821962580338
0702097089887141462158565442623752543139384253212757
3407453319116295517118791369927035391723508149986623
7794428418843345714929271033322663099327159181177798
4273789750147894332684972051543072375606399877296166
8725323470990717464054024073987653076499928272555573
3397102244685228197440635674154423398952240404254833
9769553714731599039115199581609495985121037453659944
2439645586621895120731402017735567818531957450015913
8619106408997869328313648390096137571062723478005228
2421184264275528316128586976015660464318335336103972
3374601999153889315730285882691609204948845413009226
2588377714048796551601554359374511078984718088470096
0607789076220693684073784963360963425095847082572563
3681267006429102982227999157619394123050106656193243
8529131227088307156747196820218627201948474469147750
9958737748660296312621123936262684323153391719356913
7898919660667127709734322808251984750619540620344933
3070378426798379941771882384778573049239862558566116
3352861527957134353145248103916383517055077877222976
2397920840708871158662399192331933649557410994937541
0066796880142650207310666332190372968824698040807054

1863178851938047827141225654179999425208472883282034
7685848972552574718194114111004174156679999964197532
8403240933119063192104713467023378515181682298661343
8461795592228922727247929512697119023249639138044043
9957405009271208186132542943749468080349527402878663
8624393417108857657456509859476694892184500640546563
0078576018633790396114271309657046386091763460387568
1169616742477001757012096224159952976060385348857001
4814031370011280296945431637235112508802119138585426
2210568994899518301809141719061592636934736495307154
1759066678807228201488291988205155707763583295672191
1220357704249516850618829530889889133774280092605574
8231190883191031319392993345592313428229082449525800
5239231203546840959181180376700411041242952060041674
9760555822753840278557228994429097079220373479880867
3500170223540288707487241568779150621465248917332552
4770184486333604237917427498553433628195137659386276
4032817426362481472009657057617273393219713701624994
3760722325613278742493777785892693303359640162133441
3649840271139133842747075776954377860117566491086194
2707182917441242654445981363785943440204322865897546
3864348272914836757909061246208432343903919234433434
9677277355611142132001439444322732038136908572979573
6326744778943865774890385918099259886296977925891374
7052857795461303205433036775220335508550526418524683
5194929346835243286029416899457532838210307005971426
4453901409018029918233366474407788470720216230623856
0559758221344837729629959883211943413369458344614783
5969370283268271410484814528829052616640328149408184
0243768279808314945204633401314793187522373778064144
9565756210605303373736314667499714281990742397055859
8153503666209046505844835829037062788217951701095497
6396032910465540606926458630212687402703333762870900
8636077571723127591619507653913377632919582215602395
7434293446887129808461218026897104243417090833099109
8588888835254085942276917768288120756179439690119075
6634524170061632008101418475332908113003093109758677

07303631842545293345309766615291752366323656474216904228061697515605333059925079176825022364645999570337
7476108414750188599883026552040683225323910587244894
1321492042015076366197289000406059272042496276071999
2999765156898504788208519098035733115741544655500524
1314901243989950767377911479714212766615553657000299
8064352235855946334029151965574477372577452551736846
7724114822876372680019635844862426037986498657582130
8051254867753671804496187001591044738793424304187854
6178704578543664944284385030411648192666718497525267
0736583993025400618865946300442593498642188736746677
9140102899219351903419847325760225853194848393382061
1464807036489978670865314053173481514324651853400564
0853019289907636016009140767076874864987866144724164
3842625492285981679108129205218822915194474347041036
1926198224968865018328788122865526149448724335598640
6705534886676421607699601535508232824182707156181963
1434310929628040525693801721006438745609358563665337
5409615209936844109006423455594967899258652717374982
9803717638644155408339933247328130954900909116944267
6470996060513667034017441183036622504899102028224104
4980053063939224651764328196320044478643107106451818
2924901554707466301366585027750507967666944709231116
9507428457926919864654796897698574424712025026199362
7690491868938537969774824130205607630433892247367574
7538314713417546782974962447706654093819818294053395
2786677289838848282991142392773632457160143733752630
4802632494216545565767197675193472054649944251600989
1508526537508002510756054326553772723422307196967945
2722466159738660217416890391227225471338259155322845
2515226694697281730317552536710851911358876542544357
9041298241035431744232764343271370654209963215706364
0609687138452462456633526913012207892078038541203760
2063411955394534694669493091620795819911659307574198
2692987786665036590825853102107170150184413675291384
8473908192235647086656219503198651985556903747671094
7140876135315487181593027818838207813940008699996704

5174005890292947204951246680739095172243055169301048
0382781475446419377026942493272433681252024601571534
8610440607590563320374178837147535214395727778827463
8618416087213432549823690048373821826384009251021559
9762824924148323911002469278925362538480776998752416
8277515798144534559218091235201623092356187262035618
0637137437050124625681248863511622694756689681361908
7391386116827810422466418488137749163775823530717510
9336365159207832028507487817732945679572288027029253
3039303562960965509081112345990450064098314626001133
9766003729813388131614498624607384004103873895233467
7015604765647677437530913530360277306494854818181579
8555845871362783153768046482215248418050024360485920
4248195328836784036387899563193321631831778397529919
3755242196596089650655373940446089822896508308860509
0890249651264721191096922940386605909137826663597944
8407832676362544382973826316123851271358831895110725
8099419857239426389659059498278241780923504759958072
8228798338367066100204159537645690875936082090530464
6456054983510903778477908765197460005749378268569622
6892365647368966400237614213914040885302285301422924
0229342391847460728918244015840316196570370051165037
4832836121305179276792029495844961075787312193123627
9249070877470494027720766863951289959581037918255527
5737019903563985512824029479351343047014985331634148
8272414708705113722107326378167707957042443525424026
5878491032309944218504765710476292622152637991177350
2945540414719797361893916413646795825081052536221009
5673087070599535110232822554068808184242461329055034
1124636820625956442929175740192009705746751783787094
9738346200035152023509658220513234951881288097417013
8280072774927060738429578676545651223280696018735927
9384225029823945265456615376890950037612041625651830
1073537003907029150204753714277894368805917320302711
8577899176666342571642695371669593318314176864699203
9329287314780654549961055635858785803559889382532562
7842779752774869590582901784353170386419677914076504

8129809438387688116335995347497834963258404256655648
8352302309715263896108526328413993551737005570157924
3314557133926350649126910328745743368016847088321019
8318057258996356417499479991411764640878309858738876
0126224392915251351274316311424091659579854423119407
4263914199573700819368632439542888918921590733557117
7725165886954494464905156957324223604942910611398818
7879781456923008256816350893376885360878491509761407
6727220176526327006304042988298532360410002404018299
0715058209534866786585491475231039176530439244445196
6513425148588659357253061878933173290841634035522164
7410415435261375821818187912790652821080566445817008
1884621209532764188236193739371584541654500461376347
5572691672524761025578021119822121916167692479946814
8510221108354686977604706507970232697917944664058254
5878412351378391598786857658017471735758400554500216
9915662489343277570531623439857465121125566976159579
4165500430927839580643678562017610936953432274420323
7277829212641072792733153880542657187195231461476085
1241207116214537070523460987352852535263985551737598
8621522832527062331717710576468344207112184896971626
3292124906141666241887604178683965335208134039931999
7458485163686764908868591045268078730621605021495859
1937822714926533322860968536505039861403799578393235
9268091077890548555861088592428224225927744773651178
1827001981388531607305680336579676271784577742916999
7919369629629072997268103049709697061750361784872804
9157145532340248970086518250571841390970899814432108
6327430762953464830106029176031739831629885580769714
4339567729015294792494892573053103628809298857109774
2034339038942417749608496785311587575244607210626352
2179995794483282496498179688087770356049069740609755
8151120951620501327709107803913461147510049698677195
7804672823682217588508555121873788238435502397135356
4767531284887511145584394413075616690802194047054025
0925616388730579959357100709542152424023897386614498
4302696436156975938350358000865252066344823250934289

1281594682468813110767064807271539213380854908893217
4463059788581127442534488131962175507453904692292260
7786828636587515668094475047862672273570769537148972
6486013628080150844226326597221147118721715445818774
2615869707938869559231035534774484427102772791812654
1939125547604844318093436796646334042828332733741850
6298654994600120905668609109495035208441838991634030
6963343519971372234045101839365628394905715741199173
8814206864491885648968163335519506600092884333252480
6735584171337496171505509342637189402325303542599384
3941877187420881455435435616430348910314815205765886
9444782706449109953352128432519104912469054321738051
0679418598805440128942512325899099623123240538773982
1014464058496559741586595232058144988525103769306549
7489313506032936074481814998982011182749277815201132
4046430383400093022310805472595975512167467066592294
4438571075829356865159801179019948045358247172345030
1763989149022144948902160198684151758737919168266109
8385738453765280418900933755032348767588757658350816
8084898048899461346384675835827589450046648026022470
7959607311234708701901229396384219925088768537111998
5433129372429484757883611517408335843753310906659427
0132580329543981526920681054804215521024796511454543
3197115305740995493783836932001706564102399396852034
1513173309251386082983961034483756434854709456374110
6045616668328026369760559410786005301485403212528253
2232725173232493557882265939595083733400950598453008
4486154937608307729323697805390206948984365228679285
8078158108085806495326331730564681609178514712540008
8072257937135985919602032117698516618138205726664487
9714560505647641742736841891450673424567564160482903
0981897917595674479970441848154395604702337843568126
7617715798737487316524458821001641061928767152951977
3096125795040132799512512304460717376533044348897583
7750220067414677801697322800545673449942537241384582
3677596399572285545930783851914039504744136175891007
4146226819297696949886128652985517880249933196635638

2483829419247431923558426763507319858030301534307486
1824378325227933579938356853781132755653864730024767
4306723758445557066433223967058378975019401109845845
3031203974160814952863365122483951151426513952136199
4928047761456722848443128565961544973138278595330767
3696014941586370703621756586701043035869611457917148
3445820548229597116654702113627728249354079462907060
1403720169035778923993263032726072545060040364605028
3809296100760006762109358216154880968279818045908769
9075582797111496748587103659798177900559920461992108
6218833393864367675453578236336989088161935642182109
5595110093983753747755465860778655943306224849127897
8754508135580009553618632247789455782167285821565583
4857416920557822343615032535519130694519600528949869
4046865586452883932391961240439959477905755435190582
2581270246825732160269935312376217316256389732475711
6285960699970829394959814645468124291192894493216757
8936358775236587083126261297689522140371213333713736
3657009749611146795473894021625486684146352498148656
8437129932566103690320984324524436374578928327453254
0101378735460870857849153391330184879650215888109299
0371435011496211919724372703633189011799293100919897
2066058919499183852698678005809392309173781954298508
5168466812992334259466707617776755886208012614126146
4088615606370386756461288814378861816884069210573731
0071471275560282552384610494287319949838014192749437
5100694790609597627570407425605279204037351322564372
0532009690271261787884195824392343316525246682094254
6627297934824209502732777029535981564982473381806163
9387154774919753504932179174320668434092062017580847
7830518875496124423952011896490704766018606357333213
9879373467391490808812351345513774071558682223545588
4575446863433377540313871302626071462240117170602401
0652254911986468430964157219449244602828173252536670
3537230042424986606480531127501954356523225687382635
1560617978177490363147504957320325827228208790158003
7039472207847114408535302162674050665051256166950057

3990832732506899521269752616060272524728366524467699
5469356947594725756685581189425853777257680983976858
8064964418575387172908716236658429546000645728360531
3875813863629441043134629953741827762718530615919342
6121771320103100211452565766902870974555331090738581
1128955140392716687522479984987499178502589268902148
2459578259051864454825308670960541524639164867481996
9569196375971963039809610580335461327894359818435086
9745259200054859003037296168307195762268543536417311
8174509579933164776907746402740259052553880918629368
6045867395211331046855474844038171072506636910145573
1473828250535705756613939175526069518689649250344686
6474914652615608550503791394202981994222199473543823
1783223036847137330247485594298263804065129848919712
7731269793994244683681397971930894515313010228207176
0241132296391228061815709537618452022878633617842610
3531073417592039782916543702395343092259108501080565
5918772527775548002700192946141441537672275682553321
4173720140747134788443436316759161438722559433049497
7196123384222661604896439624205326779704143080241164
0119680891010920634290380279255156967953244161928348
6641083828605441536996436593196913777787005936030482
0229130265145922961345580297182724384196876724370844
9263675507056334050268371994493547326526563206627438
0836995826335167607082354952985615834331195243922970
0398791067526831494422487587059711975013571688080770
8801638578432778181513027786831168589194611430108958
9218392897133594139288856488545091637259398359742076
7157074607149752460598639896695746231577768604879720
1480357679564845898219702887616123194701320955924214
4883555057627232344348426426011753223061822303565850
8047011011899193252571720549962926641297735042850437
0262289723585281626756378963020389847435594806121738
7390668535438453039293129881938833041183423778361478
0597575058406622541336293357809319478196639297423503
9084805932006978991767883396869131974825886474708627
9971313256137172730816533406139462568550590727545864

5068646565277682555342972140883383727882010289029324
0313242102002610635664244369661208304176869322010489
9345155973211746630090867120083557242052922510628503
0294066927058050440068181922735142563465843548110959
3207340127496949000254472079736037916466970319503383
2848355167676058310365452708576554980028239478223137
1887039652164207841403863200501687559289244248916432
1079620031371107462606935918955818239988365915310970
0423581742946007359612474329057210929097629241041065
6620923503792443139268903030622034078705847521368443
4981400664399682817772883283068082967474851072684228
5639503119239679399702278280832904039187942701256403
1731986705480903817290109382677032761818733382332992
8735425179121467416968444384160995792173492547541151
6955036329294606721879838177984886836278290997984302
1720417536252229967274325716308033262679427008834667
9931237227789280490726906343593863344827373494687180
8806945088824068997261658713437518740712443535899935
7495057639105502602348848319301097762875184555561427
9728428487603938721304909025418488426977514011626937
6139550458568990473003987622256956952852270270070700
2236312782756472091890723661453383150645086601571667
2503044253134573076142482529934735508200948111074026
4270328796135455899723876924388109759704444572797225
5955821483185792211683819202237666014705355033299056
6389961139502003559003953143148531999733956110064596
2955582149616215804551632496152498462549133866615566
1305747107306606494761259251347398672404294705271394
5870057114461774359248919999779853985891554580117570
7545841985707464441715735287088318155664906711613720
524842124067568833346326309394674405915392812434686
5274150763671083329467993079601213226236297192288906
1129439568658906746885822588883989165018835533075233
1981579035535868551557820654682183321590742910347469
5675663392485415223645371500388621789026343137853026
62274488179999873853323415250005050759944529160103849
2429647379231448519967640031204261931101839001074559

7693245743996519682211157017225000078018520076909279

0038965083547606952112614351514494096991515178332585
1724798947405242061004598407363843511382982933537028
5516415328184689878043592175819760111037188260115715
2121989928035754608388740947375220406391233628982806
6187319532355292040142200095154808807061007453865639
7258970803032798551240570967529948775250348381191484
4763960690239980085887510116129006008076911943810302
6094948065984761969048059321785213998286590163613972
9473334245297578429975902328892122887617453643431583
7531437849574608874737342587958758219901935389814229
4239179415156131539793025146413798608959887675413694
3230404870285554197809229580446989901929045589068465
9783833799449251271604941337790706486578589496757599
4050617557632934756808289220291115491864882015921461
7765449921182725498867656896225170636148321951406030
8448688429474908171412276698952976652846718701072919
3377929282443532431382852063570615807692592826032221
1940276877904292408365323232151023540753423210947605
3210171678047889604168510719739693991861879463461896
7971354678672244029051644439647829326694635849186615
0401165503213795823884640354533706750014682450893963
5084079633883393164400215576294876554514962298494573
5704556398458286539010312031199558632978985996427424
1654564022155269311761821934040528049770013958185699
5045062690832184422098580656036039665052040509265294
4916311224741224398545523345939736021584889595645760
3560112394722600290911023235818327077603181928957893
1912004228297227192768010578564466733402031860657975
9989767363004515534412122274649211784192104299302330
4754593408148695733885585311878925724359962470195810
4940834271300659716364375165746371024987053629169290
0609979797982081471471311289950851849200398640461465
3025099491414340358369556884216151820006672539858532
0356707875344741018213449970395917873975349621147236
7771510750644341932067097848101390611946814299656594
6949980301501505043949916581936434061754712006023253
3051005685661995398852109699179681030651566276114001

2393944127405040656002217098547779644246858748631969
4618955103513339164119590397189387610544264230246441
2785966320148479552732274340928634620098405982453385
7638615193364432098339181957496295052527170415941103
2941605520070797004452742655032910680168291821505488
6572979083065733200566711040393166428946073974276132
7206991377358887764684076726016450379690906737624912
3181569461325842146511243018088628385018727320829304
9320534883490802257999620893153820434587462062523629
6812240775571667614708332530743518028015646615241233
5772666545968950153512096407409879933525112423683932
5550804077953890651359848693148572687489949014085300
1062540369844024339857382126776294549192772819270727
1007595054194545037909180516361580836288870015386724
3945007027498984321855676474032471439236648434111606
2093101960182503178068353985725839133571334493036144
9170866597972333881453092174031811747752032581674338
9458264967527525203611262736721097645431340238065872
0112513451461172380163835947268752281763835665589618
8613216729989394014941251035646583363688776090658837
6967418291919203119456469780942498386090416033096376
4529279423419300230174005434322585462509474354551709
6835436975603565019923851473718492670597233277579791
1738152474353163342411729845894129107504555042885877
7627373406633041603918082687417260659615989336077863
3070199222318466648889304527152405511746120223016036
6192193936615793786736961581625973005821281128255764
6795949402814667457604557747067390220019769831825970
0293819541492759081133733236058858777871610067258359
6232604960016058991488934220473616132710075452300484
3943109899916372218862326257224723071197918230494443
5140335744766397083610698607144570069276396639734920
2921834629764438186018937668805345127770384815690854
0614032803615028038609490335348932303579251173935304
1584113326547129056739888443593082228303303222116592
9854191965597971848854238871580891403693016171725700
1557061483690681274229550279346352264500686934307748

2074666368747614762002275018155179697826673741459504
3870588723873389632912139573039946630543402891327746
8168754669502161412465503700912659179830290388734841
7613972343394556935660083801609435561378553746892071
4544233776467196313846465263157010171323583974874665
4423630277928541904501566645788185997947897125148114
0502377690261728979301308065657163121212079142907054
2150888983795453659164355123341745987948092769417511
4903117460552245578545813558670215309007703195565589
9599746805741613338361641691140099233415564386836225
8664428079403362670105226669361924674723713640905428
9852051883510036926818799746564705254506826839362640
6994422311791299733364106678173591597162898327417728
8723020526098042487577710069881962403729127162845583
5847840340492436487818337243203716187881493183663213
2424242420147187986601290829544902098739959542872139
0667769827563089167942174016882358765397504203024489
8641889636909631627012055768196992915499277514254378
8129467665083250351267168466448444547240410124528064
2178327322277604369161028807835887184371005180840179
5801410835281635163603805346307638919476150186986736
7060501475565451912556348547440616202739383503562785
6152958894681701699940143323110952872124482704720605
4602585006670407579111413682790697868658711779204356
1148929968719888032590349546258685078645156073721715
3995339107054574208447004899811282899421602122209262
4494727454056103582092642512678039819054526594437375
1942813217137033612591057551698992847294695342429807
2325629025888362684267844702983631329496605441625386
1474882834798167322881097848769413234367188334829751
3277555209811183566129984856870217344971594558142051
6760136316810447498709163649431566670016341247315263
1466469447022286028071181399281588875163721426683212
1415092317205731891117328825980525201560900415547775
9524040891350094036519708487007496783327432335886946
3126879009850231317206614113210860760486195706356624
5230487204929701679457814582009036192806782139458937

4337769312698768681171248164084910525388423933369 08
9463540923580231081725576349979969436459754448948566
4773280449886762357891730215026987964549842771233602
5239601367889026391276316733486900994658881028631022
3749553599501716187794059542720325680750991723040604
0592447593475587819231150708603864364001669769358844
4107736870284577037940928349414028221295864075270639
3539930444723850843968852757779835520831758107094826
8654551492346771164511885672238076006299878184487827
0052720312938847998209719432022757136352039888007560
9793549685072221738190964275756846644078438497623595
4164378986071667348604995364292157690926961517095282
5421086026688812876213228288701239411121356084998485
6022616743503488305211519952221309472231188245473926
0808544121534421043454311042835336107232244610950475
4903082384976233787723979857646707148507270115503350
7917688942855325655578568914111053937681230076417273
3233555555695819795161678765161127158802381737058125
8433764454963932090336308884238313464741325415758340
8532870162147846752736603532981421989990010399651663
9857816270835896248137581128520502746831434621865421
0028735798453064197217331119032520073461929812872295
1789824511177032832347598640395627061908545507358079
1658971007776402290351977055165146315695428841424375
5797570968942232887312501544659132356822348563230881
8621486917525444204250311551711252093266720935244538
5328575930785720519631126771596563353595646066381215
6991761342710503579689346925609775922911355755004495
4680939594198807691937528886502489711246859161951191
1805736623365074921836732839574906693966899486388125
4658855888383033086427922354597161408639132801696867
0609674779349702513696709492118518268068371039329768
1827934909048809926852979497855733371545681229119082
8899996496173672758296772254271826422328664001327243
2730924295092305662213469775602749713113774964021604
5186933589599433451701314743167166992535535262519182
2960685511025521066176939130589930470440130555394785

8663168437691828647234353248593887797337000237434440
5223057833850423369748670050160028663716354807214257
2742363471659825920005995273502863429413906679266972
3798730437353937957758746704387095073567124554496603
0978961181945541702455921930096405938055229427692173
5098819503385424390196223556566509598118950849558347
5832679441371943347770644174306876072873238603190937
6467452918921839273405652449125058695651761156206981
2500393153884581844064908193055138220680810239336308
5653595383285081518528602490738088819397197419266456
3416144842651154131695628352519591124428382628810103
0847554548973690254035882364831424405950604333637217
2311369797376625377898329148146768547541189710236465
8779329452455366084629871709766714521539359362956508
4168679388747451776846470197056012062911196593927169
4287820010473842269120842037473633883862747926634381
7074600861816517701247380026891028324861454672894643
7703394342046484241967025618791648972518386746222304
1651600018564312996541175982085005633239524163204667
5535001335372968491746764631941349922374424722463302
2021859547406463788211882345939408996899586677663701
1429528531270793556632378325619667821366570922060283
1025653913540119066214292393816411208069617216043809
9387993003279135193961605459067259657242443886673098
8394948040500199586995408776106691389068427993564695
0245990878656104815262619488029162203772854404310761
9152330967613456578986649276023103467080783909926275
4764500023111598815152501667563739574019057703412613
4204163590448083907653748582777525966628543162988331
4207477826120950407760433388366358024308924484034883
5410287061473396328346465783579669745925874011346345
7623216081042397622251689597347681742851273772134888
4312642986891670696316238734200146948985214230208355
1071071050255718846277856440376053415487371340570353
0471604771067752932007990830085796335058913648697747
1509376122562844835142793387526367457072220025676912
7483422379436606131986267609440621051523719848597473

7929740617724333077735380254302218943957667695095667
27981248500848626425884845796771935614666464626014964
9514634714900618867260130216748107466054111926689184
0637883531105630441708083578019992232956434743032959
7913588943793800972704425821506927979988831446725329
7689672092433109678778704372547040496937826853533277
9967817629151867127754136572668369349108729256606562
2815915263350446694975679229497645839604031247826096
8080763245729179631357063805530181795061558934619200
5525020421276892047265235195908441637059762275805273
3539905727377292458984311346620894693568462807708795
9342361434261835739728412166526019543848177450244296
8737870447818458084566985918167574593630071250992994
5590215797971267979286814183617945293811474383459113
0494949062545777573965744825041893661050156722391140
6337904426932767178357282347840242922904037674700397
1346834385546406427071026117530091308476127357563889
3444957801436719780138982653424377606720487305659206
9332816973770772050673214000573675534498089554053868
8878671591124076024028764936109146485632435139228289
6169205384742208960466080590382309965918933425890790
0622370408006699792029798194409277173502701273368468
2086738310270794793553022082277521544609273562071517
1955387489668190846802860662680526626173073955928932
4327665608205589264922811457207893258778236808279305
0500307417743535142587643209181854326694069067600791
9082134203963689530945256334022130730209864586297689
6554724865262428461104736657509041771732052323741407
5658489932392708682167942643268756947351912174769111
1577540799971999266828885079390393406103104213296468
2504077064770521769095572432685966471769863829141153
7797697600025819272394469201049660042850854700148091
8081081726650456796718668806462058478809300711671419
0784971339391499399525524545209494650784349719810361
4287781840332205706946395154769469727677477064248646
0793923519565436635083070252079824653742742569969457
7564626119873862943453280541508276209906622774358444

8627037670924884313967312656356805978534285819844609
0082502281505106367269141887603978831977318626572931
4218073290550935385624444888087051585120556194413737
0328540547572214634071373693265521086932227094239075
4299408994425444590686757411432252426167234352191278
2585438845595167978299328323642737457452544546052939
8968062635137335872148508088202055186599580340814883
2970125378123505679305081881856850573123325755554241
9605427358319447976432499228822660435558523349606680
9055029052163377847463519347497130223294939655104159
8783974016751661859360517933895039246620524551126883
7311120785257244245799623294450168341713595140252095
1792646811568298203136188273964266233216764415246954
8755816408435821258504424767069969380375857300390579
0110515414779557179169312729095998221364115981595201
4586136789206666353218394457911294294937274642464682
3921547797561573367089576184057532209850474858357089
1766352727809495354274482525113739382912378335184147
1827848818093776259467255433420690238375597658467444
9885729710015336570259383860983788837055966165661226
1881245463878074036437755829259340136451738584462455
4076490296162202292447917789014243272492456246105728
3299442767967831448193467055175708350294256732633526
4906514141210237861093296718863103717170461762893116
1672590290677122398588365964149245530812072857084100
6607616854351666353034138280113381967791228997412665
5244951348338934636181282256499053411503179141167093
8307677687742325698034291407998029191076113965307761
8040762219445151940260406347035679935388327437858815
2011080406490885175270082056238020512864218424823002
6324320559979983469262326656447019563573006795390572
4415039816423908213623513271771458619121032811235726
9933087662553440894151205179902731473868182626644475
2804067274640857238015503894189125958937399265016877
5274376974153374817242203770712866449077116260315441
7119414108348606899529507444772203362744266818471196
5636157137724246154560704796508783129001334349111362

9297558360906017594945379686150681790850760756621273
8100117918293076118629911635574502602021275654360951
1385690948154244767226073400610373342612736080448553
1214757889023755905771131745500941185974865296270588
5639173897159515988987014175869648654185324863779433
7805069893455538805052331249498418875730464447331444
5985055247398653997073462338193980085773043569547616
9828265893810030602411218665685980207253371656135335
0992188595601078815219559929848307371141617648399503
3003789882479034541053250205495569358801615459891893
6886572124748963613671862818854644786179243581710112
5518513178717745043073536450297615072923011083308025
5153495186929484971690099173039476973337895650229561
4877878048366582834827540230192303690385819788534303
8285582730067215613042476796509973673898639630845953
3099446736600527895351007751062354051809506207295912
1477879266263385428792589775958630580646504484526239
1335383426270504308670094670362204063397672529913651
8784230658396670226258056212221073354116185029363564
1616655779237766395860494693244550805903617986442755
7412949830210469698616449313701037027750848601539616
6586451285354530481559829638298598154556259248659186
3288176301101499737206920153869877418621655782087885
0289708567829701926958276952394082579589346666668839
1835881554906943683070353276320793494510936539945097
2042836730670351441963155288753214822189325967173707
8127140513347473860809636945635120190184391605573384
0805166382914886247935137940371319796687585625948294
2074632416148196268288849800968875641317790265769105
5508025432280312585899845828720832573588947631349260
6249627183220073181354243953643770564819295399570014
4554383910878449144193680471065163474031170374482458
5051857881806866288441707935660426980031632363491203
0291975370099601066619389621731876226707182631485228
4417279433406818103101838417534997349697901352604608
3898649384170852934692791583455942477874147581862606
7224662481177224985686229897440438439218402456036091

9123698959782488064463195555559308328167346023120406
6700724877475998063268452732025570156216876628405832
6889493050519390050495049587015400348542776024624858
8466667342385974454567161141984303835706397426666703
8555609645239035702010736523283527692067722136663585
7460807615994825758902615564428664967372569208046851
1746267024678766860322879651197857616442650025536622
0799720399986561469155119965918926099875691957219827
5509506475978615626474235578645011389704199350997640
6676557120850295842115591494729075235534992741008512
9491938559625940326382025249882249214444755882700290
0367951870523576276442355841833307120460124629939915
4841958135512551467709344714433092476373215011861279
8381856025571631417442644210392318412486156130470981
4802473388125696051967726943832149010465240998150118
3394145060084222913194160995009964496196330766171680
2799661459649084857174082378057131294396610368772726
9790434903189674932321665723319037215414610364718842
4635680197125709771242045599277189401630807555791531
8038863852263293491228689445871244071873985131098072
9960000540296913908632667141792364975629719250212883
9909708484680439071763198298386258976031273818102754
9342610128244583510397246172600271247264410283930603
6777543984038462374655711776604274794044711025322752
6070881915259623881035944912100259215675509990359849
0287366394653336222785601987852448078120000922672556
3043118702187832547386880440918833104825515033950623
7035345911575694871584408122253546614612133683291417
7138712079113256329969610586306388145503829307065076
4250040959783772009135428432873110669407041999325305
6831695331854406218096083461319779933817165917065487
9552114439934636910391325853497773805380142494093450
3627616581368950030951261057084123445629601328070394
8714675890101664115170393932146989030266726605846735
0596475274805617807867953935510326849129867665654263
1273298852919270082470877400221374311565869690760658
9908547798087756486559413089027045689772974195596550

1092219356932384978162258751764655242092557409257176
9546886051901000316080128972898705286108542297390939
6815077500965971737146008611522092622608527082988364
3736243877981277451170822368080610770774136633479557
4353354725066344097928989918408218150200626290058136
7815452848577595273359535974840872450053882741039998
7019521262331698628280343884972691416958629503620272
2974886898490039741471616745751141334602734497423550
5878072186655258735064125308324573880356085157662659
1008479072047704536889750719974356650630663167587611
3475164418905099495304411719985149916739766229426944
5166214080877491355367345306518299977582014657530815
7940816750357256313082689752768694913175166031419627
4122716209578299745125950736894997647865130983044553
9167618793163664040969778731171580041226555288637091
4062581788469239036439876794389944195963322773315106
2417111111758958204213822682471585586231593661531289
4321916548928211959762276658143596743190469318970709
5462549848023495501869231129366402929099667008638784
0042890442086248366177906430206330593392032243436516
0794325702465868466897715343280772170987980118148551
5792816444921354300152529961377236010772921085951314
5995246165942271641574763236570257188061170634876292
6273236008312525699654343218937450779674452915427894
7127228947044648131474412422116659008100572172330443
8700873736053316468302928700555720019069943199870645
4465506242821727117124592068124294810550504047059241
0528835740065648454724560748756247634725962019554163
0808699130865678696787553970081279117686691949683813
5150988085209582767929487854818158433903895764802898
5092572460862530061488862865030657198657936561579559
8257299189432894771618962056935467280544185635018462
6344267485715560888443376776775181119587963168418536
3912337497661237712587055753677142553545280102361912
8824660846856736084934133311957993354042333577358896
3780531839093444280492270352162230871494436067300423
1179796828639051719515750520976559027309967099890200

5130022633264738184520239976911295246061557293366996
5418267875614644743693887290887894259922714756326206
6667329080946986292953431110762432816432736086308641
3386486466836833403411741724336137908604788056800459
7543289332721406080344475032843441146117190967017625
3984282266864683881706100253649900743173847000861481
7616431964214609199373818877654827069979398415393897
4909461030806089521056237233733955299064854565477711
1323511505835187239748697076352293343549725610030112
1589126783284926464529265711611514653003449614413040
7078693714179233116662476964087635487399017477537102
0182114281421448246213204890136655231442441340428775
2981183566734855659369179625585315367510798067145279
6637458994210311881547454807524651853170218249967058
2009281703471433056490611030296600778862186439586203
0912621953745931915501116913315595473394117208613535
8840520458592736046321982702240715420614033109614862
9907590810331330659147574953943653870184306530383427
9040143059829881096866287939606842134010586681368770
0862550410669655362243076097486920666744068427555947
0825940375954543932812646151978601094092210039466239
3810002488578082153053964123626303568044502330247943
3432134418808431469281618392368201869189398393330782
5793915187659886158526588303130548206474199238611662
1691904597565625333631844676895075299258677730897811
3220554526893234119637741580704297917296184933765166
9375621514648813841726621715323622710802784181377459
6097865572452165349267876088009188070751445179559189
3207464840761990517355858488913303280635078797052313
1676769315773731879594907212372637992597153494224165
0491860959163929805153754153056010835414124426635088
4117088954264409770274228232128737818584813773935509
7493355511406244664368942045355237930229556990256888
9247247648285698792777177043958436924724006222094132
5554943292326806265100656067112487799788039988221458
6345295917166624816532287411553275264112489665623653
6271790517082015310026735395882470223528163997240153

4641220320257977808273135512050193684281552081855499
7514915110169914127116084454009076208300514164618825
5296346203608737167901905825189468389540468266297174
8668008393029512607766929924069435225777438631812759
6795069437001560625056278559143415124133940303277129
5325310711861748025772234948929252198097430895212231
4619576662075923556359726690767986661233291059527961
0183431069077020322287162520836119564948117529971327
4973059783552852128578547844286181685257107399791644
8507379463019479486010937938364054003035089249948913
8010893132270306436604092136152251750336475912552993
3623450874620625221161521345334640590731527324079559
3956003274879097386942606636143145093479579364252820
7605767366822455612779788579850905074655759995233257
6801978516473222357344466124947799906429335103202924
1706181476957105077280191727166542272028024548065568
2926562444571074844343809247355832405957279281370093
1794958428020067816670302348301074054742192686054019
7880276706177331169854901005325265807003919332218325
5176221950049560232954318807248709892249330737590455
3488785189577342825125096765197185679965291017199510
1746478143027813335716956422319340757137678346086967
1224381217307989693831217042049112414515862212057381
9892602813253361650633270961268112735445764503438627
1837391993894379695856116712668383393759855826461542
7978133179120578291237899892276277256159512584275400
0144632044579106546866732414053365861918428042262582
1688273710321538212290016053895557458048149707951428
8287427566570758148260548242022106120376883410734370
4461695313565847315846499523328897408613892603743654
5571031357309787030515797674186488333083334683061776
1996495333234340591686783886452470412755314395794027
8842161375968491828232860066928911605076118159809805
7229676116423560905478275530999022836011825568757238
7881258582934211212064353513623423335454800037637353
9228441337466447546489972715324870623432473939494074
3678490417272542657426758951827960203343622606184340

6548292910969473277581063150058025056894921339837057
1061955381036992510061604500623195895685277633841453
3870921568787580321274603112849248871469759238966121
6541007845166518759992660729902045834562796342043971
5652445650039332697584161527418689052810339622772868
0145702700318962787707751372895137492853891601345118
1479091212455428351145074766206145020787405527198310
6491319508431939379405139356086244871206328233097256
3106568067159358712039921409666332251191044508321653
5436219937775858432812272309717649727002825335202360
3346945160822872847275228184687773750722988339131876
8369026382449348885643460614702141015933553708337592
6119354381437532583680506869265160213196385900424945
0260777932898292974931025747485191547582123684275563
7397781015150277718846739673439741825642715865300921
3673388007912311266160418917228468906382678717224697
7341470030337709486294246783622917217912573978588957
2304938003585912363996896312161385831046483707963766
2679929761566821198465934155933916744468862003556896
5184061896502099578794947503421345106468294189135762
4099495577188337647484493614890337338736408448766512
8579906005691803558021757432822372098296405413984917
6861425411357801923284322356622012533756997103821037
1450536113521580875443258875177314981234159790077484
1548524687471869828437164277567966121882258983635864
6123372708731616395878299381552734158028806222896032
2744791973151341958948838419529290567529135828470288
9972904682421781158812544500275773489756610693699383
0600284424883040855689756491161569382878286204590171
5920661835559705573502183092691196050687113637921989
1638826470030323985599825852973720675968502122325947
9609213701331543690047347580052669731636628087675468
6843154412005445181096396331779963270733270078424261
5943287198367100185305221100049935858980934727278261
3245222554744663365234690260799520188298486579293564
3341058619206357658021349497123815423332633081824963
3020386361806074300789362848049457274765559689769047

9630772584358960972355626885277176950957548546741563
1893654443452682522268733165858336717410453518601689
7390037051140387216074925657286694144634342819342201
0787994447931528908070441678372085914038078719202046
8714895404296577827423277263006754826839257204742895
6919167600520523215382114088732406797255883699722977
0397817477865544513393695280467309791987546440540501
5355984214901764939708993368236817978618263713774776
1899242139647546815218023565700846506246212580093382
3937589398535255324737030726876131869326125773333729
0274919695015084048179987728373665255064027193614777
3259880890814946394227307546211379742525447857306562
0616232758456633687160410545655582196322844425800161
3092292561169521705856174292971169937298798552686573
6798162230768594917332186376150773517153378053363994
7253173790467038575527223738278135885645323766083898
1202294975179584990141689663452187860835838411893138
4728325768648734746219535389978008754241505867497801
5601593113654055207095080352550048121231237718152107
2980032310175918378625405659625399485447107620238523
4083415014218901838963027669086460628899731583050006
0541661052112618332456308874942376132111738323599102
6715443333980903010767519215606860915099297579489847
0913404847760372533164866332739977457417078705885849
8903647825050060756527667766673018142798346299786311
5472471904638130827026950271552434583777132888840113
3228561232764247580549141453340043073513682001671030
4896740791322041732936558863808199024025042475897990
6199739449424061393859002043745081712616036278391241
1472682090856905268374225068910991937677220777687371
2677015290712968226158437571496653462961540352898069
8498199023815881324900728284203166454586451867787181
7717727792832125269683229766412454967397152787968043
4765895761265338524573915134381378450051873859153296
3414053689484439722550801196079269028116229367043437
1158371953865778600341946713096653442535523561350392
6374333559024877800931675855665020261424517552023105

1803797924160186816532721349074474187926304637935701
9572546568707696490256283113949083065981392587716575
3432905182988307442209315394562671891365093277852758
5614188691505843112818062116453383461456499861027990
8878315999233208349703499009644828973619972608413030
5016138343757350335026967919910039457648503139889980
4053476207979951035562800942711980771413862537468942
0067112929037940211050993128817678635571212882205845
2590232988278448897285576764337655132098372084536519
7273566294540752078683774825993769508547458537785440
1518668703212703251083788575535253274224674561655301
7529469704928603493523766319377581531269112157125054
5649366284046131575493234361611438689415519179552116
0403279413870405973659682877235554936953672492603352
7449892888220448868443652751589546895588589071831729
2891292345774428441927250527684755038702706328297997
5853882599387906789963676634726367997091137000500451
9151507050208574470532031134283753039645068373494746
5152543161640695883965696047762481007698125762324027
6563247145586781166535633573841332037563285771114579
4773611775891097844959748713454994540500897494312370
2669160022779621516016443144632155674657986969134302
0917375379329537363102934825941848515313457700649437
3909764208589573177423145767288296790675029922315257
3286983302634123352631634902064904297082100632648815
6767632425444687039213337678948960012513626523547256
5170222559556998628425108866896847107872600167332422
1562512429272130805593262213072140936864354996898787
4303526768849221231834492419076374715747446252159745
7646635724275279522289150406427767865661151919331918
1783056716465304813810106667342459156864174457688390
6241920186541022526697061538909990725499842854841956
6819245451974709306142275315129844530918275771513611
8167303580932146032258472352811825504706062154262243
2455144689645726938231665552509598950410934253743085
9997971370042588583403044972671096299697632336077767
4373479878835673010286471384545928791637490145406647

5193948993522123624743661317478304868846315160365922
4357676276234466653958979646879055292390270201075721
8919138214831626852490495848675432931241826413466272
2820936532772839766755726728973193812934194305723962
0723292007186386746670306364601331111642546802512288
9430533112509853860120123607044969978521095859875329
3032771622679823055107676926800022074188490301650050
3853447597101830167378268194361241656963925229474103
5743185176583656034123276433900956511863260791733899
1262772072135161752222552418296124339628251823286968
6254441186238123306403453315560164069574723203836514
5663557498734411685994161655182496042597983926781613
1483180902534507164666442670262761185976491324768295
2727805703223834351506367217706637637402490304659096
2859602719797255378001418201998101813981259504234866
2483440439211364872366629202063939628845314483748901
0260840361484073120067415622915966963669408360326433
4149637120985454752501773669601971461784645155559941
6726373970858649598779532421583282184109391640528356
7907068642107803466075719798914815540054200510730096
2796234727249970112217781656798449194332226334150338
5675308244677341045503274285611574553874214000719284
3017744731423009836576075155127779628101472205306681
7420350596794105098046656313637782517247091409925552
4710368126705138246752117200528494295219748862848985
2778783562106004878127114406349908816459244518980104
4293570832904722016072696604619842607722478310717143
9093492897379507505647105380291618749188699463530135
7293501873206687311501731531091296794862549795815121
6822075712318919091383383534471023659794480804712338
8274440350534679995291335460941392728446513911908360
7622657983981564246382915999044162845276818935327913
5674740322735150689098775472181557499848834669462227
1194343513957275609331877672157428578303301830422251
7049632971612296836752748983294723150697497887414002
1926700672065677292131084935729407635928956182812900
1109784724328303846519743758573685912309817836030314

0528230081302663133041399253399217941576479853470817
8636113772001408570838639437703529183497403741835116
2237004017318826939326308750548756645290932656650302
4394453622727917080038157851325290236510568096258917
9424001638714961212169469925442398674726206005713115
3878388338830780165378838775211593119449359569491793
9405788488606239594441849728928723084955792607211329
7712137238869698636023682916222546471088062109447913
2399015406678160289346942821550627212605417982917817
3824891997338295301682667906177801353365047188633978
4273533585623279185357977922662703802445696829686254
9118748685305497857965989184862186237485563935321563
0489928348655615415406495122104661037654818060250676
5491340332738629416911776263813834785118364105699966
0949202044895026294461684668551060662420883145374012
6877947813985957776990398770739941703256531935561300
0528143599456061261608322358990324895659567527576985
3245744035607612886079681857619771788765561985237527
4472235992720206023891671879081470886706830279398976
8378023337579683748471679204205611884614435084238369
7378594825884978259521414316768984959213193894128750
6959614919327114703587453366081494375469714291952903
1019389435637183593749148307423044934029596281116652
3899581106000938622216226552529766106507452713589497
3344740728167392634862226297134715553293624445799465
0823197908769074458525370345709007744081534167863870
1480996741240038378085234273977881746910805353305091
1944331387304083804306750603068626953212452290166750
3856318585859317437769494157410810572749444384001399
1522952924016806674584246966055691076975969873109507
8451825185768980009394286371219101669807885171057114
4669507031273706962004730035675368235205815249186823
9083974088092648527045816800639153401349373525471509
2352704416192657211004234584808532239809308197012158
6417329130532589871715885516842060650340556996859371
5915621939545955585570093477116811798359958427981955
6435636530938905094196464188924341766121771175457371

4429402729377177659183107443058151531596094826350633
6557238614139208130754146107405127413481388906875208
9651754728644348902015018720183661384172807988272958
2018977486126338360371109414086804414638189975514419
0511520140241876289786882338665288749564740110724599
0553799217515564781980918495587675277828080382262981
8043941563979561725694090929518577447883651594478720
6826785963694547637062382066962023966206659210812781
8321912746808145303142177986735336848938082668189691
2999835199422321272638771597576428521321515883717146
4854288124231224684028390561577968199897855625102710
7062837939943190735797975362287371994784521038316866
8521420822019266723155810117372442375609151489386343
6665265794260371682892815806931590571523794802566191
2687088764750695085011137025788023338180190302100297
5975592681821635953507064188571900594974446797417420
2521309472461919502772132372470257029616316814647621
8464364465179953587775904809172469556739679455373497
1032219369455962778937791938340697253788415502062958
3874830961954204615469902226843474617677113197374866
0008793544360730243366328086536847335066870740890018
4703067698214753133731542862215155131814095414979724
6706763436976964583092867952120199414066540432666834
4081968691862291765441036492080785729242338875506180
9836591222653797288411120130691018576030498329532694
2141884259428662146952768806320825719648671342246985
2641941902223624118633913028417184472482275572337996
9707482002437580371792180734202080536935740618765664
1696077391209098134947021207251972136996423442093054
7846506923744649042088873263022615635791960630923699
1602782364930003449747123779455951240858239709946570
2753667598133047775050505366345747155165583727731007
8578178715303161327684892535760784621147886035180402
9765696058486717567636659308748016099927950787178913
1042038494789432860847970515042833265245718864231983
9993285634226860788344374530927289314609254429906078
7111736766959849633062177514884899337787867859785265

28057054866121737921355212470239532560819067885280383
3242296807554471743774895014302315014696122549489533
8362756944869304674198022922555065087429772758076095
1068798271091938371422909682687285963219428367272424
7744390906003680485278454385481995582874334418909552
3099265929588482897771967505439205771668938552397736
0925820906934305789867423572953120514850903846524931
4006899617373173581622229445541614935787147750627037
6192498036384400160913611713729557661808926386467940
2793656703853057799129885739447837576390926794433365
0549677074228596380872187039958271475800044022240421
4003303590360960548004718847304678286807740989832225
2624531680320340844351093743194993802990812417921108
9542392709654258219584858667992411578844781521955749
8322258335674226798960098032009354865108549467676713
4053103434998643497580002152868358357213659782084357
3260466126057046440920052064374880868404199958540869
7477316017505390253064903620449458476440882040053860
5715251822177935180194147116600865329482810600219159
4469278346037982926881867778488378271314816066812848
0874790430234200377130896464781785599461837510620688
4413586284506303464419139428937623547427775867690146
7822890700609268325225032463995333756672899766025424
6597951963260902742615157481865278192977983681101331
3396516257933184194070269649889513869239612612753695
9206022969008742083472084083318841582683801938335897
3224335136412244321174979404766824167809635203566415
4332541509645019779105414609437498159904457928380288
8013356248181407261142327597289482414188702595745493
4254722746989976877162316099322885042028070838100814
0918873526333183584207407484465733978384298053471060
0237421998721176268333490920907386533795907492808928
3030107207550472450851183334676304759820661789998004
4627448033701965550213204413964236745069537087816973
7996937906163784820116979627072127035848047948858065
8308963231288673402963848241128765952185362411256969
7491990574782803262998612317247930503236377058456987

8577453161038667067555844068240891051181842902580329
8514096157331538756311438547792152983663838215871358
8240820127783840973623264758443526302816647560799932
2148392715632124999083709893094632955985992872843352
1252427433494379023824949445785164936127032642339094
5448086200283535262617529818355252978804650281353991
1284716128115341446038970316546773952587653838444574
6110351561641809273346254142217903310714720310599294
9538959584368857734894952259821038315964206232730714
8371669179896744454184189037251127283530059298273937
4737571099277652356370360647348724784839684203742309
7589988743878765428415935659735883450609361299244925
8746769154280459813281582587299911030078063159248172
2052213206010771492336601003182710066727266488949550
9423368979354810557964237715449541371774079951775014
6669546557410080155793417959830131871546171383822033
3287263136997808093756281698575352925390236568114358
6553982842832417000516419900517643835120057569334304
2180293152368541142405986805738973771720903281648623
8395498050084360235358582554618855942442612928921434
1478982670941767604522251349298729974353338206276224
0633100488377452738118872258181998221942829367666600
0040379914870018568674554412319573518712379305995148
2159548677010540478202585390833356406182622252080286
4866680976315610713764189090236060395412455357038066
7535676524726680367517673845564696693596022634258001
5557208962384036477132142966921934724207987296298616
7579674608295971274857067903467039157865858111273825
7432519039782995445767430575299283028634186242545022
4962419791639827349043594113958984043489575783324648
2216524972533181193055585614050310507654858991552554
2652886287528895545773678742029770378468475636642494
7670848543073541328403616349134710744683298589809511
1020124254488464307122717488658696436722375125740766
3807577968593851823215803790138851324567042252785387
6610135195682865233946040200356733860252055134753079
0074689452614361638124660209433968818299857253465435

5285404636104131214993769216302614831518234694209627
9115494171946607206655284400443565753266414389342772
2090557518423691208034737988670796922839869375088816
1460738382464200081539367400188625730736953499730836
7252810149430436456349752135453195195003507648237036
1845384975636163397442943098863871989880818808674749
5831760222984672501959183717870015464719437744024587
9644193433052737786174502452497071499070005187269292
8345871786309174843878550639754778139797614710297480
5258930692216662225235373449011353986266028141926476
2937097680131872041406668762554595422292493849462711
7755017586202137887676002980515741123780955192781815
9082066363654035686833244556620095160463375225688255
8545829200193063815338736565179457437025887562647321
0773227646623152269937958253816250741193599257543470
3207518963927929216230912990259044551217209318966179
9346949541502186833770152207591130088868902385799152
8263986782465460887462785262268142473318858857241665
1261959000329224404728408961960264923773072793032869
8350719950917336222069042662113793573787896339821927
1111792437518683817576213472922730484110905289312739
7566546440159910892056359539550622684903481778340163
8880747758591060473586645607299400942000631204356230
8116498145514965551300585461173552405216715566604133
3475875987920445592175677563283767227690115416492112
2464223603954033684550113426542474489895459967920364
4242966524827350687996495015740462148251111674016381
2882370549267667972800574632906194561799730944487237
4670630628346193769263728437102506294302398387471804
1127794451518210864000155847579128464012873995097762
9770826226345882505207818345760505308157127681646160
1267456151310391071769738455787322413300300055347195
1166901258113520801563037304690809309792735365856491
3574711350904412759076490299193882008262173939592861
2333657297066464102705878385513189346579626859330479
5602601154503596771014005799333688900402207538482513
9930863716343366007923712406457617650036410612205435

6886881774062530570060230189829110915340711775171244
2370364363715890220116231710263565013024399121540427
0127303916604348528921717678005443537960268144769874
7940557159937783563996621006692741927146810896204073
6111634720258986246474408196120403368752089701088063
3542844369252180174251211967856991105833494499916830
9449469847078063675466677678253837230405284892911730
5480298931061328228524301397442127840108229799225637
4991861619095395092292352403872656334962447446903480
5751356594650462503096250111859963630240365418782445
7074024589488060507416839071505803242418375586267960
4489403118420715618426638993005968351960880991550054
0819116094261561779964945557389362335095602169384530
2940741535422017008850593410802153774416896976552390
0070011310946928000344435606360766131030272873892742
2665249899098159012376515704327731921850284488111933
2011035710571944438712183523225548677264408667340454
4135367403990104641792881141327732957052332339987800
9160267002892904670034550632113551822596454563655802
7046215314706032147678038734544203988775731536419729
4374658678276336231119864674608317162495938051631791
0160217431600363721351355065556811627671648322879623
9003714331634809586892438471169048307896510059110496
5015992831438312018932525166768955897310518020709156
1282127947857682315030996548701378014203423508621889
4451130917415520121250377976572630511758844557918166
1243191479349987937188974667677782724332922702482645
4802849998567554945269468703275037839400366514426856
8208130902094905789962210081407736696556627978958759
9381603739294081898326023119790605145978038449412185
5073472344404641363331714829781976698669655140051818
4541976331055635044884971342236033913005897971734678
2373472329230517388505004636025681998062728258112455
5915860150184390904098641809717100754618847739349112
7357112710753309507903619794617087334466480524178880
6067731106455884142874312055368645075413123789205016
4182455985291702855298234917568151981749535650404537

3588004097369310021016197409940885723368139890685230
5802152257830798584444988490026722154928888612925028
8528135271737803182076280866581987021339186121133602
4618736264912859838570424605478859944208240180919736
2711751540474656341180486288643987511052601860076320
8664032080058809812466828727691582888514535559929721
4513431881771664556450266633627515714226121270282902
3587031467862427302335998951338331069080367912289759
2232090053533983610528084879743470505105124297994696
9587732900812070797287965358392324265767339214438047
0361706529595672993234416869309201866257158203504592
2274601133491784768678310636302367243553709325626949
8230726186313109105016432061267424608679167037793094
0669607135447772041240171387152541478713374566022914
2745368281009292055889007950848372326787186595562128
3765493043122746445977381115639667409274991990309678
3157044379273964166675109789264093117468241878846539
2879439142807191372281945062111996049420141675675141
5522656932859693990054101116477675292564944042879583
5710036845090703458019087499993092734233237906647410
7462898117101040277883382145098316061371850584279038
9539496134598694553433217338838044229221868482471011
7148515834710609975786976196816012437330230684469271
0557893261660012959934985974917184503344610562408400
1095249031129151310207353660669914250974416710891804
4279263850255766220625664347056888812091343129654781
6198453967515482108102441606244493185873512142860108
5815587151941939765526106247809254081424759646627019
1943785507186983496876926575171350176402003599383530
1783027817671022024492886556546201055956741577115904
7285830165422561420054826851371916276898252726600077
0336835926768927117466145886443256295441705121686083
7357165976102782388486067014463296368213637303317464
8717632014278800674249348568445726886782552555092500
6154697582885492108122224766822902775116822369502543
9873245618612099967380501457521453467701080259152981
6042122311632876026457848920881444254178235178772946

3684916863787103355988029352879751316600965034502135
0087861481652756934254915758254478587897790042101592
8011354809715815493253864902115138985775663927058200
4783308103193586172095928503098371977956384664987334
5549013365660629589933126670354255179585895342556852
2216705720637316682093224155465652870620820268533260
0866580058396609069504970302254534936941843479918148
5403175216153188936016989829712382727329618815135404
1870492734852626566640813648637887168029974341992184
0452670036155802038750040963721886553766105646252585
9676231120914555806149237446224865590525941467834123
0133648812086451317814505464179416456723857750904521
7705499758332360916182468663731199597425637392431936
8360663346878883664893997708709923975176942932704315
7163405058351989947721259861246595675803136402007793
3287978651130119476790122849334559372745446777306994
2456260202388754930902233573983039664285659923462394
3430754355766148585186128446617314397997597768447092
9792773827647093562794945093757497580940229719554370
1438592212160580810042397438533045434671191438712266
2709140126153844627736610886518271556640204899738718
5384279740871780398587857487216892636293407937055160
1837140508771496281607873833623355597883713608096663
1521893228751052274037101841254829712856895416419492
7943850639454838617154528632987007434474646146503414
4602561936493892557193423209623857284093622072055176
4698253040064322875603806977314699966010186101840908
3474528089280983391290914925830365117302996765473925
1518450277244844953768047638864019063487296774799021
2485612731663998442736186230885517318239967881715818
3206309699648514729573723694647944254825014483727864
3035426699644315398152771686798446857777731767242149
9306359765181359539276806871032304580251915603646418
4552722886148251459740929971994529105998334724104185
4202720851360543073574876227384079200167634661510906
1471910813300876924398905054283828587174596002008845
7644825190313755480860179403410944189883726523194071

8313705379983523443759548981321534240842874824428098
9888047197105452923399847655171775144109635033144384
1574283608079013413016396157944559087366278909144275
9845229763054393408666782643140163757170561881345065
3637288873684577300189754353864153639381737629018229
6333049441891940659730575385121339862756462498470327
9184151114912113525010468511900896117079021888918806
2488253842283641190655874808838120731232314134423335
3144433609656271921082476403927206088886262852588519
9283013330589057652728295714261949791649958943631773
2474958095984149163996087240559405897409518518453701
0842391107823544795389772207975226175997379931801766
0258416783458521545313578584209699130699520991878609
8861244401060741198637447153099351033428616375680948
5035927570474426589679566193382876884746673876270357
7987555965494014662899892099869716485407230339888393
6761101330378404511307837997043311605332621995442577
0307103968439752796919730812802511262236007775400051
3085974983046454049513097048034261383540913445405641
3410146219371605655280444840088045303964949297382686
5022745282299484577467343378675502800997560510091528
8666465879026257768957124187931583948729738877148353
8424812931191683160601354302997848368635277312029030
2971077830277473895813465194275616066742843607020400
2387686104592077696567626787819706560612033973047229
6548137344619132198858923218674391232224152577419290 7
8225709141401815695728457383362291885079486832949330
5335931935720916763645955813679923869635567492986511
3248271394607316285501241323117372648773982965149234
2674132247288632846021041366966644267728104149594302
7672387634286066448079048426771915985645126086187040
2572744277451430790173615156177315157500598839964014
1880497306997550669101292407530374958155784627683114
8373516100826421056868786356840858920119268252437039
0352517666900923840826467526170926026971040704714815
3102057397997681579182981289235304146491987593615632
2124516827461722779681573302532557352230229683398277

9941603482649856938263973605905623213929485507427648
5329426710589699458926421441196000845353331145040686
5373131957148434854150415172347068715966588934687947
7616050652520532551887794276200067792917428629514803
6393715562492149219289945067840972054346001956298474
4096748624653637113020873814175483338166165615185111
9113468473236553824853198785818181450105386941315804
2894105310850262582815712311114555123885490445347986
7002570776217413802918927623452389391402805293096864
5560208707475029630568566687239774998591135620834859
4264702238540331396655512294052067722982107716988749
0683123218656678892534843739289398243039270631046016
7855928753060178702221330681129914256487266497168032
8549439089540115982149377017032767620929876361529476
1022963864003909946651742860527160651172140132509592
7055929483973612998179810256713853317710667503313178
2787325121501327837748650207033135506227558130481080
0294605171798864186469383014272229157943503979917780
1649052827713029562457091027849444590050250012647562
3251401612039820325502752696951967074235168421191098
2090147345534524738516054502034488652611948457946739
1032494601754606594913347356487848126818801870735259
1838083903679072721987136512691128737953771599527413
4266740529860582672777608419974697064196590299595629
5605560002217637882818096465842429443116043410154032
4161837112411833413369086040738182186785929660060152
6097920430269051432225681436574696554200716104926071
0551621362987930055591213266525433447237515483179612
5640786777424307070087662202918140655021360191663843
8599988612327515290352987034953210752896906114040165
9802188280376805348714902083087191780477531360858414
1065967519604324017985891535324434233629910033903677
2618991404668102761485787215430327575243593205031171
6517034302427376082327100989621495063849691002902577
4167136585044898207551353446944194185199121456681506
8435730758765412713166542356681222737222333875877673
9362287460341063681565186649322813423042420400173053

9139580450340596804482575134055549046416335784381605
8868602279917915156776991848385745819789813620129705
3878672660688951885072200744261029707371283569426697
7933820878091270526740228190344887810682804959591079
4308881004695618358720934323233044096976123771968962
9821199168787098339611345262017695940034586738339782
3414732191382499749914065896773447434028358031537479
8489967619249699851822401768193052100224551085786038
4690568766364128908975155436650656165061922181855860
6395256352043478945916167981232360574968374823438905
5596433504292754772607193219830825380715538852177309
2429941314419026355802110535798866536264151014646071
1985598292895490755148136944093607176494262910340381
7182161439604176985281732820882103690612597704683140
5987658981426853170031066274257408280910233116815975
9586548561781614606434476519302871734300921800100587
3954834544037627646013824276340532946558088847737466
8362561690834309727006997481782424574194262607770979
8914229003500843321192397735909559457468156646947811
0101036963569468678903093375711027620866070878210655
5537792608816413375293915596153910623815381314813176
6276032319888049792167796110491023883227506396471007
6965247441460982262594476729758448101388084101452159
3298873538518073069810094601261678689308602437849860
7208280266924451098153916959736018213287944079223067
5841298484903656303436908701425831419899105413985226
9307546695019990277201199343809980965719482859198767
2414559171595955750006024391473464999909496228073208
9018531774166215707333892838863916441375939741779799
6190645277409657976928253653487828864697225365575452
5228031688471039269429944017744150565410839248185380
9722462655146890300020782127549450527915436981754966
6187513478319186574125583553974077373341601561144528
1501716175117996399406119110086304704795713409531 58
2791149697550614252596618790194047452875733438892990
8002976987789308698733990273247229361876519329728094
6392158801058120917330356069608825523217970057600041

5904488179392298445358037974607129470760820065151683
3656451241281129400207911460827424310952033600285057
8510317812969011196732086099490036742606578833236759
9890317467841882737621128226150437357826128239235832
6023506215025388120503808761075779234110200663883596
4616931815750428606621212402530812757970025787253844
9057874024076775176118282802205700768033173143733222
4972089532223176991483092291852524674905187716587192
8339145127222439107688038146764776825605199162428994
6664158685700335897100581727461170013131727201526453
9575067017238873314438527194969975372458504518511212
4532376009243047239954398933276325847419966021261252
9805661849768230504120572683027889590129847903701012
3634777326720390810711639303289268969858980276042853
0981257919573240805314535999506802816476376786162049
9087220505717926326447013802103274475785095981537692
7943735399559906920110868457276158737474149781321992
2100979463616836883769800680193267246356331393619802
2844660290825749708876116126193917988989141472036055
9936908838930580535936933911450316665837679068253381
0154946336850527021605286589896942257096353454924087
9532449834501523023103683349308340823516829151896416
6715750476290195346765505045433189157265705149877638
4149079126728380317905379403906551343242579313304132
4948076088104697312495453454578562643292457539754436
3110660436528940344384293413102992185638619690395362
2936190101639935285350105729932771839446878649027719
2411969477667967432169166174018371906560463900076521
1961148350720755592910178537877056954207460072534754
6329875918008302027150297749789152839894533255407195
1666575323092649513942114255404511537786456966234680
0500105576656862225655975320006948536438622303798485
6936822387490319549004916657833974366986091833919987
2371947258845288725401284645056630547236271099264278
5702458292237304220010398925143760741811976799800496
1158488903136574404814727769349793351969079124128680
4950501774453583056740426732857897572640251681129144

0172893889390600786220339806661965078580853482490794
3715105931869232064049673865635312813040791072222135
7665482187805198585300198832071946026351214279937006
9407085655958724681365543416712160070267748292362040
1452985056021224418548337825955416419100110698441606
1119361341572843855737682243702736802105490498596516
5829729445551918241516040655118397072027202084640204
3930729863001390554348608057272087118125877938449849
0437052921037497010016639981519494762949998642849373
6752536317521883313087108880797883924177046278893607
7376914701380205788950494781158875639904502685755056
1741605589946250346009210210935213094767593435082242
2873652738883742321134710601092049395617317488537802
2731466288416038867881534237539156003774078668328693
9848348080670719236001585719202923111341735102217455
9411995983544456137561917963110704018047448038094383
9754826744551977505936659329500786951398347929873388
7810177945608175447381355918082998124982315003735066
2654337764521831661729592356655050362988711935601204
1679383725200771593141903519272458016944939389396886
1290001191170558851515798078321975863643962234115591
2478451870829040220207055268885676776757208433019621
5790085294712798233970767046678343101904313793909567
4184931794875599199055140961968939225573319387182240
1654043894242976165912825960645565767896269500675457
6610574970349472098549641722192264151810279891105903
3065391546659670220214954542925225680119973223318629
9301288977260054888305180190736561784872448961557321
6482574547538161434670840182571136375339420168400511
4429600303082324276272440249343956105593933073782790
9395440108058510853811441266551615428095286811705096
0782891078997195299893421677946200201699898496514405
5336949093144156637478982789278074171170977983171522
5227691017629063753678298686927280589881500482306979
0734921618995536707907033479375433604945307207964877
0163350386612477167988940461720890124337095817160070
9412249362154964957549239133890521379284813256006540

7087219295204117451146578362108110624285418078166194
9418458010948482260635786400953180380553175259088246
7441944083716975126383772221899903516608185037840103
6964191489110716279202497884078503577014056163478742
2640000635581746899577459816171765424735821336343905
5746336700418263131436114181603329667629676001679942
0553340364035181666054990897216378910193314935297881
6209395419682065819064283642662413237059039256804645
4636658820270576326491829158713636046873638450544974
8883936255634469025847997339724378268667920048942544
0222386948917200466558257328802332499435398108946466
2938621067826158527899573711364825491949666999004651
4833047867362138961073979915734993372656791738205067
9489033578517034801568487119057331503073236481575437
1927707826788849883018486465398211832288477459906823
3974621561258538266237228319698602520433786277355751
5550720101605978712174650697369997748850582368618239
7405962823899061718458168239669939404234038027152149
3416458066506094255166330604931071671973308360331180
9122267282616536491778154547133361395136935789707903
8129100817208674970566963962750528372743979162876486
4557681076979043877788536898437300360143129486649262
3107727490370595621425875142932668782847880787628478
1862459567816866258302682036445978850982950892525944
1721135355859411423995153512414886100581148133714476
2057489372241692219063913137828116201787876086438886
0506587083240869863946519451168883795277463535977800
5996422511881275601601791590224752139769106798632092
8338406008610218467139811866120510377179678586471588
9119197808206511049872092732936744466455522783331556
1279846519883483619760801583131790674551241002086775
2382206554614559397489786921632321555311929060258852
7663367002810850040367760202462557785265266977034006
9592157761564764259634743356471602185512767526489221
6759980154725911835301779011021484702267325079585252
7548423162615892967491280149800575414928943724074644
3811161096891279255334866504394777647016689704662497

02173347336807947732108619344342264216085801303502460
37111089416681026503673752140328243429336919373726379
8916898338751375557910265452952313728785719567272503
5232725011490885240111221231522395356681417563608279
6889420199323245274911500056806619067100730562113125
6401829504993342817836112004413870830454117779908282
9852391031652175532217390838633772407026085160186509
7547228451197539120397627596019905383829494982268416
0694237074685285118596668768797229864602757184502991
2469206489946948977483050501351974592227789428049458
8893626617557558501668491138034540655338404509254779
5648368531413220672805295322177576799732975000077209
0302585104432463993406745109433124358732359862858793
6132286240082781507656081553948158074626359271910622
8645291219548991273889934768984063069350153058083956
5139440243232506662242987659239682694103113083415119
6755536137839854221179219411449561916538849185979570
7643267369745935938106687751104593905615875963496617
9567758163129073143952002117162416023638780963299013
3247037859386552914051894854020217477434941180746937
8036532613139946208945557490603976970949550080264968
7908339222077306303315271994479489978198182895663978
7262359965053845084022616071288706919179553483412729
1556345078392909554962176378094144760392676202991209
7978526101261615871270753558678860701332293741480098
0534901978765117235013926032028283193813753046174386
1854870731522878960438016260215193217278125010153660
9553959394826629785009647214769630414209285608367553
6971457674465825137792689300349818767309536656154094
0181812138145157189348521141376323974759124935970455
1181113512626385218226841422290576167233622174163859
6959428555841526603258173569687347871528253854686617
0831715678979869207966857938520386732793631311097017
5090915363971577478546692389841880008598498493115913
7283608134143739359840877268138815763164529904003317
0061435515871321048490860408630995545829324985126941
5089658896620196441030335698575787971913718747479419

8902398557644542753175912782182067919400959794488980
2165519947491076702194965961007134306836058843980625
0293333929180504378749584578395597521918499399156656
8420764440157702637349736079804389437323371035779462
9495956975286424169076671625974311702801304335272139
2098959101629315031440616760330448713108889303292520
9532460424387158071374113333496752189467842043520120
0510171558881014072790337090139060829623257365434902
2515857159734383964325496828242539277712422357476365
1474626863042160036737390572809741518026332536477880
6297783650316962478876630494658904139814536834964837
6763797240301315465398808417396936142586717938191404
8143126221184901047710700206513634019865582459549151
8936088602077543374425879392350378338502501167727052
7206099193802611795015938187100439454786426117842514
6172560272380476328699796611311614717459878855420378
9603048511936947192108774950855386289958503777799044
7987976819174494142900098035975024431445721638798732
5933347484330457394081255124793772083829886046552487
1445206812031443287202182491713332412389413032203595
5727805144049560581365140986160079051007464455743265
3939453916473024455409044935918995009300701806172333
7167982022317326195486552606537659624069393912796408
9260841611488033064649940540746459731482403965686343
2159312994393707782160027095587125926393806265309063
7227159030214454082789035367016709815238983270423840
0163481012749869965938331877195079369730835694367288
5650611025094535204704190595424682019898279610699732
2621366281673631229890562473777629237557610116540065
5938523727391506690645767946392058971652286497743450
2284140350888083093446181683666737157251416382605626
8400920108841378388920760123000864027233105874037792
3680777236337609996334549142295140263407406950003571
9201635541039052403529072732698238914645854980607121
9716833951774684572732705783001650434385226732890660
4188841138250834136137996198101697052736772471979593
2611776223445398195320472440733945521293754849964621

2828779894759263556471124809844788635485768440400796
4540865343117844698666996315507153553467533182789421
7490529540847002365155937191300707653206101646242846
0575445913627408024944264302474203872311368140350116
4916738803397962812882370862631477737109509252116696
6772840005669652355331235727344712250584933421000654
3804671533515249182317846165102580881801644604999405
8848908523540861518389404360719340671292367260694480
6459997807772492209029038624407445869701420590221988
8068460965151709482306000964657014364407668066372796
8756940713515678279906490790632772090445245032485757
9280708225203262396833548515860693145978385283616945
6336186314956895751252054275834084337654387735755811
3323322445874169130446233328853040341681853276507962
9533256719680304662622269392924233817614740621769438
1301816446836250602887868826388486228440768649870505
4382292580658487040403556257325674952011143193754775
4077091962079537181488250616630692548318888716236852
5155484810807574535347566506910899890257900674711653
6104463548486258453296011166055248123665813348443402
3033838766947308531146822952900928520339707297247845
9503263973047096928772755667410787921270676969147191
6296467784842643755418575698510816587182715147495503
6156500371320738054521273860737134932832995067948381
0466721696131667455649838641066553851957114779798978
0405313153130469535595601250122307301351228235903089
7413153187246065767650692731207540535622805396956670
9404554331001699200934016030151096700708703308962865
8695481117116447222429564592624702284383736316827618
2638145452738190115715089522016695558952554622067920
7427677769230811522647511068244339134165002524040693
3025859894563392703641944075397812008218213585045547
3521574804387050965377446147811346871656555888351972
7921318504550683553943076050369009362962387836408667
0129443989385444418785088158837508762900011444690128
8812955855677523752165986849343242206433312659157488
7295399533862251753806821534505112447440946911045116

3239958515057551231258294012367477151579026785466343
7832997684375181671324663954643020788768384514133131
1200438362720947289667797439488890340393339347714916
5560987844062612358122351295492562595858319163624484
4911239536576587105078807396708810976691210513367202
1088490924983481410808514719793119832560149134071181
6926537717669488632567738508113099293924279731126967
7458349060895901467971121975453286879491003849694060
7005645672377105349189064450652802955744757018578593
5333025343731414420530358189472502565974199223900855
0333847929662370767233463629037599500875430750257963
9741502039884959037585768291001734410030163901748485
6330642201175201778857984274579591250260727814703573
1188031082540722337056184039814643086670132641399107
3500244188772556351318020125185808148975409736973081
4873054088717373474428409192916434244021024496426392
8735739824038105842003734586953107032797985022930946
6268069017927241787098062523829754926742717401093133
9084400603177150398839717565171566425066140863519754
5734597473285460352577081637908053873158069228053306
9866107176170472318941722385413267567686410850693619
7728508808999120593229478701723659957911254740490230
4217356116439548935384403866196678322738363099111005
2858370782496250614551882569385163576399303075590707
4091779176896009094216626863994593098716651875276280
6121677655991791029060977988760291131293858955350180
1828282284251271767414232794377249083446842157094679
0104913429397381565935133360706191251839663489878904
9470872644134458081024139652538971443908917775225241
1780220168874982430162373290165423895888029875750062
8310445539487277012493083152494979747126764811041792
3690325649790862147591407233856998568489932072281268
3150370989919313076022276809177596019419663136065337
5426514541707897265642169499127767201935651871297423
8974200774227060081833146868926602940980858395534529
8132643374294839713715782663489388817585285996432152
4684920221705042364386297153170378612025782854723968

5501094726486865273393613270531709184960842867973063
0043616542134626766101017003598757970699862232054 88
0264185324862925109616879659807695389765453614545744
5540016522391424814892972938142790625588597012238728
3489024057385524642344391199345027206577171521049912
7908992116992426409704094162072318039496941688985426
5615303280722468255424581111427009573232719015598853
7895755711619245963123390013892387272152786124203816
8148964678214166675876691828545852443941373067714640
3734330940413644769293578325756754722460492377254530
6631226140550175638115999431970278836561469974535618
6625199217747587896680220466677625977438338995660390
4036282986148270213861905360663668457915145149129662
4149189690080815398786558385378115703426603443048225
5013197866047676271115191413296063961296795675148556
0535966427176487338775484216680732679344682737453566
1080150860574339919862152957878761118559244723527131
6900900727602292778572040739492840810280038898566540
2155563337562229145898264058171848809035219592322984
5591916946392957967530091549871090141039887383479249
3628931057971150462061769010546893013669125649607645
5191053362731791560064596482747654805723188947139841
0986013028648665616266629576250098178394457435203937
9491631861623245084104361645553981702333968280754081
6067678923505102764705204099569714819307832159932255
6225791336901779370937542504178257657070596223970542
4120671641874246415575661781751832110091846264871776
5091190334572307787317880484937765442539452471494240
9147933707351348787631457698510024967498296725718389
5783784649478639854402312145464070231609321036055594
6195476083184107815497585524494732214389320523373497
5829477293697854244733192165853383175552249465848759
3974612031368176924912879005517840370751610615086328
4334456738495665891503494240520078974138322121492467
1798085284634286820447027578369882865730447370179875
4988339182164436320438360275261130900446100374779902
7949076124599381124051619199605965139007790963429358

3119034305624356715734095056163628748278205876154899
8813622840063195111952017808096749067049765894282031
9324591324255961711416431669441618015524066188633117
3995068796437781553809722292974598686743643437724460
2250082170131493699181402454209157676839550131681810
7403428304112686525498680326457932318450295097745201
3898055358194141019131984833867985554819940171661584
8836118148504918646756391772863305837466512519099517
6278621822177837392160428438123658550359877683167988
7691766786403769600339672806404573452597520193285403
9038432784751855614753102263637335938463979994511977
3456856654467428389581911035087502447542075011754726
5579343040641644816400018548961723573698765002091463
1244006805066561821132029336859567547226664660246868
5420080392607405662982996622827886673306450655032884
1629388295625887554096946807070658121980508578924056
6782073019132006706036421683649165256313537762548308
9594842053609872295558275245931500943341900907815467
1122572710798521222737556158326103023920531679278861
4118329822645975576534045423461049029572239587533119
5796195022894605030835697403198482479375038973982795
3899206897189129627097026816017999488236851544102490
6345037933046385059805006893688509215960924934546170
1869664702253261938819301682122648436877795239618136
8777350178398760142879720848366410773287642886925358
7146978397261888103384503337118151711484715753572829
1113771363131977821202456846096977749219637968473749
6910945644214623527467527261285301800789573543032753
5505890027607241710824277977227327359006266638666960
1352230256085097299631537578243379250716217207440633
3137963875171443926623811455393900438867851842471758
7537903066366609326888319312103232072351409070336054
1657508220906033720166811388503146846445191695043655
8866152125950693828434458152870871228293140727555933
6996812109903515910164212560711075765634430635117056
7316743528958194754952161142519311002589041345289007
5077583181812267074861693705113747405144794546197953

01747600610679345348377394055213592998818346545867 98
2587586043173404055460118233649355330635908322439166
6806121029285929309962757245031274894390964296332087
3077467150077733008339343158859637012443376957769454
8260771609767086154816667942389103506090461460441303
9687136894886799835087804068064381621772406347917811
9162006295777701399370934394432172497221823195212537
9413260275336745685586088441059085112302706065537968
9486119033343118293391087619618565414570968938743695
7061234280197733557389624076816315844335848770733607
2070640126367241684125509830095138195768861512464866
0191044190004053873335671201528782626114531444001949
0501156417180146235530033460802176758915614799503714
6733274581502728112721182646692255544413188398595093
3419623985945561184947674786532207149201414043587348
3891207105258163644906920398813879272899285398846067
9469997338628784322251003743280666592641993060846936
1675677484717995553822497857465567250556748949309600
0388111650259900059937401738666047062621238852848170
1094671013876835220025370049094466710475579000862748
6998607580100559897397752748532074183466193993789997
6107539930251144261568920485519723078407582278483831
2358647816828634723970507033770155108037216863941507
1758912025235200309364453816100089088130502039169341
1591082327549299699784135448332296718754241822465520
0379622790431069770241676548293949761640495002830983
8939426022430461690484355804747722403871786693491539
2885786302298924314368417304703015701090230606750370
2447200332641348728560100321972365652015909492934148
2621229998231733207306487960120379727647315563630376
0929383734234682091833203420880375831996892409927493
6352907356498472392751796483564603811318074452718462
2345859794497228431805274062500057844200483005238238
7510848554266486618405878804641206810359198983960987
2713115064108184549045557992760943542184006717645354
8615105282475682962685981806029377282987924425294387
0854120731025294049832789179127749003152175521488252

6034714160181953845417671180625218368758194154070367
6815615766181720477998692331464033613803346520401842
6158039026418253618572246844860612887368699927202741
6268063766621120692903461969545811364344741598714018
9211660466226582661590542069763943593123662045528756
0342165003473601194342256149140320157941711851715427
5639651725686453846709545248371305948985825597452775
6437837209390373760644875780538089666661399183963055
4346351531548185886779262912725363426288985256854464
6981449746189241495863663671981400650685888608602242
6733798812768796940649702991545245272132575428195324
9173115066208586652077490952965100753404049227356548
2829570256906293588816904146510697177724209554461302
5854381786304850806058990637380905430695026138424222
7053753545985909932669673321651519481725345327473336
0274472585272453947857870490548475863311571836633235
9132347588259340641521039072871936326796375928473313
1612339781549856507745957426301925013613442181778657
326845949803925741969699987645982495670940955954906
4514319299753296990292901811334684918939731673374047
3761021534979028013172233791279986391471010573645808
8249640377936691442602252243291822035969479652296324
1504625930376366432840865616023121610990271779794048
2442374377242175453274369030749262617258880652233261
4106033816532093232026699108708475868198563990498575
0117619963690596992541043687532918190720041575982623
4546672701573697113335704140320937934512660607079906
5586879616157998493410954090321243654431087316158637
5727374817450178665573793984866922911759920434224760
0648597600549782806294187391474966456601976892636165
9828965655744580409914268909472497067352204701161915
3600945273632530666440201002320187322781976148686634
8989132734701444820324293117841009152833330376911197
0512525118902970829429748839813714997780527816493437
0560432600535269810691898658868981610889926992043454
7815535740465293822554757923965125781698498673421821
8531240734311529608214112091999940616010158219127373

0165017695811861903668977927569046785710181059393731
4388119291474944356522189626028632586636519017453659
2121863876810777420915836469090916518273980753103066
4998062448492774756188452973294727139148997268407785
8977868656048723305752422857172473443666741818123270
8415917914786168197800328752421946480119315937935152
3941740904190984125780990903889407794204672704915345
0002486042745530730683647222075893082199445234721452
1842825620929143768143878021396936399786022126322109
8220773571441294126406537652642854348297069365511680
6728306148155350067799273428746717408366660205372922
0484840870252301258571791456966579523963596862706459
0372072680587943981340066767014117658125223348104838
6867717405879736896155996217846549730739340954660431
4586015404956157764561673447212642746874614083020638
7935980428496622472232550466095231768172458636261128
4833874076507582889567745895887361095197420723212623
3355615193382471760218186838953006797502120769043883
8128356258145012500711902715651443462706481495059019
6996139040560790673726072411292447199477028384835318
6332981761999473128331449159687775042903279770347619
3806295512384013842291003576769296985959054439828926
6660878010344059670559052076683702101595219513945447
7311874107235279456084435494667607928826819356764666
5891613621404247856427926815625448506316685268327524
6560027477652412794270534193482805462670281599245391
2487393755809401259197783465336557356259376680876997
5257460270216696492529983775381969384625470851886151
0475226475136489188335738189169712195832681004195765
3770211921182725665088966824675048666996598850420411
9161533208988567238092601814281176234479558294288085
8136988607372754974912130687437421943014866766259691
6951953688569117493823196975595579402179388737225151
5997140544728970839551036548666281965036848684661039
2745568414363590177744038756512080771456364877728443
8331563572496836756314810394389680921192823374145076
1863056577773779414533302578207887933713552328193066

7781034744878814533022767115982463014677391318064699
0171453231819572964017138299344586652810423902920440
4205030108503722889707226673204164075838352116255472
9354209770671865262076294672043633643273505663204112
5212872258941499762804689148555075973614650511764125
7930283697453203604650615692381197213251056180613452
1343817681732762841418102164413417024917558436568114
5479777519582806628442497503791747723620420424505736
3060911154840242699564336003530143666597511828032803
9534210540491910305673142107541302357934877234239073
9593893004368913281485468821970534826806406133447603
5746590906564609800971717699575204375563545216850442
2439082780834807215680031213805137447557386633580661
2898256952535572326630739519540636979469139273919509
2970992913488014867607271497851682702690507107678965
7653044046336262688872262742931202875173849755421431
5049989074564612156955425357015201474626587358829154
4869367168210972638770366929876270333269267640511235
6591787736324770461161187528378640880382801363490414
5085131829558738033657159237554043740366009431266249
3744735018369408835012318057025510894184693666883038
0623604361689986818150462814543993708439467237952819
5303288606099631547805411053979831739104819179879337
6309918240071636953592567835866999085256828346179992
0448492158286825542660666948065905588375476747789006
3030763977320119162644193123137332822364181193438305
0586585544982998689911466813117119421891601793736025
7596321853210486874992047347029942678712761333423766
8342822565756501574897202803431803206244849572309739
0057150931453891843449438283973734515799805156259164
3226270141862062369447593021410508502812036049109939
0536847805602166270463685525727413290604228995635102
5228475190703082532738499555330449578033090259275311
6352218098898826291159803371125701721767669045456806
4922305157474689155717101756724035418935061128887302
4043144319869586521866760733038549036027746096354501
9525296534070301597032409851150252930588656719011250

8328471496806489438130078371862392446817902162173569
1228872394802164846473775176818942122207103565559650
7979484499008271935528147914340403713871720817609031
9885645874610810991059369177437947128793689503247741
8648580648196799561464367082487089936839513150720305
6530300788685982036072066991673761641475665428771935
3591044521169167692816396364562978340737064788274061
8340543570774121613832028737879037858589327455361495
6446125505055478266874544550908868894692718598894942
3449507483821850184341303236200466807001917504592628
4083850536431267686980340268115807098103458986041308
4173550995694415179917543233480633073261339139979788
0003821091327660145496111570428580281606751623381353
8630524295638330950320192878164132492230047611795358
2495605914530004246447880621302468978655969262925763
5740287994013569211675539040026996645560256829723695
0495909960271709738166616867800483272929905942810296
8161574696306060610106226717021253966747951389381374
8538953517578329451364260069361377774400056699301740
2011366773178770446945060129226074969110615762076378
9334504137336511956003290783523596465326574749743528
6721116298075675851083850160699896935867152259646305
7009139087628741649254170816909684730954886023329828
8800101652962808976987723196907421027009348880934455
5121881883365197843185355960674763507221611787287356
3946565543427614068559412425912141707811630308010192
5876219938098958943050939682518277130330334926648853
2956187952664463190634938964927974779630581823776065
8035402279669108180932267251426142876850855023403639
5970562141251513762046722444289799785545413190137205
6002945670634038242991780751257244844635771525893472
2336845268000504905794076592611834046276469983532198
9331271173271051129387746272008131726320867124724783
1036953250571166846693391999383834165301330924772942
9358470743882634240070132171320972827494796116356678
2615071566282502125206276389751566585134060455290109
2611263816622423668992710649041886270014421128735923

9399825005658006432506075094589358500721807293223082
4610825818587467133406733443268051547565276112290094
2154655618310341268701722865416226907574466637402578
5058198390337266853912834251138897761037015547059697
8428438121104166749642361936898406356138762279595452
1514689288132823943963661294501387816765909292508975
3247166912368348275154607827280445182434537596550492
5680648532309992814514753455092755527796279414245574
4052552188253239556250857211799635301956758117744367
6255097319751155651484513728542407490483808199558809
1516117939219104261909284857816139051482314744553101
5161591784145999471223905165969441039727295741158339
7459390620500775807095967658989249614373434878156377
4420837499914618345134117857213565765100288216534378
8389069722998862946226868519562883232057053789421789
5936784489997283808225129585737429565314870430403219
5433016472204581397905745288020747285881031140858798
7472118632864898447340531146044004800111896474216571
9993115889037205900314664181261792091194088785006677
8010999185461909348926691850911899853281758015614963
4425272742021323083262625378297537478084964862057473
7729805950490214555010896934806451097433300599798555
3203313182691751915919500655848843655165633523507949
7441486995545934682666761749841931923239198584931429
0703397249074104335315507161857712844527045561514872
0821787901075809945990781903407684845590803485256121
1244638832388776007725640595058576145061729291017634
2106201632276313357086984161133384335191595521993423
9104565565973100823169947997845631959834114305637058
8841358572324394599937160845154432247333257474326902
9321113405536052609072416883753354361412870234237825
2708953823576585165964173281100423670224794265894895
4200246277977930989345076021362417137031065132759759
6134899242700470471725333264227742623475663202251798
4249524982127655951139456814127007662315485158257359
6335382671792266993633391806685509480289663970163358
0306357779206151292995868181602332782682875613869534

8983385807840486775650291623281022321262862370364711
6725909111622074499444703371546717193557960340649572
1839913243157175868905635782011935622777763421623672
4756676876172206101521338942884427554638039142982934
0970637684665116996845497749248421623454987901900590
1223071102488058804992082026734152133452619482271804
0452465922884429554190815449214580890457049392729832
1688341342995368258674641083672846833132874321981230
0745130314440076835740244627809770130021421871235118
8443907814286457148671303162743814053388576038852946
9050776911243964028442753921160112468485588590871112
0433136983355129831770447543952735118094562627365072
1342386754878162350966398758989078105843337220397544
2049861899214534394570010766993023334045070623558228
3219793912107163390447845740649596837368359996102705
5981093432754571822658191627376408326491944633925414
1257047278930012181409950288177663218498545308566679
0070376265770583827317495612091752324760127284885442
2298986799882662839508678945771438815809675505801148
1632951013417845494953248474350246166826049086054349
8626070769522068457693088810199363886020569081016645
0518721815005743472746924004566280135244242904323796
8476832649959785995418767960988824064974266522958440
6972977619146264897831510287400669792362033206040445
3547944198199894752531957170520855316177791641077276
4613921571177402991355550151670979661965240622229141
6099722702986540871469079193291110174604012065208119
0334793350740933967335438106677644251621091064097827
7403934724092269713998641932401833676257879516616692
1148570844034731109857718604143806570305811408965227
9903370706792777717324019606366905990239660061747596
6027436477233417132112564069573043007387189697608262
5377840761689045650369641210172605511050631805878307
1771045716789172093155510051312624885074971257088277
6081846297351566606413813185775692321942216169819818
3386159109111212964068347645414987489259676939155208
8999743483482510977171733477484904241570447657365745

7752887031370873919072182505772993177204212596617789
0946278107374893967275333669497759775761401339096400
5994959912477424058226022767434791404365977500711749
0687276269345637528762791538610334809869295742898490
0870413755216037594678764398436269591372899723772007
3145336673352699683662926085657058390388235938311896
1476361331704366529764436074940164969717395660023248
4753127826135116274516934985914972842578011566354011
2574883834495782054756513486752535339309492987778566
8247383224713634127815663824590502307339832536083003
8730248396399541840286629808766899600543606746374781
7597385932001097038400943290848252148607855800720302
8392624814842107356768943650847829172394313537730783
3828620452560793714347989024775448000157389116577894
9114366003679354363932067762630211052152101359215469
4544997058887628365793340606132391102338123478911651
3396064823273430276115853432570782259674566926399445
0654928199905402788150706299036213505045231732501367
0624394106180766761378661414563785766044207492422926
9770349845012651492167176963310516767267848837295490
0567865723978442763113473249771989060076187591408960
7306658215138449134561555571151084513215973829110011
1428738959946162393850583603232344208288145043391507
3780818631937207838031136417583734875197650793352053
5426106483964680022831803234766267824389703438282856
7674099388012245584182825061658871841917739713484249
5575355361542835106244832821141075660967969951048252
2794215870693151868682289065990544542897676834078428
4686635169507359290059446525171865184120454432711637
4524591249405103177497432716433047861042520357940327
1210384808446436333013715290149884275237794983457479
9875281511965159434402988721491706555591849396373620
2353463688821813143720813493658980418852618146166856
4638974283915059558370394712383217345273482136156961
2832630851650382152956508420824710834855636841528927
7977777563665982862156502212507461167590321165420965
4147012294283854575672141738992979980052464181684733

4818021732521988211950351846705828084292711592599970
1537509747900795093033188056558010139808122984565468
1618715385799281286103766006844083084983972797637004
1660306524614826604283117793289355874205593451881326
4760502991798338271637395986023588785134609267323195
1588729729246677097635098946770140282292245909364931
9251931742859694152366564659951119254685886480010377
7931630738794443787115163435808683855098036167341177
4679552505415696325551756593001109871034383335920075
2031771871321169429737366650481616228139743691943602
1979331891137147757928460485105745428328748179328722
7871096158968260415191032160358867794747925929885099
9499049216484971385441255339167379832698374244874127
4547096206337139047404836079415752936336390448179541
4111734936202604851201230536055436888793862914480787
9100525559893282527733543592470673999726922739927756
5339725669671524096773073517612992214202555065314295
4908742617053358503331777460258915289378865430766613
9718694743502046969668750567663782500133665443412014
9780395334094843058804080871386953158989175443230557
6148445992971287256207647002380883308255786375448067
3545385987638085847636506952673628098611990040267981
1302046101019445980359274353057449296242273773395520
1691580053320669755130197311337039125611913305432979
4169919294829390557267957952817726968793291142577732
0500214760199698152202061524517942389845818526827978
9797440541656480562100833028605473010316050320145740
5188876198453502969999264657956500190448278241738609
5401791037025463814362445250887029226639233664043519
6780035665454436425036250795296489213906564841401827
3697145021452643164662100373809950418558874896308246
5740733426300253099497815591421287519705271001157101
7163774988337879565686539344266119544077414439924874
2320604226615977147800654896433496235839622113531257
2898201355984815657020457482109173737866280253930704
4655436889747490658477949409598122447874322186601530
0973454802469617267129477879519441513440173011883236

7448720447806236637003524258617656838084857368856902
3709229088212227208341700897912925654354194078268993
0557315191699011807050189588596474048139046260970734
1954494227067754033537102964836062032755617310215914
3468441553909565909746549927915376332947359960200898
0264079468292536127789832058536058561861189451406113
2224271286111668672502570336348863158571739711603288
2123866374918745662604295318985208791769279853205968
7082732060772049087562497623985063918737880636107372
1159075733629897056657468837735159852306207834856672
6739945019724570616041702865614312668507975571689508
0673861961335761307793385665294261178995576128220396
2786163036697657292746317600522624881300262015934469
5993323013044439085141440696129490951435113240437281
8037803052701614745966697318233961655755091009447720
1123130169927082110437284835364658022798435317648760
4395208073622595864078734581915310594558252403107537
7215743214571344778184309844749991891988572348815392
1904520156617192555429987533454002381109748520270053
7114712389797470101585475386268028132317028620129038
6585144236448649286552358171372029388017529410102252
0285691187186079645010876529842532971631174418650752
0187692927464210613131762030850679066588256061616245
1232983338560212251996132072858164070209062312718344
8482230078940409243916327452408745667813673557003975
5142780739200343533132128228732868954206188188329141
8904239295829349293059481391941921015430324356744769
9243068489549520239545645503065047097125895308641001
1569687327280575346288820928949980376942767152372469
0745276874941173761124890701919879296423247494837186
3913292204033404762728529048057020058877226359767881
7471579821421764176643490054851532223351305020047426
6084260275811143081158773578536814136568366713391740
3160705650379085284322678216464359301519207099123433
8444761974897013637518951684986306334712439357199345
3325119243402486722685096712244422809548620967064207
1460389235934010684400475369638649751359735833109378

3260019092515744319203761229377089055745436284773845
3351375649811641233689229522888668519991591629961786
7208191837173473070068228127018606502753029833901466
9995747944610702671611160278717065002344548526553180
4615279803013588943109664382224754397716230467646353
1802819964493756623711511605197875870834292146749800
0152971410916725705796936154875615717823583111771013
5901253539556871274579972017592606546190059379608978
4619027221814507235879548427149913315039620305124110
9165044196697558251321818617835694275540614559727053
5238267357110231808197208539486018220548268986636028
6956681834866852454461440840695218263328048760444691
9008266967629645490754572236923327441664913195564864
3994588993387758700988054331863998554049323067761591
8285928743896057804640984208974055068329611397239222
2269790364697767875517303966644741574726584654780659
6395648953581955700357971668912269469927151284486477
2273981174181488663319289946594706008921131898942967
7196570486185276861342368815000041723800282976700527
7922765540848554333486168898497387186788618987323238
0042400963864067984351716251126972592465867872110705
3801531949577164948506298157989469417142820421641655
8665990728619849384917548026958461964229477931498122
3836415385570380897890076139010323497179696325471965
6491227455826354132341424364357459474929792785696077
6359148472801212182057123722912544332455660534074849
5181446769058959806952003492300124986619376210850051
2364425478264357338213296609669731653535425624730809
0288177611337397206298364305054086194062218385024498
5475668721260067633974373153257838354874824409780973
9336148731020239045338094741597766456031376811062989
2140901661232700390050502294761351885912410647065603
1298014608889949278623547812337075637352432127180061
0530855171703405360330737630187113669353217698428260
1761121860063584896534143606709141997792464097211427
4495469891463555482864340110401472230084740058971939
4255567755784403993657012637700923304017720157097140

2261897254902499639625668908485897750415713042927159
2893380146276281780424512433461156729170872181169866
9587131261066558109715515556963348198442249377278998
4900169133406501392255837452536445871131537749642845
1540053640421859769093297900720083629620246732239658
8374131752905866262694467352104426037919315211035606
1561327177958332423894101137830862454629514095817189
4165381882609858136255070281472074410103228385697701
2126679321464728011596824377110164058829203812229822
8255056488502000903159908409669801425015995974256102
2026318361725571114913921443611018533845568286070315
7664486057682509194621585069508284940853018066449912
0714286759898668841279064653948357153197916296821916
4287626912567216947228774367226907463108534195440611
3360848716300832207815371548154354645837025948503616
7470707302758499672313280960162936123357408505167586
7047032285987247398086473267940394913937913084973632
4141390294392845766383519842186763466430180868960692
1433831960422100947209843976665228225443683042220547
0143256511942687039719929342480639118311471596792882
7659016065848116137229441994332637690168222434355926
0799088203400234350990585913929771576049472707702830
9575842707091369770437135759020267227121355344973304
3050674103705440773459584309227693748403956382048859
2647043863621119935422550000256114497085052705009162
8929604984739830597708931412041983770187000638580784
4261773612787580995515950384666207484815725018212354
0825433798202752568074574177955302589468194577646065
3749932699328882053951518474085147443635682109242501
7105258634953457915902871522129953659591580769737340
6864938135614834686259397425293493723922991126095352
7890147415209461916937523396079180058881837850668857
8822174089302237670789264905590178357330290498610475
6624088404630174420916460792765924033500521369750536
6658033405013128071242798970006273729152590701666746
4372456678743256001955542420944533633439391956768045
9087133098344796212966325811637654664808900711247402

8151564343463108144532733971543233454492686193074483
7353290342429022706323445090815537951237757161049271
2179818535358509941553871821944942708611496926208222
8311054505260176469442149847969162493833508643899797
9437257258503494733211236420156454459583232102586733
8212082720110236171813291628134769361336231630005551
8541699451237030874717693475306380949148828231811460
1484735917650196830571470471397168054171207608541527
7906148408154352770547111386619355191825554967376875
3756559018915892076743526148852937632107873123875720
6484082373753032884640234882928758674575174119642595
5473254712998878441337697174478289514806062975015981
0410965179248737254241586060709263434511221805151576
8050395232079839070384455908024897675124281118683112
2353594495362332805156428450911638253284682769037798
9405609730496598216774794290605229064271154090759630
4907500457866948064424079141604249897363962238355342
6568824892490309401353227598292267618021399446801899
6032067580342939794795005811235633986972790236701977
6289840733198614297089945534090372605768223447488692
0102976488719461875862934213170932727776916518710024
9899666585508381133567896174811392400806920441462566
5382945223391551604594824870277524026560308024160840
6335831049915931308539474269073234720884119182024847
0757329115072124544689852685531159985111793938108020
9773473817493399988897385653699403875952533621748239
4715478348005894603936659188928996217521047163046603
8444412373450910382932483894560012983649341732042243
2165642758286269646298705494370864746342771185293824
8043935821601986070062171196591832591807574493655660
3190574210330697537073223014044293913607299742314821
8620571279724084322974222530647847022287728988917204
4541364168301862059266947810650015772730346839982955
1105996427519834420909942982394136155832653884068529
8375019610026743279608322676608982437399230458381935
9944552036471095908251899166291593196398638609984425
2455919444237409807650556117505789614643591992556426

7596311962778375128746537965238665256816957003269642
8785072018471660573792272520952321099076112702419491
3962807481696514943204319306489969675235237780133601
7153557941527226744043835477478346154171361080719710
4730886023745031213890081325624317204976886802282925
5024334939599178274676959411554430985036164313163944
2573259674614175806634244925060402322028029469873751
8293127065541377829886195848109350266364520750931961
5250820150239512810696188302083787791753142473780066
6136610712556138095687549097630224818254733695777442
5013633346505272962419515030700162982343399091000615
5502494673712884321972353399452938067223419621701616
3432924793757445914665771562668285110792098013823602
7790852490861406132382047029603361421239678916949923
4232171528883297993911294665253847268640531873197932
2677702760178715157112713190221683641694532904519520
4615033553473446978714152244708770708456141208314980
1106667165207465553671878728737469049871627624680530
8575835228041919956073276659175695773002879207062873
0635622829071093172254124410289965621943930339359793
1272982490188505998207530280581826873453626207698842
8838928963551772699775527089280711983832712641649813
5850566092974668194143322037636016031702386454010380
4193375688745593512383982760565299379694631115736236
7658035157564368020797909806973589289503336310275094
7847813570006470316765317984374893859319928446797050
5014658422277826960675919777431422848983462296380957
5224532923435922351304412034590961012744858914050582
7476775911036197187481126259602724586676557147275493
9412055513362289169042638208357399520615723946451744
4989912970211065709594498556475671053910136404655906
1659117790624364595739346071857771170611184517001545
4508099885004059315587509512061655456314620073873944
3327439156540822226551667114981361350737399548933917
4808637419664809327817100263955025914047775193472052
6936117215529192894891128512312563105277097234938677
0893098856247973589327598084634232185162498530503627

3164555086002448011287948708902187528734539294131614
6620881482680861416201591554912204198602598488609910
0100893552198600427434057310121427340294759435672697
7628542777276759795406783215499987080260583813286902
8183862100061003376423791980019442370442063331998932
1465169743345769912822318261140907089861441540819914
7374743368164498243252660816669669733619695332127776
9129772635798430150971081585627795241014031219725399
5009854837006991572638174933423198417086784859633091
2936697738356288707840800239223578231036122923133431
3870871375607264795530687856767876140867978538841839
7508504683509284144719683435669245391453969726703865
7031729624183897923545368707062910535840958625288172
9281692471046395913765433977510330386186905416278540
6796718856452314625346834643002030663624300772804183
9150504839977463906523527004766822033701569158523770
1399054126383476484663841179107634163945096265761345
2483409138987537934888710844082251502479447198768883
9920035737926073657685493015534302684384838893140272
1966820387276849040650786414983954838623991443271003
5484142857146636758141086857568497496424920588569843
7459468657448018493422802795823563775643882682326241
8776221622607090451984593267734773501828543606939352
4165896011745073761140640689598829299446453818686066
4747288899190982296017892978047357912479632186915368
7036559529444334995425160580908304927359408810125128
0459106505047664962675822241339331580270942092343548
2413754573059087156567567670109209505467111783773210
4759766979436357024999172477640990996184223422593896
6846991544188772094653007034437183115728705732067398
7595782140794073633423608496383233359179022771268263
0327694877532004680184757705394343007951196667752439
6159166308278083959053832295112723382075507441530788
9777628677166251881091187535197330098637717478618815
4764116022239030319559678981537333325829836004518889
7341313857960092977745880614240104591501478206797394
3636291993558227602367510347827556486627121882555285

3178253586010351082258145026112047470924017186025646
9006846173176799057349011007272876892619452758833587
5822194738432083457863305280755745493828952390059845
6827291413464323488171485846067830594826014535968762
1596708124955323057637815649344565257825522938249626
5751170749498866876544103827053374089892104036867736
5604542858451695031402232363375702641796577852081754
4892465570992408132366557869852645353888011188919328
6251925922213556676815576092763061557593066264739260
8983278347680214605557131593915751341973623814379449
7888858119633437282923219663203357801261130770108572
1598982028012452701419440551082110912628261670708062
7427106208073756237473911790188063039151918982517186
6613757572759710389811462813209411824321201157828817
8755591908312841658981012959939724582828850347090530
2829222517897243132948789737432150734138953199236450
9403300894441779949785506114695615294856553112229462
3526306051560149924641646941653581717906597534647747
4751894490338863769454910133847537957012434353283214
4929827573206494600570358797413728473085568500241405
7806994944420965645454206704067126917703420558935459
2147513994658656379532449893948072963453595989773250
3522425366975520262596619022391137446470151563774946
8173440379453288595394926777638755908647972478088680
0681723558531314356329754317660439253838540575689974
2985095171277824885406609203265598281270195713564281
3925406613098387251913882874323038507006601992157068
8871565313698645866717928365735525658627188144014431
7143679374810596913709321216806242471623729661715393
4399533680541456986793897178488910159860309541293529
8746169109192523809742944551488558318499498594795902
4263664255486555633149346893561503414883748846312946
5300565980746769773601945598152863062761262676635577
3597675811384216123733145970872986047138417401248888
9187971332736362625113113346765376292998408903778952
0000853939976358474428197985767807210489634599077801
5426697174567542617238402203277336476397551754916653

3844727336853524966915629769248343626504746198823359
4559385423887390136417704694539579811875271215977684
4251179958071694546686174982003891413675742529519923
5363028640995847738080667594169715580833542623996679
1371918119745650950142157414450246559428230866854834
5475504175328090492439697577338340692003659626982380
6321058421168311536080639600030298348698625014189519
6159770985309893415906918264474621242983607543464462
4346418375891269938235380143849573643323589088034003
6503562945098957173182119387536060403375025733118252
5704669344702757566572460202434358051761798043050015
7661220685910711916447836787755287555765149553864629
0031477843352420182232281862889836049955671009471307
3625117504656828874933451108004370570085720732275083
1967368410921087264603086269870121760440904028069794
9111839643660521843149230735163158890444548056797832
2555962857732503063907386237033313393794864907355954
6851796504296661825531052645982451735375632502837394
3551018059272053090105142909243352731278690015151305
5874053955359526490302813127589266878502683802775398
0878419695424447075982652771476368013445122704646965
8616014050005981355426600455530412564538564899316031
2805596469709727717647751362379318695356428514304456
3012761042826494036797480293301901344399860928405868
8100268883287786069070057522291637884595429511271236
4162904172492592870335181217772457206347444140710457
9870116834865389408608793464288581405323722052203338
8278249160655075142849889502673047338689414665771519
1706700715462849334130545953261782576247455778605289
0705568120939213636020794840019045348227175019919985
0351721819829155537394052447488160840114286829854219
8154173994615194467566539911086257026617289157211616
7086128078644225968780654055240840769809262297989089
7438864871881241215285862107144171314682739415124540
2315271846416021045875096948375943180736202192937400
1190274750271765673435942473670661803986776730560640
1858990535751230023066083708711686914757913363840862

3253843492671860661390466843378796516790070950960770
0445453556313624051358161617384399700872078005727973
7476697330001951815716651448215634062181865695556952
3059973209613527960436084916464544009853402960861674
5334380968998239436938249495509471695416706412839424
5171412601916247382708669769311806960971015572581478
5319462574614535039762605546512543502145241434329824
9241381217713203403237186715206273592816337441031547
1046396311177083453501544825945760866597757173914576
6095877536672498405172457008259582124548521961643789
3131098824817003842233206328850043254208492159442373
4706930124693333918887639562414254068324429613965686
2403164128403058782564652342818336566518958550369909
4325953094250608082468414255314370373906651098967653
9808735874993770334589530194951951702175226452073203
6027546394131137116116222899004570802847540361406381
4770890413618963981467936090582948205418580676537456
8452152011414513584740042491714183497910852675578692
3646084051029640568547162911479605166051298601164218
9235847067744783697916343692203021988896384901524172
6483510873673134583953442150224626155336633614834786
4323356138053007496676208428757530744847676122129520
4640268425615740219898496340903610921847604312888296
9266380818247741624321491805174576051652767148595672
3605854481485440213290753231193378970542137669259894
1462443758062516153217997346963717964554733535492490
0140560743663106474176679985725698630252839944434637
9930518242598412579835864959062789069536518549591816
0282523112964886245466247414987802348961990493472819
6665830714757725320158683554707783672825756239924355
4639377454477973382960292239539101363924842457990895
2046563980451217891118683684611173674956595237321588
3192224769664295949694007368505028403776890626580850
0246717295897639915288552871279469207921220672013205
2809396452263227086822123147580066031785861841069345
0455257909777395661801233418741794136221578605007833
9034591252524540485754363277089363734813762506583880

2048457015329172877536585102843043037380946458279446
3170347696134486828805493874742078360720918196695607
7978078659557407096044302978609309079720730749607010
8155868501594809753435305272934116171483174733966100
5175217002301479901086903452297704877598405728629894
1810215296900745556597243348361850377715055086033011
1505317616416961877236613812722787109865173452038573
7873021661272225806239456342389511827106389993828193
9468089089171426867874887032369742369825435588883246
0845820284023483626235883643493266480175590460342821
5172195639634953043008484166217827151449459540901179
4488525950494726556911945792036793654037611386749376
1913528838976548861213181153030471606196867043644288
4349882255233129687502916354586542686608894604692937
0596495128448974048567800358527040399356260489828221
5255571996993525794545261740743270859891130014290671
4002259432746902189829951795534427487181116421174293
4346618581259577501754135341180155914001252399483939
0176617521611923006320392693503074408005564853217364
8116815833020317058976467823292023818247676449092399
7157484669006626848707926979745436105026659791718725
6545818722599567184718953968963563982739116945400828
6772208397356485201960596067264555193429252306818637
5946771657471639851037580102664513149058946532011702
5939029809267215532618811216870595841647293227196915
3732331551498788130347398894896254271704631085200203
0893497607474809695231438144859523362875963931570347
6429000525190908752531657407924492946317614112816050
0604332367814921702447181040900025235729074509400875
4190944850132342773779028203768777598838891022428946
5863071885783632584011400402958201515727775055220497
6717418065229681281445359630774683991974361507755608
4901483048815266226168875496803462833040684967248845
8314489610898411164185347246790495429842336879295028
5053562273808693036290643496106389027342398164443712
9896752462213499807179098325385375182451431822981701
4980471474405208206771734829307574244607177847252459

4618574890493050965097953905422530692123607017038379
1085714698302257724852517384591456079105588537047066
2930526861151496257016777566050987152218931199319086
0804095893272714652003159911804363740649554498522211
6311079240253084120858743275083025736046735590502428
7620096058217825407072535881942427482290612611565067
0990690000964622866691935026133569804484990306069770
8791796420344947066473435831304985932397059589076521
2059389769761799954609902557501292529505175646332819
3778481798272892162688397915039028415489284840501018
3293430169393085976918820760983272888921135516982344
5644473332530729623985792356457676844465574078818475
3280032062040912485037907903369696799857569854811754
8118386688492826248933731346365620962364360176047562
8848255746879835231668920327581208311926727387077628
3087919441640602074628031822157640294565833974760879
8691752555031704962919619171215072124527733136375472
8630499003750245934859600321151449928406621582574367
4227447550106391222421889039120688571499028125033222
9301019625987938312748207951457466369086901110213105
3057387506102876258248047297829759703788665270217441
1246083737007276409150371333361497174009051602135428
7018659906055371259092989698875726878000679158690910
8457407802739901018725834025027067523492790845564584
7233838793694839321219370566310273581109630944234629
3573358743954610171509748417603259483536217516712490
0482878786934431786340777895613431476530447271015873
0835918654422753350609450045427094382959523450061795
0815499122260677053695403470872316703773580038588820
1853606077407859203091001307366861533205143094832971
2861083660252455592697326600103297614111917437427678
2789747510302954650308104060484212662927492258713195
8043783255838142797282061046716445453966782750663376
1195615471808114106637289044508607112165066033989238
5555376753205387993468505034921858653621561162560773
7850783368139484509250569703460543116890143456562307
7243178045128441499021187993090482889896661964477429

5261486897574573204681701313930905117056129681336246
5657670327529978842563672610468451395578761745542614
0794992788515941932345573064585536376766627790456519
4687523590507070290264235937689217411258335714398554
7271796933471669904524573857657346363234020958011223
5447644417233019968875948411158859193880265208241262
5415775923953557139009940619257885762438343967082535
9850867717452030647712597168716292719811087226407167
3162031199505749535335078557905805522805676870940035
8862145084193945110212966418030102507190041435180262
5839184169633428710839244701121728427303277470134379
8411733012446913775974881728083780863283584806041092
4220865767728752209963240080429944929304986884989845
8249983713858916691314115948053797704200159706893471
1183157338901047464798780815652192644112417536626682
1681770769432381466336419486790863825847134143907867
8526625420255079875005983442086433532034033854071697
0048589542381941646320236449921869693519762514875895
3644751634449406416198941671134104435014824843798746
3916000978580071488654135135723460466234792972728314
2415592080025103467895454275219424132570402630697694
6540161354854687985714429486803039101844108638904144
8112375443712853308239937328366819623130295691856958
5662741137703388585366462274719316725033611040733156
5700765207124240797569995015171682190064511788702874
6352292980881877100729033972992256642113056013757759
7719013994123632672808453894003190961542149931926133
6411225553601183662732783852674019875478187635393573
3949284710295825287103809975654397325671294875582247
8362680745273903490374539065811519419572645585878826
9618859947491839526549635447571365041228605931178327
7470431717021755542733811316446122057779146073657791
4630762301569877794279947080006669333908086631285203
7258042871394552756934418643828321630754249357674340
6689842924817540762456348435859978947995073584089721
1272601098018591318726985820436021544935373422820998
3215127996754771510867255688821989769067943231991850

0345646597546894209085865186885415650053017704347774
4794386727039309525080717481143880667694404088033700
2762289229403949546456869467365627651215744427275561
8547272970923160771008330327612046440019501088255436
6611838401750643307987896018495725640922702645363838
4878282644377846764678894521861373535543656377606476
6781708984505435511469123142741416483679764597496100
7517515958073916479931951112693660164858482939073318
7973916908788195618674358831373515539061314098627551
5212954448771045808977219105827633989474849102783391
6995225437768143940741866823756712442332351482346586
7649964519452476330870536440687141456832606639769454
8019309437100867957512398911906085980795618977028617
0467140202639004075521169679039679720071713355977146
7791845713611149407966712462922999331477634216541227
7835748258627534990067921119780603078857495469732841
9646448724815492388450416874884403265627747095296067
7481195278592514810702849070915018652287534294183631
4061123708587326294024339098198386808979186012746208
1893950209887488319592020920204199143110243288618404
3867214698447218058827477611885531433454775949915708
1121547243048811098267530850187927122360672654247225
5495116778349951376070493031367598922164576741761560
8894882995093114276568187950445739072603866015581213
8169555771358420435493478953642023389746495492766853
5362013175286570550588094409771668261877850356569324
8837006191686881325769898892153770164294154770352800
5611948422472985187487774953546599864731283766810231
6847838642080357150661043760802759009924176268410207
3191041168647523492506036456386777296180495366105618
1453874745364735629557600768283850026539033822042359
2553982932841939449420500089988724891806211287049039
1300294855651474954344575244822871583406545423074678
4942749309634700151432631241612982110976577978086462
0896364347208805915536423226483126451502160519650265
8472066706130120493386969922060721205507484698313250
4456790361979375114510561099405972270610245531643418

4231593135390530727736431563672645801322676377266861
8633479296099412432700177449539823043204002544985446
4125821814920560149218878884850042818418295383774717
0367919128937630870104272072105279340761590955605690
2878841543593704129448706773764271263821528379114630
8614599688138515495896934947775300910908956450628879
8749499879189773300553955499696722311303299622373574
3856778002884728964213358326697472583460615280371262
7432217243252933925759249244741154505859760314253954
0190273271795348245344711818326753377256883131935700
2083161783185793469555062509874148085073837262014483
5846400353369127055621195890629893555627791783990885
7649562403797143088390971110264225897462931766896722
3404012789249994400301464679220203830421619880771734
6647351646810982056584456771849874997429162956952777
4417095631284596102690145422333953644324790898827529
4511632099275307499379381898314756794441245959637262
5788487798245921701280055758027161475792168773028767
7241427839045907391273929426788438577192396805229940
3405334707357333451836725155426582639619999309836730
7950372448686461149373049576129415707070662032891811
0723891552753833666381708043033055670727526167741946
3060608701555657445230874655839612405030148057910614
5816309013148988618821079382747513047641248682780160
1908496784421890111018392559678015888440508539397683
8736014191230960600868884840958759093979887545712509
8902772115401440192622172796546564989599214375691142
9020204111157924873126508075595972847278699682789162
6827869491152476745851922811086526770098192794353379
1355350051468985793823713882735357271717893830142216
4857171410290069972823532328848462192811289408170797
4024421905443693038017492997032084340110873321114536
8423599993209089515669085964915227766722963969422424
2341883180352109911640280643484735443798320000362808
1909768558492561175239297741658204184481090895126836
4708462474117396127148810193294558673819432508612653
5883736855992025907815396082128790730606533354490598

8347470101680869637317737325829139402014992329209692
7067831675968147669667520807748268071111678492935846
1984186261721109253221606852538815918559880627768721
6438183516049553852793642109031310360781624389147125
4331653700492954357313119180428419650321615780904252
0133124856053203608591448176717054976690739276489514
0552152060925413291657024918171664437203666136857876
7332511082859038192154476652862252463592191750130342
8439331954912297279269763953174862232078789202684450
8352041249612609478597647640505134461645779957974192
0939413873112767240579405410462075597015741827181915
6561183209665877740814676916977483766418386081752547
8596851524694807752817906293992706525231293089486645
0147904547107615155399799638954573549956164209646047
4745985377240519452042446951551105760149422144598507
8562898005200150144217236030394428342444870888738111
7569548458688101588473050820293605143013701045069902
6815819894595150453748572348798071613236899889286204
9934134778445964826238868934495641306886939680162402
2569609725905836903591447830464096918458265475815349
4500776516075819566412400743361172866660578064097906
8506843839086550136603417156612177413653524666292403
8527316315842924751071820737619974257005266362873424
8527697091241463260439116468257193844749765274127912
8833276475340045879787567219728508025158776549056558
9220471496205252104012016698764453817493876153962176
2099818669564349377520407867045502757495420828343826
5272799066404046308555601579354158018388708877140523
0057777479101336075834637081603741403358166217105237
7954778031082878931138989447958611693922813772482097
6622315374165551870724300378446838734446101858764966
2483023535529068562088936705012914503374712977082985
9205524051878310562165213224612288003475473255932571
1750916176594166482394972913352370543886272408483705
5281700296298090586508047949546245822401357141584900
6733084457388181196744514291152143427939269965562257
1986439196067753038374686672221703556890842503483294

7667841503678368198974756263607605617395949590453787
2887088011965418789247653078202554123421527555465375
2784851164219300210400696015444285500717437035400703
3593565053986517857070065820998041957449512358764478
1248610414755659045005537787454257327663137388250201
7264896961612910104812793263745673134738711663877099
8454206702700296011172263762727267452101811865591052
5861453388197012899119638473502901740007068063146230
1308053975457760287247409909750691788261928319716472
1284958737918035495590854500617988132119821142982583
7530563509789912359405448601790248626076719483518442
3490587190753372944339592409185352802957201038394209
6233427756208747723170117527568724101224530281538795
8450963485798391045192131863495690303912564113905964
1254019436292949491921353071715344840968420710642282
2898118640267635071607969350470210023187810985613567
9106593066007897762553197898250066315156298335443916
4449258057007801061628032102201957047579865658116809
3324230005606648967369487624261724101199075962069434
6859404061641090571155345453768092327128723539990744
3591953983973721101334949165692714299302802031477456
0505446618457532305901707861564430381970179149567653
9404594356060660507017393893132134625850381073195038
6744835560194918228051677305768502688432549589089560
8141995075256518935540226366100848161462038654464771
0712187951630698336202140746124327385607330074172908
5341371467646620823326530075778457801275507872627830
1618250870084281877453320787977894296485859758192842
0499648296204273540733853100546993954125461947347217
0399527022703577938585129683664406788983746565636135
4447806668384669765176614999618543989623502371767110
2740890919399583145146872361287138497242069080950442
2367855102876286804251881635840261629851807211528332
2375809709057453569178401938160024396543257825045445
1969403488409243408578999827719151230309473805540308
2315560818607161583433955617281063733110606015264964
1481568404319462356043630174317509207713049088685604

7273855173095380847501448651760497567736078331544722
9253433596384563021521710498738592531020397895814333
6638415592992550894460278070179622745210555067119131
6326265279936696359892383006096981600167088134430020
3911771916307016377800383003711264182414601498704171
4805662603384977517560384954919157914179295368495970
6286329717452214945264364000401168512758738794348666
8837572288996134299386640236405894482452912428201658
0047816694184780636604784838176185651558401746037289
7821585659083148930651177919857231716476472418930431
5319990881549971377420721011833196861968494405184713
4805103762448875818172772334427215708740008524939194
9339810308319995228854262630851814551410496748964257
6817203145774197605540116651437193376372188186506524
8544503999376609226777707479399801422580866214987191
2470138746989567658098163424079513570373348683994607
4001483818809109122785049874225639471028589036178924
8694551999049171162317092974081951721636250623714208
2526119071317918763800373820998941551673629031054940
2905372537989577370008817865887049043920910261127811
1129355194227781921470074363737351900832947336873223
9654947632992827531835181262400711892034565885546809
5030263042519219879777374432637260928911109005219868
7553722810530557641476148246398586371897727977619373
0876704264970441151141289006212611388530603736722959
5816117021395030074141766112848332886016736716732035
8804701580247848646398079995767647079233106445662603
3730736158992261787526742601353600729527851144731299
2791452450623340290963971759232197958011191466992839
6066090546200798237452122450360169104115632219532972
6954455515182840945395507867404093718532366038779628
0514312676627403643982398318817605978528134663946956
0555811845893980705011357516820680694894385529135088
2854473482815176097154238595221327308613220412923307
7656558692790957004784303955681994015964223359594550
2281056543779954708624485590800370695015608019307214
6489726731639389255381599586170220591397302709006385

8841495346162764260442837389125191847819850025498263
0358510063410344376334073346990312703558308011243548
1586611646799047040995479238128789718678010981520497
8806050476668203662967250936939607313669337472036724
3003189666204997506525201754713578657346438364676379
2685019584615106967653639024474899543219972319061169
3296228778946122656683064351895616389957450772210945
1279918853244493764877890405442242334709795872652176
9377763707960407327897024558295386961641263246575492
1044812238089336380764949967196348615540609091415949
7678087746191227708286844605325727180192324631999466
7678926030359795781182790004713940921719399874076000
9997069581634479233605106166460077875593954578581356
1911797348472427564395444690018537030388994721998308
0588340583672047301059337164585377733373826188199786
0740674509062922554889908344358447071868346283907929
4708011696863948511850811643813460281234771266500937
0864349808106119251169983906911241127884922501074046
7001002498755498035240563673715644048348352246102196
7504033422680901960919183676975391844888981030769313
6036846490751851920899807967484952064252815039882199
2945016312244419290790582175500212464156438780114736
3534860323181769769925688623422436511614137835848603
4572918456459731014580144233910343661909121175141432
2400433814714649570280368174781573399255278767581363
3597536055953938401768676743047462011227916421387901
6271782936332025735406498294315950055471411440663728
0225906499077297215318483539621059207509481870223099
4825467297018483782461121232263919435007317776200669
5405996037403765441062871924178666242088214648278279
4381136197446758843595640054338403532921278595748814
0007374970832504832785755627877386652839778888430937
2421959455553543681653293719886363436817907518019612
7350934580410944655172588496802612101534669727381161
4985607135933184212517820869764413662803131959266299
8008004999380179915344918391418600149176520095550574
2943903063235405129132722852895331838612800900242018

5920683012455768851599860366708821440361799497576930
1015816840489636950570719416181922986998546181106643
1624215434388574483433888071994766742309255414252204
6461326333021925412332655422517900396237001187722488
0121899739867454023278310111558138887625327523466564
9792845609117583939724769569229581647917057753460630
1700752743140171314093470826207580556365138365779777
8566866723143389971585405128381753951728867267924335
1932427944640427484557220427137038338428243358563142
5510375691018147458356414678953450868562901848676091
6770619988006592305344363957172825088842599663928971
2477574013093554392473324079182045553817663723533222
0935125401324815902989026429742592787894547500654994
6921212466620662608214349320020805522405722398438304
4936328281922845488932895874022579320820777499212636
0859118902964660983895888100292388204924622971332292
9703775199313610098035267052448685322637447804638432
5646301000814486291219945715644790099446084293682847
9652744612765333244881103324202517473862223449069723
6755378803399596664030721437536023310369657332315237
7229874426905668961779628232189871890962510009667381
6079948561699112561646645110175597163799499487451735
8658677084047168447477084990976997944707107221646280
1394627545923362533618584457602386869037340402342997
9586621889714158804575375650785042610212069743697096
9214409411346899426308743785693181269940438714594895
9983500248233904352960214104347987103068353429770829
8332077518071528480882833701459951573669234089522074
6753110902000408776677345208304107255415939005257603
4609120814028417742037713070084243715867293605934806
6492560890600621400735246220346694340437762973714172
9565773844359400547535778336478717385883199572538063
9809960264540532720562873151122978157160862957374684
5665423600829004305765331541271689897462285133851076
7066427511106126351131616092424346656970070932995217
8094757112910514811469846816362099491211915737590934
6546661760859252499642964904995300849587607819636004

7102673940074073208436244200346144947032423951767853
2424534809937752802805259240486576284671412186729974
0770308393074261740145003221049726207151701671849128
1343898195596976652192019081370003672782082548084440
7705488949773529007409396411521592813098524327606824
7595225330802824831409114550349648055883363143637880
8692685098402275435610953038777070121200028980325179
0104923250778850259083263414354851687722009982947719
9602305243885580448738331603586172269203006710187874
2606941786040706999025075004932336269299321409465809
9646531428589402950507599383742467138926990433088969
6893171221522509329615283223140401522472006962251931
2187694832867420029173664435680663582105332804176726
1515623021242888554325019347747162804383419392847933
4861222057352401287750089411533192309765082818111896
3550249584671068452926448455574277121347428588612865
5316929049767845662596418247266927783440026607279474
1444083706645758512376709200048312194034837292085600
2290277977308648541474618588115702402873149042337477
1022051225421059966070353839506482334592609712442501
8477808097717267392473194016550396078670123047395765
3328065850834841252600889465846731191572917457422417
7949335497989190078206213086108038655162237581246483
5540650724329292911545092667131859296769309032438050
0504705798029574716346546186859666334816473756375825
3029528629237343678949466010883529136131466638328071
1983717491455000642982081727450617511533164317850686
0559958041447050514513443466135434913237773369000477
5758504046993549528650821231068855209618997124393434
1116809879539624817085293385576090027022794974124179
7404757178358915130101948311420886832494331481780233
7650953324299552859443683435519727318517804946558350
9919430937022568193180246844883186770873093472780380
5915525023120406422360947098853016075837054367783741
4644116053421760045932648901251010943688839493750108
0782355178495090823589954072647462043742888105083964
5588426654869276504840328312182091922525491147266406

3622030793455066039883857248600510852807898750922954
1520536859292619877916892215922052205519853201479261
4710859188431868657411596378993112160998976110979633
5654449027343590983289478678839749661090310579656789
8848146337793780972063390584010251514800188040123088
6131297554207215369649270580145527502161467808691366
0749812251621634354851006334434910353420533934694524
9747695508626569362762126109157268918512773037638397
9571593669306794929864838663384454486312760855100518
1894581104146470845401536291458227457290153410579881
6935405528318673603825088352304739202146070349331910
6727265127717991413837226705030042889515491348029236
3239288242207957730865773544300787149940796063032576
9589926239205966013855541803419669893243670285104942
8362179024244477738192565459972644214164588018332426
5738561878678959852363478423536809059927119995597854
0144835966062156329516389073004275682540019235888320
3001911349752328613006510978769294553113966895558347
6437080703315692464770566831708735818394804538875307
5171478337119507687976407728846486649791823184502315
3810551758734071896253879853762143717243682071046881
4205249239365508704489135541655626666713154192176366
6664454613419706588440480395862840116136033685481345
5050935903141657252576055258698787030511931907553636
3454402515452339582336587160381608528218500714103549
9360846627840750612368387644621458919217791425930064
9731241854636559956751295850203082278410326137151135
1983906124159646333982390087897387993371672887203567
8948696127846918029884000770240461416843570964611623
7948458002436915344623192970276370458105466861864892
0064818578844704613762942820614210703480363084411167
8006148150239136901396657580309396590441591251209695
2973581273097903258329274793996928243831492062590293
3092860103323917403838342745711351398457747832024488
3860219680767163316534986730347454013999215982211167
1115125442585376091673432769990400584497434759055642
0630769469347283565064991077767958926507983234810840

1822448911707496651040618759184748576972103674326815
1484201956972207473448220879286996083629358763165389
0478390048742747935151528419727078614321965828417069
3904968157725682805012202075109236346551769315157232
3810501802258551751824782234136431788165206850661394
2262653446972893597805755792607832166738898066125192
3880001695343252262005288995610861328799507595735547
4955247424308328701918030564770827367014344539375937
6315501750935185158798505069463905231958292523097514
0740525459563140074366313653185988575773788784190261
4118689544565042160337064786252627247085434175977950
0692649026951120275026309543674058016472131592679453
7943694975261018446608580007831271913160212350913829
7042587023364160464484612847091863431519865365466090
2676182247471801225355996813235223866795597369258068
9597530371264802448204581370182034328690863716598337
7571085833010368743465774110621814317655633962100638
5831674674809188717077376535589866220294998738274442
5547924116346546794060365024746024561895259093499667
3595122712732052401111139185230270676142772309222947
2881013319296333997818550318293443263220646980101295
1704557043006325015625043150836576283267412300204950
6397173678418891005142525305249688625673873847779662
8662164973624028836930961735437337313667758358357017
9447145747081249069802297768156882572257089498908991
2066055402520608013224504825120813756743766198679642
6412986059431128300054088818812331334729747312126744
4431186759432091066818417810265573353262138773747375
5345782790415115931321851828588777441772296620469138
6602093496942560536450662133317325961987682588594298
7942670938011410541539939189608363619248741879169306
4635784480721565839931203794011832448020155671414285
9863727859846970740475543618234815879601141366235245
4806720345614756939780122895658521622561696712983690
0853900438345103629959118876555485850103824200072825
7383191539907527071272412720139581995535663735512890
9069762510703048920279452591556500485530466782494374

1234171084511127271602210941931455401579997525759706
2357119650920779653514558284969464124712726275202638
5688988076835163458821474438221218629646612627082674
2263505719504739238635075412116648302620097127671004
8982671613245191035278362070128544446701113052325226
9301508703063384519741892706340692454588091376803729
5753346806779350468647874587707376219253052874813619
8511385757845429291076654292834071343502250882818892
9152445277839874721900813143807204302072454679317587
7846662665684652886150768840312232121873329607244268
4943391394823440048794731810015672345261770257670886
7907794884357597613518177223303246999116657596635615
4888040497320129756009526598823496022332949604235511
7872785529240802858776678063370228800022181033547073
3819373826226757192925933563709540615727780572441483
1116404280200117378104537735697122670282206349312690
5454981671849950281156890107968802900112565891790555
9234547229213785287232821812125034518205954809677227
7660713101778603438829573318206423587236993410098106
1723496382468683009133256534492346840680693901747353
2015044406113834755190428877540607758191617811541300
3992574037864435913019078461131559812287433383309853
7839728020727286892320298943973394819817757139247491
6946466667896123429109370338793251233773971388025498
3507066465501643565968531458265075610705724702998374
0904860172437641981422704314803738452936874360053639
1267265076156807813142004122411959494039297701634281
2407872080852166630645923056703207098161722529026960
2306308129260227978177435716138102919298062250972902
3421402721191693803271329832008372850667961628159235
8286686576099904415991587203941836822473090366949690
7177457012177714467268351119457992357643789469356535
4208606297301046730719822766077439302309468861548281
0951520215970525502361564783557941968817555609138577
8521922259647769941023057003836674762355069788231896
5569824146502867863189211312240606180938604883264517
3084016950142082157326064292228869111525025499360218

0051041932609690738474883239140241558533260115285070
4306609322444248252414417460744488445341855282414037
6224643011608492948664582998552054267151740545442020
6280703433294106933727264297066891081535848401490945
6938246216847992970706873787740628928325451342260408
8371872199038355526747220941231267659633815572502448
4611269592746975238144932164044468989516224006928370
9586273806888336291637240041875958645520463746275695
8305079845381534715230664110072569236293492763936710
9131351270403054576969966845535847145591819283771246
2583521412445812027496607847718105699457043360508168
5095894537463079518395003471725194844359949485446852
6812973590190690653598903356956031006447968263120457
3191011544496555112268417325152277386303081359758217
2290597661376276417661634769091250701619995537596432
1155753439662168827108403675119299960008862463219998
7541809436005308741339626929982076296680154880905235
5871868076169855606043417539252404967999646517600026
9269165516987526525067739508513040805388445060926010
9643688108942026856159073899850990825015494639182209
7006484553663668996858922863821597110767179767897501
1542808723039128878869961867893071982867549919743262
7093410002685103259296326881852414895791244327648721
8014529821857497714292943666893783643822296585911141
7940518494207357064528325988459122593949274445571466
7765151145692903139525375804633758579641581163621872
2766151898241220535132242248825616774245967654125403
3178449883841113760735007720085697277626487365278338
9163614249642106308834930890332084877942864404972226
8286185994668934379485133485590582820696461239464975
7412236283139773381908745580331675172722386647360336
3229422165312776358477306315342973727749231497394273
3916813987517165017810326131745506377657709788695976
7574221493716352884568741975606953600393373320206516
4936970814922958343311631037274090986625747050514702
2723096296228767689369377387123902046516729559253963
7570422968270137215978496180362348043379063970358558

2422108823529872394968853577050451828950991937634747
3203519382811144209254673933678229258496532200800152
6318029140864688824845131277306267980815595476482850
4172711392640315538504058354882187206324982256830837
8114474967278883332002871767003436969011263337714068
1492485609922237095518547384457179054176873927917178
5165931986828875641818677687684101914918079399611289
0707515885455246994765082176756772116466364276819985
2241009652130504650834072757448496619761614608405906
0741022144497622366799448483777754612007723490483468
0479955363749200222871127185135606616822912323921800
5582781418581599044062405262546367710924438317339847
6328514706535000797831849158240117595659005350270640
3840870712334125490430058925526833691094710188568063
0974863894076607300957659146215650501976073448899217
0674595990100871084361834433277876536825877230822334
4345432927526450350827510368852787696779247529054351
6348471778306004342306802510416927098404294232615492
8313761218792561598055496179934882234043222837656531
8861257502640780451529541971941954779774061554624267
6714336470510868155403216644525162310712260123089371
0475506825086032404639086714436907732441343984928414
6788284714793329962152726567683401040035380272025275
4184352253722834489706581156085710599336667745198566
0472200841601908600007880595268182669131384173418199
0559010582359290201354051461037292598408924440579962
9261209829819631751214535332103905565185435755015506
3302303939832174241638876581418283754957382481357953
5376415648600199102097635339207548493483886928161124
0904286051688972415625937579583189895007292459604679
8855441909311233298527184462335677735993334804074703
7205328034706976370027157282023073057696141487196642
0425572750491950366323004651395418140725108930709443
9783121107332766875928037765071725136087557193764856
3674485588825788766133918449302119188229270901950001
4684234363699275319954611567138685704507441432163390
4078029992349058144656520449623351543572010554182060

4402629891318181183565214801386554846503939174422768

6597060833397337177144364986798767187972512615063197
2885540631915126103864962031374005154413457210449044
6705194515692273936673249768573916031054311481689222
5159257668780953462048818191351201284162581472102780
9659799919441603443841823262314480816732189123799465
9746944737199473447546144729069858854201016252306419
6805972372284233732451140654028375526303970784269804
5973210194934570025250559594698145422107028369067651
1133800827197043045192478092511785627291035262791697
4258068402627307584190214628440917898442561629867390
9100537002978855069858231909288554409934577175078784
9254791713785432263146556661535870211691604317209842
6523231396606548893085383019683401924136867142069727
6336719758147787236143301726125552055830024756665571
7115576955972073111366164764921241000743253672111718
1902698647349658130101367113732229307682146982330958
6260175527216725842477594432158344832516677179538137
4496444065673813832664575174490574800465056840211185
8982706460025498422293207274982206974769806226082660
1109618485569352118066229299403801572610103842829608
8987460656304679850229299020929193246177160616038480
1422508869123734248586441731267184746634512149080855
3212758118947482517859401758447229142593819966479033
9087510366666154164686547007202202254597534809842350
1364848574947885547459213375096376782165427914135474
6700628127674422229911882799813572690737854690911015
5681042971730577884247638537402679741523709462463943
4605121639245148210507976032685986609400873234156423
5356676663753122763480101077045489051032405795659667
4130831157927974809669543472607247410692015339209093
3458277744733961650093575247112417075063503208867717
4121934951466676387126373728461160408770630158376001
1153364921219020331806850032466637922173797926804663
6376195583620471957455881725511200080419911461363395
5520113003942391597535667411367022325519019341764557
3146948220222522954833329774556051373077425177097744
4659807607594546606593103127348715553264389083383702

7738220743145689445864371754121066049337745424904700
1490395946574594377123789728005842939621055267503956
6321492376729814066350061202360793595072500261188229
8817024409323060665400972298243353765243779941591851
4933904172250146997425335378050436214109357723549008
8186907390269514016697214836261672332385111763897273
7557228116899199803995729374603221728192845340933258
4411696304062383003542688742902118242985645124579520
7347984627346195104552425613736299433014983937291110
8705299695191835336505348442428480463104765220834208
5936517309658449613028583365481954359818675622029416
1384321163257685982562967201828906781124186868784597
3133871404187297007119198677012931062930969498993548
1392187592880483984513874075864302765615714733518354
4073506255312343664960053109148813032384462652043513
6290262659333068120511588851043585699456499388695707
0611192357275711029419692899418765543688425692806386
1435130310638444334313396196979733561523242666417066
3902403623655935749442510383568249385430663695656283
8134975783190902427704990944514111124123020604342232
8892597491438231575907984423517784541128724259493533
3309857109787553868398175482521217518103082181765051
5556345242994845390457756267446578551759148916225117
8070532881170459740597598645830080056103223325384300
7509811343101204508331904228654685877994401229656165
6389081595922394034392226001019722176573621711635990
2575752900033866022749259421935596129475518513699393
2496927866350652388267689411676389089823146190959683
3861231185867149575727057568639974542711303659181838
4717808180841752010065566213047632675249466820599746
3936880649990764134258089308204090236323835178306322
1761720643363201104034409940915840514687832626047756
8040712615484260346833880268094044793089193734239036
3864402549244811573907631680266467996790175567187064
1363324028870508745716587139591642925361440259784029
0871343774441758956655811307376296889347527173711313
7790003100827293871224879869141242800845027251225463

5272199201876242350802784479678433702368073614399285
9048111125419311013509593127276611248889546891945335
5636218793299748845867834814486269804703990091628952
8836891137461673156170241004515775700160377837033957
2539261354053250114657419224801218243169853035363945
9885750411326222400541097051949162925743792819168762
0492675684777486034513839694679909435305586150042794
4483112063090593443247000937536710056044926556034727
4778079078363950451876244839269798731353528434846388
8133230439792774802201529619235548600047342730950786
7806253076736433322824196132427056557794522660656950
1618926256269482380056950336491504711624285796896829
0896905836375827828389289552036322393096042046228781
0637658111782742762532340505564585343287393682881055
5823020155443402566605611991608331245276376749383219
5233495631786258422134166574826288774471793387032908
3625039959451591113255050506004055193310678540221401
3928930774710317833410559482918371307692456979522064
1402279584670586885775784687252894700750851850045306
8757078397466638450736019535737073785356700362585582
0667372433386238350663573235272550298747371571049578
2761939356350167592836742308538385735127612882617320
0948780873129781099401372087532187962176750723431042
0409830054307545430893475609999670530062600893958753
9803004781681712719509393419106833179242945159543851
1210909172267321832118162279570595447822536126829509
5486221723123631665072572186345317410723976503826472
6120658627233900281950908857409600576878745913537314
6151589768102894467346086010997367803156129155718923
7969413783512316274075169456949086805441787597920036
5546926344461817737322716700344426677604609492481205
8797064658497352880792415315639324344125789177935725
9017863758256938063425044305458827958137410756022875
8444781096542765176706222370697241979180215229145483
3956256228460783866292668106052637667735843824873378
2586428850121233292472070960759606827560580289356980
6800904247794144805224614080192982744535914261300673

1539974220398080445375011953972619448484495199114028
1906600326280269709544871657792318799155265031123235
5363968668453024301596197349226235227432305360422337
5606417777316238560718050405591823508941421612604389
3732003497532633969317683963974083408761489660534676
5002018018095017901524757381590514466840423320462519
5859647960289531559077547204187516804221048827397952
4827374967425882212290841842673273540506297473956251
8609784580701322766115173325159280125006610307845655
2279339066119554676984670693114454341653858729999117
7255340758362673072059819231864158196833440775617358
7601158541062885902514859706630256627278188193432044
3544059426417472874446384970887538924365482695425677
0519550050304857396719259369183132249952382903951907
8750074774192883412833931514360545655499274003400051
4752860117649136881975405835181301064281918542772397
9888911556544206132952148340397465595374693176797126
0593262273578893694395281403715549812562291836028970
9884483519995909272476142927384761134273942707646596
5860996751230503243760925283745354475985587192797451
3545604092844631238388919297638722450948694664331645
8586117087884072595663446488772838981448069759334634
8920920847479336641147691694830437595998302984485104
6697116761180315756704307486831913501510866268148110
8066267644488187637341191046301486433951333169438648
6719125305119771222605129272066288828365146022194470
8196954631978248180623064295919343803191070756728911
3034166693906837665143733400982917691778335023511377
2378070080134493751248050500732197381238516250632081
0496669536517392868858139774907719078245297941956168
8697223167521045066713791920865389367499863148664101
7004445762995894532195598663958091794185105600868613
7283520554464203327542845960428433290425439861323590
9889616104911040370538129550774681638755768131629919
0250815702719076490775784360216977227904172832152483
1115467377986753486330097098407188529177293638378968
670454493802201715742430599092277663302429744170450

0745899447515007657427829025938963776349353159478320
0225424326725184230050688202082549721292140563372558
1581127104741570185389317829398787712798274886567150
4632246643855553484887868766413556127610485824205891
8651972343533639572360994808168173974031190309304510
0043961806912344770855514924729667850895219861090165
5248983577004961920373861655837363132652345982453199
5105916058379000397996951777069795246952298367607431
4990085485641433272200349890338404285312421123402100
4981624291858924222434957536080826030676240154695788
0647296192668452751116009590537854831636712454848677
8820172520546398558650035823500202150851642366350078
7760715198561452562649515910555074772785379815406327
1681951117541022889298378222545336756651902512168230
1691819839425989997594117695875215092794637661963360
8719250944468590221362185710407220499663863587129905
3055527028410467726202127446674545215457723708236444
5335706452251497567870517194800500802567824607818185
4875838922588816417620490361129043128167483207669348
5228577606793484645865766909520661008787906430168167
5339162115205681084467894159123374146368211388337577
3261402746246664515098921044063818083056080401370100
8907417193766295844421691227908932831982677233321279
2007680916610268387238432454420686797560394826736582
0673451550426763088181585580393191617327544295576436
5162015166446685575937259409711518783692173299938415
7882766022670925291474618234186636721931122698423608
6854520418831280199780334878756588271058702370961302
7479913234822581376961217046227422496101673375478801
0634085514866228699569152716263350178980339827364487
7428231538148290013432567888322876071651101159587187
1450105783476297528874339274581827686000452775717724
7880209896564385294238817387999187471297587411654133
2391556510890877525768628661680719210903432650301463
5427492044749312465201473231975770361204501174793986
2821525354972026717908950519508590979562153891218056
4878996785730688499639032778132294015207305052020960

7654688325251752641013550651450298021343533658755591
6168367494289441214804865998225611198797208669302128
4995016647952381706495642735155258468462892001418813
8574351067080680393400649398027820408450155974311129
7779680235822043997189182740819520245230161759947970
8280586003288928867410624713286909469466843904201205
7741066994621572385110915692375927855868449397736068
8001922934914717580163165254747272854041117825003154
0494164953211474507549665480231018551844742049336821
4986786738789249575147069024662390473966302006615781
0935792046362771350394697843435917414264377803119021
9547522692089547968878254856457071355666985023275461
2398823027868901317573219171678493016365013895529131
5901174224896073749460202948512798559245077379356908
1333869747037415739655825885737413490358278143407462
5542355164396890460779583866499664944507475657969935
1493513597320013303372081166289167650986948072944032
9917612907230111404658911382427272632968517609428418
7939802605239565406645933007896574157669708671920929
4525393155274373528280048234575091448533542404918205
8302117366937604973842272159134835520404257616992806
4285471401936855614110276582808570403423258758656946
5009202308498308267843272774869277955040661110043597
6476867865384955027225588542249861365736317396148099
8936084740971764954715433406237073265869752764592133
7429714124207051122564494990791441924448328393791380
4206362380029700282062393180756192264853973949330832
5197087313845811985701710024913652522719868408318404
7846936429025111292751766993988620343878917517267572
2174421453857611002834874321905243332939928094883906
1529406411018793753941134303171916852635531673529853
6398677616350661196822472348640340968103858504606729
6014761310740240074265340243680379210684461534197080
8088122076807110643890588634215785282582287743675470
1408310031138429853493528249358504036234592392668831
1448077546710591069406547237346587789419248732364370
2402024017517597046829540255079677309844431798635413

4645953902276574320351884311155975463523304523550552
9285063631152408117678464547141327981340659274673208
3545282276634305953054734309381750377872317646489096
4910442558796940411705277662970232569524636717758458
7882202846053053054661817220965509818396754440119356
7027827210833556064466575228098058628703330757873821
0821129210210389616523122848792032803766538720195820
5185618250144240962836609846332252841101574173614882
3414602873774984914984590462988903284925608395739914
0591024120684557624803459047235953688392746190063530
0602528684434546600578574568725002887974078126995014
3596964947641493185049625552169444528557452948072324
3302773655574561042966088356974253073611130310823800
0315853873628688492109174918033905482850546278641000
1316947199894889420953086446502679610071504843926206
3311711860861208187157327345513700141867686921967048
8590103624233151677601727174223305049530615852479196
2889675944421346235651931221942844071430978955055985
8862442017362826073505116384635711187846834715883490
6233425883436142747938295483412634174168826028887355
4781665323832154076835909681205488454817189269206203
6709653605640898759284508157427478621941367169943229
4586933049535372296799780048719102034782473949179717
1976726253884952817305136008922705243085564874504121
6033551514931039641863808926595190343501979503662299
3710706109803283472869345415135967181506297663744309
9493236957049534959141935272715141873765869451223636
7107938011524216624494119800446557212597535804639991
0239327977083871847534799854360306617746854816743025
3509114142253635633838835435446860011086384802147927
6315847935107106227221232890003040085551034607680665
1955216865671891813348506806937700390687013762540552
4668497671014459934961842246898041701860235101964990
1337786508323436869443782163920218850859965813111265
8894829177348973611017473924599639788237708464359325
6464055446635456450627816507891000060357895847141178
8114001470455657924684793389859044356095193955983048

4119070859683908269696463010331291954054835663347075
4825597162240957245607899522751374317292043294421457
1757021119664927329504589243511542249892841829309680
1627909990143910006516066465476572330384646243951138
3067770124161991030939928940398116061674207603382917
5930656604400871622422948641209276800548402635460460
2128643129686176048676854587332459049044581544689760
0129176759371782877815746344749744317472245714178249
5933265895990123263165431809785578257753550495717798
1990246125777284944635352521703343393134364513830425
8981514773623679199418168832859918715721722876519947
0314248248787596411631008191276660288566473109611646
6676711893676347212909630086923962342070886293386312
5561287340653467909741993828816638506027873280048505
4491820199337871130829493390254440845445283110790671
7557815800938675360038603557680639810642357510997331
8402806850770996662013993830215705749043949115438309
0561583001130141099139583290325869583767619707448919
1132631216409818434725600876621115969513338904625890
5040210397168875297634115653016376140172417412754156
7757338515633409689244301406179749753717442566473493
4545792469679930110261943376364486675375610876916130
4361374833047944122595261241514898169980768101848189
5780038325763387804353118569719951740840404292157320
7178938770895931557022742728265816149740801332063433
8400868902388939183331420045159614008986112156292436
4169472543135617018117920055959285439183334369451919
4514317154135007420063905060716765819058242162028976
9684409055245447001281886858143549545420556689838786
2448320752121741094873647410915786040991025855586394
0130915517560027976586538407568236397358347233648255
3352052477229608819305408210532813113048809262550440
6239557590391944317141526527540808952986572630716340
7322439503799871488151262446063128138524107414359794
0857665785251694389927465562180588692066232504493702
9212885127459270213767483429277774127189221554062683
3008286009017921875382948629150162454072477788669988

9007098050569250698581837390135076394365418723293031
1836581686333644779329007078574510611399148091109971
2792943849391367015842418969109386238027257645216678
1402236132290468097724877826864188122877993432976707
1638705695119901896323092458999902179757131345174259
3603251369020883958608546750477732292906284648172031
9507281491180095950512581344757157055467848443623776
9148191799840529066771547929117426693348858567088139
4734484862143811115954754498528040574225069514647200
8424624472299321579226467217881452064820830276936119
2173852367856740962791085234229560906326667104802881
9226371390907376844480931844498969929310603687009530
8727241026213678870697269856973400228036350475523688
0049558697359039020691931283070710617913556767421587
7257598221912943186457454772051328945878273757752103
0711942554517528858363344593580055240089312099188257
0655486639231448086237441459530559003065809529244042
6176752542067710335536849552465464357701017897964163
1303867495239610828903257724489918622996519842327827
4995958934083103129326766102724108073795462818807634
2698617617033100934680083065183477132870542549962326
0325011619282730212993493413979963426151264850634734
8327591363178682472894449321466061847965057304067187
4828415191474168416908972295616067007281977661083114
1061869274357335356462725414379409813463832286624336
4133478978890912641531873040605317247738103537082669
4106544406226612937662336325906138381976319401303989
8558209482095391860248199098966486271356649875701497
9534006764478586820491076270969775705491508796197691
9441390159412568326794263957901827825455689699680322
3178158977822565761387167701321480211552727673111734
1653118922038451473699023164776475938806016075537483
3823285458259506291023936001654473539429828766820737
1219275299907369744081134237202013142340450525931697
2775590585657003132022465479086657583451266545682630
5546193147547187391279239721184692177837620063262466
8145632249687765602801045500696828152865754259823483

4009264710128328430643025697652644694135027062145831
9110491793194143493017703487026333460168710291836083
7412532758776315402230395628877684472328030214298729
4946474939503146491804133514202393881524875937371592
8383266003540642895354169850619922353038292186104869
1825526902557498893197784720619919805017972262247637
1814459796413713846207982090839588211870480155631852
0813430516852374158421747814580115630583117678770897
7094256915360787809113628435482402282839456580419522
0130311977227965959838840936358355901185742704482665
0563438421625360498340854389040285430492689930661353
0295624400202826734367287192620794029735061796925410
1291753762660825396822180621641812462173118967334743
0942208876706063054671699631418215590292435137843643
5063226209312068014316916384978101227107150216792472
0562634687067390887567539442448270838250878826535655
8197441663577849241763188148362164412222323635499389
4299078409216509961123532512011031423383550654938890
9275961105369398083023861037130475667626055583092784
9435785197856390534743267450560490940770859188651065
5453008198264452528174087746547899161511634875393067
4193023342758365049641568635388344929702698830212838
5075011730722531947412188603060660866565477142449026
1810915095583074276082948581103520110703303058709257
9512353389221572474247856716767883589737676177211751
3058693433553676490367437388170444054369873859392664
9447153503815259632505530020490676462445852260521393
1219629970578255032704577980448589560702099021534466
9537068428452178214452386671046940810750616731747856
9718979427878871605283450429022122439834339069476764
6413145818070481842165818603669376409894336493814267
9991971798152335108415706476165027604538632999522443
0338938448698694879649254150996947664654689769216486
1423923568275318509654088133533236031501845430918115
8979530096088238018229548364665073083461587537390946
7531125542610804096592752714562722552735012176578243
2201282217368454018834091125636960586741735441883030

2910422954093121391632516652716281214020039348524629
2677025335567801321211000746875104927292150677332824
6340543305146411016945801523928106487165119374849628
4714520592286022164309227073291131485055146260876549
1036968970857811251399477489383288426598056121218316
1153030419155186977220696407051706558307454295568020
5586064791985611372083202139733715897180100934195299
2408631804186229969441590014844534898620679645053077
3611839913611440073059304205749678639537317291277835
8482156291480030842989176364034258219775859099886154
2291506488185489391946492442442707346065366282411043
8068746346424452489217270080026988968400977049906879
0618994349059987912329939316865429778734053637410860
4116821223280321443018450796681914202371152463948220
1770931427298845543362729864739836067619735520646544
1486062065827311631505717724208876556248722268292666
0203714851121462441496499951520297356963468957934911
8486138732356737272362867229569307022272826189362420
9144683434264680026607719950221650936519124299129890
9496734886916099755708415385632669799928723106696812
0038735043570013902925124556125259003322197307426285
5770905192682914207816015920684685189496026803241645
0232179784549350036868531120074399878675353188044734
7851395431773960600553611144063536421618126021263865
7309469059325590639721949203805628767882874309190961
5406869152123189362789249870124538829074308381607486
2738967877453782363709467889116034175404550891697540
5922739404926662391204064685581911120898761302214445
6972568625452950130411217199809106691154157468748584
9599866199083983781712633106815336413320408361685895
0592515016827684752401634734415535782918949921385910
9112248318187172194868838657130404829477742444060109
9690156383523782761290959674815107225829181329982874
2705498735967748422085620675206715994218091188510214
6438815360796234541509213403430061299786868257960293
8497659187127068051527071418619605739031888512824338
7268376641020767175712661092745822335229051299485282

81613592546990669160185027594016816244982303860880 13
8242498830699639176230939613485459945178006710782255
1932420508739012679549784401569898875079123165277472
1289479153173128457969061288370019751895109166908506
7985676051658730747998414456849249212797902881242410
0884088477837315919361320454826303567474927756541385
5998730321590540762951436378852229866981851496708348
6052744652644981763753211029230625903686358567994891
4924880896041265110733890966344359338441249583269826
8242760102098965945460477365239871180175186513673393
3944944267677542321191082309423728968680220347439593
2486237694374108666028201747655631949510872258892022
1524980324583927172427958667737838065801661372929777
1184466502501152581240730709630180406940749655684993
0872630421830015704120137922456787319402582131094546
1093699749222618583746519732322036812782167978548254
0385357995868520967119563235803556744705872326862792
8206036487179366605594411753901808983779028084344224
7968708856595271612788363404327608000580450948161376
2417573420339477986303223673828425719406058354743783
8904464154065790999931856081243084635339679917228812
6378879916003383378858845547198231679388933628377732
0641146617025954208508851805129008431630715043310753
9545541990796465090231644288344740687197189354667356
4945681235430496129888453760929770893199421463048220
7660235398243215761625487612842541210616113313364882
2454132448975453566265349142408224913402066175007211
2694306124966341332187855468029459124917217464103818
6713621514571649507321984896957276872688364311053604
4271571542891366815712744283321333976334300050812739
5671874826350462103931432431997549195809096136562570
0243201665807095188730589919787067936829524404853420
2386527588450776292787146342219301500563804572513175
9401228561135543685420108182371586901339410613178329
2979204109913274541054713071745330047233839565461871
6344570340185560471813815780898159470368494257868219
2636363447742822186059642474579472170150025817333302

27320473865732153469489387723270055569460064750620414873316115686228526907416880934990174691276069271989
3850556484002433703265597529368602387426960159164509
5650888549020898484988762500950696676359381231697790
3451178055406673492287980647697339913892073568086405
2470563221867382478507164461143098095494880924737318
0071966058926964767438019702682241032886561113658492
7486696312680737684133674344484354915694758172853359
4401006332275230578399387530325507216780079595410679
8161051287012353251890481593561721703239862529923914
1178571556013166744954143136232251343090938117138972
4401951230821032789265836285024029590494049294153277
6864235979783872458059395139056795560489022429600734
3177928261847519339555645937150468756199757974316901
5357214406687580868655452485004827357130853173124413
5792320993581940084780355302666178999034001645004709
4910138334017722946299265642334547881046056477140301
7078618144781116162479625532392440680096901179092285
1352731479550364501613013003828067884533563391135173
1410242678439770712448559939264307744563271909032088
3935185236603305236997613131733483452682577284960574
3916715539515852956333008194184226563499761732066839
0993072216558028540545869711890122196602406694608294
2278143215432436576101750205912960184367126183329145
3458847496239270976581693940402863874677684859859221
3820703486498305225048439168577534548195511374947189
5212461814152399021533658511335253541411165644033399
1014817160305596064373428680323910031407094111326232
6323998396995376192271369735001483985838849714481681
5171497459079590177449277445111306272842013163584376
4179209332924316744005546123114291616263976070663560
3860065608371404383030041343474639895484594511269703
3377583290553364122253592752497345534833372325862570
9633843342035966408972856443178958761351810094275452
0374008880107278848188959516723126211891250019208737
3386133650137343488640425225200177046207266882007044
6561847389682355647147107826826300452149043241055363

5545255918421354196865113739788666309750081425643198
4740377591458789590609845742607885786320231440257646
0524637248392024315470427111900318904204033381700098
6422864341807477777798344255598930892290697457018720
2046818294167524913485599606198009894844748916628760
4198006025970012736569393629754093208594546675623408
0461501354582155086320722660389340137673057625340655
5169815277785599299882419464266516768776119173622270
2092278336052507704807075907180343633570756382836596
8139953907607270681813656575919866837510546115218083
7811919647554096709582495601782824567273685631218502
0980470362464176198682717748478222463490327810885463
1415173718143297928832562499371156297157373901158363
1087044860251030049694691425838693706512037704663082
4216489443358000596868730214852492879538242286100073
6420364967914869424254773064472810425508729193419606
6705256450640960879002440406424731141356609900651467
8880932791384938464806546101789056276456355644526787
9731766008564598590457594504529363273229140340624093
4385163140252600210208532500280314180983752338963958
3076237367334254811893427718926930339828412036495177
1760100346751920815833829363212820663131089145602014
8225230455288294429174005143891311827980981984843229
0298386962825148739445820391094065328018875407720949
0747861179157700171903879128063762366174401440452070
2292452320454057628069657930850203981218378402067202
5012026675295531308349435347193634177273406360262579
6031365119785548566937284640420468489277157780434586
7761008528960736931441334648737735250159245211976597
5459087695020605617578193591077403625835765360080893
7653281370843694390227229865322218288437400138825811
1629715534575674032149860975542868865798743690094970
5097986093770278357223388331453980493989210171433582
6189674003122527997303364571061607284968264026682347
7045583015458557482717137243584709948613726587130254
9402449573855889966053537090338925114540555812456929
4137888271651990004376107967257280599874820479895678

5593885849948346965194930897814997277634733058570717
9027093568227576306393049702296633955287633799130785
8593142078113351114320121026019873042167062601435758
4117977079045808380884980881666261853588355924200630
5302464346289923082030708064941073041567597710077523
9855868675945731744767094556842689038531128494988018
1447745665050961489899151762992416428780004741385080
4520329530539184097689946319969559127867694931959273
3662054309181205566924621527407866514323526592070708
6787955864168604527753575020748767143337706011912940
3158574310767777952135902613080828983248839483209 49
9884568307672417592994303402094399322708275483573885
0741991713694004987985861942344627960841444735665203
7928295317016335118153029312723025435629105545863957
7778022116588666112693357407294436145574905637200712
8254481135578340290160485176052432969813550274714705
2635429352648136623886958489819516790476124747446800
8477258871394552736710887847508425688259839636830667
6476645133082342995384063714939655126025964126916639
5532942221627797607874955291748568842182486374632474
7783244929832354402571567607928674259528494338989676
4343657548230757547840335036965376873654980223987801
1920354404912882683594195397184364725540905314210556
6632073204638848382768379261055003805739537940215136
4136624967493537324104404348623823362492049535442857
9053065452772650722034659290443202201716324235831378
3512521095764152741244657762616754360947097433564007
6904143622180682993551510913855657373411948903218456
2204438771527004821101276120814078245264988636103832
6508480852529514952263554264606718445430426533826668
6100665577169517144295655590542368193393871753203864
1155224288474087963872655996503545316017872842995906
2489756943146572532979956564427538102595666725587611
3030863545950868484208170230903776010731371062342933
7807454750823785605494798769021390566558589286009199
0456026032063782729076155397038311018008449011214811
9277796748391027288205755978205350883461500219034837

6576463110568401425042106378331650979093472594994266
1704520723269101718680689315989500806239975869483897
0524161223017172894039046699849427213392956812616100
4650902845621267573941439279503195865023504811047168
5635783540426485721275402638812871946209203813254648
1161703135867671064365876605516551331133170227182321
5687736219584821685646528460697066190543954014065106
3097333651381196333165949030392164270853542280497980
2671491189563642517489134412142636155478089214528367
0822169402598711263211438852993916963048048178929629
8820112380749013052942492948016114353302390080670657
2137816797198568613029030129939944512498469010019891
9360598279169730514759434649602883328969660815056345
0566093781292361334905857805509456421035309073601958
4463712165073198201564242201326845668774183233102473
1921868515643412032717030573066078517538509706917170
7917252855117436278713016009522089202424050305756402
1537273695926679974781070727937239123557770934682847
5601076301279131199539176281861594303820778398243261
7319663133362063793496768750895240236424692319045416
7386235836048283743927886654775948590289204020193959
3770656732119490991043352855179871403502030760557820
1914838828809464964820842417669924567583122624780703
9055765314126326024292243620371953291855471809159644
3185685205788235010309107612806044570442514799758960
8880281259978623877435496599049296732208449724434582
4350368978036518490995121422940156691745341683830903
5284779643067608611599763678720495505795636516693834
5210212057124671890236358379083391190802068995968969
9018812232185525286934857365188863016045294102817973
6080689549524036066488944683485357371170607994305471
9216487594313141269759525166102522909575375509509337
1854490007290767612634676529166464558037153306020553
4741620555668380872331011456706082197136019911669601
1772653512414405109362036010017584053344689875653490
0244758018499028511290560362815437279676288312381657
7437517662456404578370496485690904281846741434107660

7549841146574215334379628252377393517758770399425521
3181690173990186164214135439277973347087659736948171
0103318186376892728376366023019205919792959179148224
4163940318041477900282857125177644841059315644675363
3092415797021262648130428083893377067239822865434173
1736481424562966180793136953250911287546949801550317
9945166912284138446463087410279878209558773461766677
9332006361614129983611238785269844967622494946016222
4198481882844175972508965043238838826776211538694490
7223140800386409667479556596033658655008345015746681
0037154981215455917708285526905878274626801895484098
5480647767322593083364643266678951981323034384780554
2571189332448803371027660806642619768000401457681926
1412342142109083788260348803987158967469186812759503
5419040689672781395132198842118325610948747352764866
4367133593683737190716713615344289207252730570778056
1606591615442358910784646554736956343970737221781859
1230109443692313952203010113674073457059526133029367
4379321204061599708906812035078623541278054168265823
5374259385696643576271097354086523033339574924977199
5346662569428121211926674888665256315169706607240021
9396266842825154475614963579333658452377240996873579
5322759190097974155172133484533357868142287399385190
2093678274021559991420456446438381600099906505371881
4849381608655035722706417743866297516789666554999878
8957217902623090845448064651856930925569645317224108
9451645426796761819728832958413935133844596041672854
5739914150804959446613534398450142761805422096598486
7109944082508151323925213606951062673373679223322142
5995230222936409047664596154505594842048813114413172
0464692670497597490599351169204390276051574466773968
7080324780406343777841672502198884943540982821160007
2772915050759869365684722016941046189444582618551160
0415494510628158872485140345190055563466615244737496
0766113577874837400388629388488610195028128078179274
5034958405752928452983890915764913247310105633314781
3464026504626291567537790921372478289700319632596891

2513302152465612054358376226860928203077741687004590
4352635817494636724551789784931750675390464041603363
8472405464980750039300245766107146606057194951091402
4823273526691221496016070897220722054628810038730762
2968906215262971114289273463392143785758381679957096
5129751212882470762293756572134890623618601418995950
0029393433011746330033297290783402638252783796053000
0473559275468487189299720656136533751537477921962495
5179692200855731479445742882259242287677732128859806
5370465402461993872964993594356323021311084824249501
8006757189398611897262182430778317833445857036118160
9413976344651627256582886168782130134255890738184057
3422275279094401507963350696306831585842595975834413
3931666799730480514710420516213562175409048777330227
3969806564959009456956985365843208356206159345292542
4189291617305222097935246571227066400541353921262095
3741607025988131267956667461709323717405236296319608
9365298444250743022804976641640382829257137163603061
7625967249957176153695852486644931720109608534572342
3625450385444144127163847672628333081895855936476 00
6163524985906328874450325511377681813053346646699501
5477493242098568659350490106211412991417730998045997
8865399855599720886527297388216508774800198668603163
0561230114449331935784076334183313859772732345270212
6526577296264884620440503237750927026440915992126524
8626771659965913245715413925400153811699661401449792
2059852865463119881458741918733755185509581187101969
2417664292423893754945163159477245311019841450800876
1556264407882172093511259342618446830352107379400041
8382893605854407065172644916885787285452650728104911
7224129415223468484489897349653315569393268554021166
5594490751531039708324623445957019685643267568038544
5193586873351496819597696008201253799008400105463352
3364189127960544687635703710651413568371551244836184
9192509499414144624632178459676671911648776744489599
4644315839584871818846627420278441899928803275124496
6696486793458941329860233034829288762606371364458073

7134010172699240031409996289875932823997324878713822
6525474190348822177498195455707963780042780145879194
4118907707143580110302662454293625150543461651519860
7934238562390664551545908689970098727578338564769103
3468638899428963619169533138310635144431946929978952
1504273430274505489128224046567516837384091737414843
7318197118822641196702951400104844973686883604892628
8540745371246015784688794778131708392027701850083959
9401350787510645356146154845035346787490153402751409
0183464567541976045483308692169390248980675092299229
4071550692377787826669912301589909380813372850555299
0599347167842350786739058036553895201811147715527516
1383726656687055032514568315829590653570060806572699
0227214337914923752422195825551552739047664151524230
8413093279355619405005324441453950610949163270387153
0370152810088754080933294790986591783965408974119198
7143734113651271643824052441584288769757149771141471
4279508295887029927924683321337051526756439423113502
6287768903446466363218444592171575879241131996329875
4131201832522267869678996413293411317636653889683205
1191636223996373640065062421869198223064419813515321
9731985910156362569862181748547088883782202161710149
1243249216532386557690852747254785968298124948068606
6444493519183037483665508175542257335268514038987865
0300704028993343819723019614734086482834761260730198
2226861441179898436755838915900846999131405413831939
1816564308843497882991517174297486490963838967343065
1712173602754537578343113521721501826959291493227874
7374257245721360256626384138952626279302130009966196
3003232522013138218844822215385253127676763048551870
0683146840399268185487653840563848319210024722319166
1009134395076785551383704821428491510169897533907897
5632339921778903880476336813748465168922635716230718
4064156632492410866923967601216010814456092321337429
1457844880612478637738826410208618024951305733883694
1585087823197098151586711709517388028679580151067880
4493390248068909905291953284466968288254552920787080

9050166148536753308133690700480133882858546165406413
3202506938355963174243658840647261575760099347841140
8406299823664823574855435335905053612627428200187848
0529530447698632263662782963274163701153111823408178
6739876610728127325778513921138076815418944404176329
4630490061864780759891264283257299873528716127741833
6805175637941952440232128885491177415065311168183622
6989531900495922925083762608050033174333856378486749
5822310586318894073980761449692017917513933532988588
5343364499791300165712868099995155763688357969034499
8472342604194318599122046582749564413763677702161143
1270014347716120164648321329271182571328791058413578
6193118937459532363102391270889013912909166527192377
4586864170364801203295328751612012917060959270907773
5616740193911744124471246014178496797282493661458990
7255008243499708909680963641689156962089845192562671
9343047171456304432399815568869354337262302614980035
2837166513591216931783823097964852220628541884734869
3935943843252998753765119249233509919666893106839343
0992917742911260879728304331663875840237022011217239
4561144733654127633402705845417785774852486316499991
7048547694843205312092927399866107526313197643376538
0295221416374236023722185067791138872580576777554374
2535744238997963358197140322779356139744707119461141
6517615151238823627905648863589472686057334479728309
2570943913779516563058538904168168987692580836506882
5009361192610789112427098812226934685319851706637174
2046809637665572893641713249386443340528873527902550
8689950937615120464984509308482094646064177940775927
2735187506149345281817675171084502365204423677681513
2674319325109519200587674918493027769559654098122539
6357711694671126060236069439457213648076464990166378
4374841097357300987497338721557269597603311371288315
8380306249032383304861952114982623586673336359436081
5330962043523180699058672531667967198977573967198505
6332039162769296127845043250930278493655757046636650
0504235380700021043379054365452676915632116230708153

86679328752803991810222879667549274141381460065654850877794899447855050889491480528826587688444566272939
08196144006839830805240372569506411438993318116637701630751930445002156616091239778765007387433986121377
67631637994991603580029425394150936118287792564890199706361151183343773228780513781700685461893977078002
75450470574657440961151865016887216781580801618548641080898632223340991247492258081118532699879573620303
63601202486339705294671240192398688862699835431092004562279169984416988321201809559455053488532617542549
53638515063118962530921766529165824315900458349693970687658654243281945647647853735525153068989109796668
87810025698390687113192142541988664192866687545377244317614463541566576287140553536428785180176621964712
66899624319482731009397417104259055243749687333036596872146018899966928025823050002504947431957887361774
39811819412948004602950542736866923100793833552779750038359156208164728629951618141511824304864955870476
42038715770645275872367708081805904084235237757757540088468775611166581392511935673190940209614289011096
48995715039720715905042478578296419139818646456980690038836467983679810123769463228883609158544301049426
14876035037990344177685999567599330502253234765254688995525806289317811721823163795087784593437287864151
44638730344373009824804924955400332234343780588846442656516711172540890242516568074345760295714812822409
47846055854321075345638218480483756258915751706013714681964216897177605495301992469530239019961482626017
06284818796357991239697001594441468677531856458312725471743944500821629828304937569597521339743912031065
21261696229281178721499149753725472129306870385087565061502752264203373041216162349639788099435270804166
93322721835932242791116573630925466496712499296143990750000976357102050913521721026748783818004596833106
99995590778255464891283674339451582478152805746103415113435638105663770354879809403265266548458248684582
9067556435863245918075346077580169583997580648813770

4343413010596884392991793174174742867125143313255177
5956673912061135467364831151979354208615472228327585
6040177338917321114186236520276449176308259943093916
7130160355550663966400637699177448238398404527277641
7201122912290611855951275066506498354602961029657475
4375906955550751018593507588378946923408088442421040
4435178451018449469776602243257275763372138266732831
8485104191372257279900302301951988145702121716572276
5189190273755803239808560285417910869633038050283058
2541552079221546755099881660712637966690626696222880
4105219437555359934786322393330880742944031636339297
4318457429479645304848126672429605547789372416592547
5425727294818303585240798970601383390192188164731155
7850526432810671078304253827862550735751441780943794
5152087696441803929450533710695800288060095926155424
0429538493922366928251865457871541550543768172161949
4162439802366970176458516822418482349498561226122052
5940694686843355828800042360426716492051983003040063
9082160494841822317937800187897147326541391659694146
6285446072011663385275508319000328193491650079157574
2541572642210731921314240334532607660005408832422430
3953642851701019071959689962221685724205827008044884
5866160085246854711766440337454171697257316643571299
3493887181992917594663181328648661847971944939975673
7688968894218741250518128652265436890284789523945318
9075978183784293738692711233245224027048550113680713
0149912753906376567761950408247294577951614725178334
0582936314389374727744967896513648114070023132746198
9829246746688558984335554557604277573762760912058048
4802719998495539526723426269717512224621696901278008
1098444425535915175083609503915428095603229402450125
3444800135907132943742644076571567194141395791922713
6541009016170299103199865705258409257998434141590790
9404761270738304781789551094730906625750499363514227
8626507467665960409187273722547794729752121567356857
2609788566464575410138193018478212336515699772263615
1699971162308147353868744559534540138665599995625362

4683462963168340979755045640501027726108378785182034
9370533279213894340837041728605351627431809719641851
5813346386178640580852899326162674792028229007798061
4873787741173358302761312079602560784881890468622896
2784390204998370368536881027711277649164807893994443
9040925075904833289087283241424493352364630816798184
0710198476936630553895402873674490337398474907585950
3560607200358784504301688110214426498847297177449594
1058452003823012531660787088599066303712804121528907
8551534221312154711394784386313780256937275762215857
6792891521263000669716993847263443082846306827831953
2124797579764030143681437346152646919895021034218176
2593659784532667602506674488467124609668661730097066
2524501769227885205690673590084316041245928474805668
9469781827677947558847010378813457163188174944289989
2987167278685725466553677372831124446176730408752350
6505439172831961983050563809001407118907712213292090
8376622049304967945786678841810035863738342709135352
8093942551819807934978546448024964273686684335521670
8089658368204954333674108689934436229704144434396109
8861270196212361682942389358044719702097389199208230
2323146423505671064991136035773195638847191819769880
7100580670387057250153019061535855590146412669669192
3629059503854868733761219137217442714461555526745228
2574000580347074835030814855107153993891683837311092
7502058170195123131177677851844877768726804213939286
0342132389958185132776669365598180645571558540380121
5929712677686438428537188809179099573080680102254116
5164298547985978012005303253225752499741035431088380
1984322002357567408273306083966425559365951758305161
5349713443826367990287717942890826604259749491147707
1422970252511258606039005296024025458262483257557575
6121613127958128121568538408559162780570929372466124
3672058886815767907693035051789575639340031112592979
7253577754420056996640220213834735660376916954413189
0619160614468251813160176609861677232066349745604650
9728826595127539727333686941755084872178893886406183

9846003716868968509185229834465775516415629814905434
6762684211445248399413123038525805184868019570889226
1883252235536436873645834791469035678722598255759755
9602168145222994919737926897880657277499602061831236
1912598422999744591793191907004469955405843606932673
1670892612617013398426718889342124987556990243059505
9536766540275330755030949068933593376555732175383356
6489041072878037317426730961814074515620721259831472
4574830121106609479787962949043813324270611655269733
1798122204673427121226641381929147327894366091827878
8827641461469764222050291144484184413818492376352771
4914690697433608081450429276576158754217052493940928
3863732947357784234240779548820953126235340275503505
7102890394313368148199515356610924477427046999116723
7898955163904637498396603242741993113944290390576059
0741555533250655482151792922547642545087189622131135
1669933045431200007471829800650638738989262564890543
9739686679432127054939232744427995717336359106333484
4185146953429812808968687408323754080262605943298620
5629181754412222900184021005925843557050011626334138
9111647224103293543067992468631553900279513923299722
2766212995130994097950530207390559581191512433304040
7885249710925372417474301388303179701844108570451357
6815129153624429492503752616110118373210046518961467
8269724442617804346498440708181946488570155664729124
9400183231574748921227215054856761733105517328675555
5513727525722807015844443069091168420794485271927516
7523884694520140584365412441900688299574590054357430
8056155946524228819312720329234092490339765471465181
1131459251905725804935151124366891654022600627554580
1761174352814314894999161467189352381443368464042767
2953871675260813395098759620778572778924985598283482
3289177208117777338346633424188492279198052968309263
7567318470970487223370762475836981471774211480432136
3094148725478149263087466784789524555610053498907888
8984592118410223597827373165212801974854134198697705
4395388749739014574122119048053900855254751776918270

6070979677126597248844196823381058259943529823826723
6317342426715578296460105068310046137927489065603076
3632598102793661123570622546093038459223095699447464
8995943528035957281220735900214846748760962849701879
8980716170871860113170396984354371096843517664924795
5442027422470637716000357522942761374328810277374243
7633846548182324254585865229937019088847477674002680
0100967319726684955864544676707987877175135398088398
3207327701780462499327861888076713309254338928428954
7339980467826791459819674690198306839892263439290357
1857330959662853884503112265863257014951784436813918
5358392042964337583894923848122175655021035540671058
2772688757513784359797904449145269179059270350881467
1877816814014900991554621690142569780350359587247391
4976161903348045649916980438948284871605733097080720
504665480348755712333122248624733016399867137951278
8679864381025542560425357927516241316245495529731023
6459199301134961429522185316982957104068514803822139
8883769639075814255195711993591711187550877596254773
7751359233870322994013917636580370678440085956246876
3995140147145722468543401280785856430439394706997121
9759406459214429010712931914057426533471341641286645
1075856458812395114011779550807321636781016037343376
0157315563492556593937367165619355004588107322335593
0244825696996555838830534131866761689980856668282771
3235687068122625484629821031317607718012390587255347
2420474152001617666021805882461949664874606456387899
6315071244291538840423245075603214047762425803652609
2051914829101027157745924142628710445597292699956501
1286066885462748757187665066969779602823041381108469
3087871684835790925462780349442642545861204599719960
8003166303479589928945632553125425344317399853694598
0583602867451850810533130476475285376287457097777675
4130771424300232534740409303069421829116816439039165
5747003216588110006242071852475797469805276517097274
5153025094618659937285040116814957781425963612401478
0968378688851125147127622317915331487044871205793776

5503040166429701507673885504773832888178731212460074
5276123541766657688170101149428992573491012356466763
9625806511271339649842282450273056693708923735816460
9535611643437505650299631516797453409382353289997025
6271802165624362551179869721624983230956587196829602
5466806500004671622240239665382418550573065814460635
9055955496419820113969652844393015739310520628308914
9218063428715535370059003504708646096354109788410609
6566034365354449061700707899583180561453390650477052
7431566417609721517919489283364812866046083530718819
4805344217730423604084119157616446130913675931528639
9233963870540720547988479579886169962180644201535353
1790044765255822727670645936350572878042757348558987
6829668923357242880868062463248979189584169445790029
5928632288829382800916032460635320238223124737736787
4061794090104138153305761028300884886494159225575438
0946063025862676989173184615639367995770538251493415
3048349312243480633331688402699767024473274906189736
3345433278280820774402667097781782078312085726344560
9470855485214265848910184957350318664217482298567340
6328525642088746642534053395045023102754493429019160
0084415039849520215325600163927669366477809584126560
4290645256806170955852041871281481347434451539374647
5496344220538926109954944289146367538957876055842484
1902583118172885021588583788068989305945215392813746
540887753610042351014931084607654329094608679691001
8840384162250090888837411244830646123775233176545420
1486843335315258075266734685821776462554798771580037
8073715241248408692569070492890066678114027979163653
7433395145161411107226456176282259912478849099284872
9888705298083065245993551737408241133677579727233568
2680337811635754998347666990823837814554391621551285
6385576818804480343213210293942162408135480262308822
2419123119256612052550161349899775372113033318398096
3621662471863300454136003383305305793860790211293203
2887174995145272045062395800542689317348183540808487
3470938873886190102136217565025959113514421270210116

4809309303749416729159362456878600346602240033953522
5142449151606042997647942423498377961134986659102717
8292993124652842563039824406078979869734206039517007
5318757863061612647145001003208115941696043578824718
4765644413513385091862089785746218053947270574914758
8858463426651824146838798580804824882080550201928877
6349020535515856240018934643324268636634117403031223
4237172519433745140146909287147290589015061523895622
8461520314617026175667585862095154776828356254506431
6264374580760714178578002758760804833259675887885318
0959596581481600806521298179195843985925295184458469
2724664770760915168517463856471876221491932156230101
2713105284450748107002650244641785872435677992213039
9652183827914184826528162510461472116304790782242023
4842176235064391415298230055436257014763699588784501
4405130574818011867308723759652465204643509343130396
6655856293925156810381694666620313639439917344896713
9829815627411988262723510952512218217047506862533992
6753779784668099954204214991134354070102205692681999
4042516658933928498565667347332579493437311852048961
4803989074911035031394683340777866936873547549110776
1126947029878211256476568149156669190955293673205595
7727929512982090715157730091788477963374300214580360
3144778906200771554904764805424270513167766893532669
4373590972240083159402375838831476354214044086042129
9577016536729659902048363057779854577067028190587186
4830613825275278600758790702435874381211840974392908
3622398912538931673184265955965954849588145573527311
9107469179828345743767858841443980936994532326065725
9969758824219293587914543693491043904226378176755732
2498731475695701341604149887297960683857414839097389
4482340813010956583016594628071917844798263328455536
3597451218726033253578179240687001068616506202782368
1063429287414508137306095489089247445066227733083852
6810902090813243131511849533596968989039103170864371
4328426824906481266624069267042133671604705742655313
8242434220856097935006501412683886793330241865462230

7472025320717893805031719917873491122677621899708478
2360811160079801956068213562664092459514059419198886
5960527804596189879889781246044507520439119510874257
6253503374285344359966802548772112938561910122074296
5112088842954685544392519090409345969976232251366844
9395939143105822005497761651193236335578447738700727
5167088770183360897310533306167400940748760848572870
5025403965220624984387807914212329203806188310481723
2109300172019148512553721123091515719016127137406536
8354640345909017516919077376312212625722658664549644
9591237496183719395515196650457314903486981404538174
5736975898227511377456307867235621697668252392452981
5904675621852858455891453014822265840535952639098539
3717661971015878387327823837558472442882536372621081
8665740744020130772139829403432459845503488984691608
9350460604602972075243052225050884840981344559999075
8681315475222046561457662331404526326965352792923797
9293739291903748696719646307180607247847473965505170
9747509660626023429037646844450582247222021323863885
5714513234749820349966040662117777297860644341467636
6211106480533161430705465164829466033764356209291270
3240902104351027595403970654729979272108961839673150
8470303622049840208056686992252465953583737496013314
1790035370052464560586947767468897309441369101074420
2027223857514205225470819089634991421944317404112374
8517288954308018457409103199946112805564334422622895
7934051736502953133602831432970249365622011880738379
6596866938563040365542449375719742102493740570193694
5138482569860019532151625689714018351165254960063601
1031202819953496704616097462082917736572997856457130
6965165001622777885273834072598355967397082404631525
9176738042297431785283156494449545036956401109369855
8179851150272119159176380048296581307898594703973120
6537802032276034416297329698012059066824690143029493
9952501589890477105080253835157158428051781526426400
6574584027917036556283411551999177884995168857898088
1405096867648256081575871689213785261434824019867008

5959491834379968727493803924399001240892926764545234
2219220757841378533424878833535474932468597801974307
3891678142961050444496628579719868493170449749155067
5521279111615838057069681965725948152986672133352580
8676789995964951596061098889306228869661326312175575
7586483262796857114153350264721407950235985958527648
2031363986556281374476129381484774020225044508591218
2509562477907388965494526502037078510053115696562202
9849937475086136190439762995990313119008043197524580
9400009145582174545266258052740399813169325700182768
4217223180514068200502309296617727018544600370433014
3234525164686645610521720692906036720273335396157708
2848882372881358894045344902269945387687846572485781
9287558201702633487954178253554555754906825006997973
9505950461342053430926981905725531230338713302912850
3192281356963960128385345521159435691409774601581987
7741238198895866727111427295831082708986392046218034
5379455624339718562497127140957925961923078188404535
5529797774221068893673388554858810722367687848593822
5154035440994822529059283484780136013500915158114532
9214423539629875548306801558853165966563394478186222
3936580697312612851200242204741143375961776991063339
1463758052035815470306229527721703514821238530493265
7058446113196562254523852860644493309181674712723697
2720295002626694938676067390723574413268607147813063
2796252252135834514617531970735280901135431467773945
3471024006089517815558307572118215079950055883774454
7319261292538304998174304023070223890981607615311099
2888652872560217949886633964206182092131604317680117
7815429663673507872419183405805978019413641380162857
9298183208684244469747609594675465935139271920270860
6667009644691264257896976005037528190966153756618554
5806264352294921657908349843792574319715877064686820
0181823204868998256456334050138496655764029708344806
1878669689312163513396687831362351177497941993054822
8986619040064715435959579227544277779439667263372974
6627797753573196084347249181190210119294392590380260

2648417484447820051685684303466144125006122544118553
6036696829948065721395351334078869245327059129149828
0174112107188413426878788829800210711931841547690632
3213303566470428019983416257261051670413116849386770
0277509498844108513693169564448607593170835467673690
1777389429731545511459227701110360843055771824121223
4032928229874439864464019195609230001439499345306044
2579969384917723978161494511312042048686379167525306
3490066523958044028984353925557848458072200332029250
3465974481326140173373348415220872649858367236488056
4331283046930530487353905968489776941066248996816465
5101825562769089233065437474773251574823464207618269
3720200111288490837408415666378790491771579162617447
2533569211027963136363961933383031690960585634786515
8364104095218542189253938453651900094568218823512196
7853491290747273345761908795277007145342964288577789
1979700517737331894256474677870595141670950151254363
2545858505909277772235744136906107059254179657940736
4489401336846212597403776943629267107864806916569414
4947649627554797526997506112392906590555602998061827
7579232119869045159059424907676014494433021447538110
7886168394173626824737953620485786673661943401837539
9507887357076956973633489060966234152033032736644168
4091559726750606818691954289729554967800742088808731
9998422933180164226391830114079597049126719567266193
8762353423067783745037399215560497316196545379184136
2376013666098734374056156461634598523847828523319730
7913701982509058532692942864012889661556236653366808
6796762690219338587009470620408502701789450516817868
2770319342784307016451931313911485790961696844160662
0928373208333878676414883913529892584818453086699758
8412889658670242875568773123590034961649957608292377
5226893655570763541340826557724889024357548539752579
0911342017983026115347451748939422823882771044974234
4359228203662147297399136740367101215970943082487534
4769801066976990314194078502080100063845162203542748
9532856955258016698714012790945546584468531729766388

5922327228023922957255162170439537798680918870851195
5501483450065354205895881728190715946327770613634760
9047316518417732001776274966861929830048478422225166
2526812410603171436519456728348892810958904469510765
4103618988534832669434021847931347638061335551520236
0217636561827113154532531524831850160025503530023509
9811874568401397841324504129248995106356188398860593
9985186066266983743068215608935364080372210569221706
2106540290334689571523900667996984398197199449488473
6379926562713791440855451262773768033692487909647451
1063094304810474408259752902764930190996182867206680
0838124770828042534854515494482673351770991586513972
0744453559629062029789651482279964382284624100494925
3809663171584947464968773242941714860117757925464809
2229392563484734484973447687678972551867684457804193
0104358838478744984719157546612527742106519834036887
6821770985647989749664179637585327608894833993789803
8693590500388591500418224769262139163222115117073297
4075729995059216141953417954539564825806957558191410
1054740858366976388974854435670380887776225342372523
6685862528686070111207376644417703475923902205402921
1833635920768287468191635734436212258468551849117378
2814989331732943286878667341277094195061406784309559
6346611830093772355931550084081882042990111253625495
1586587987779332016060230253996395820888578524640683
8930603148815511881851063392801380688294775533868576
8700628738187175509620230167890829577299370394812355
3225117730651413774797053438937964519477623368044456
6103772824237746406531747191228550875257012485553034
9584254775511192310417412608960417645304738448618495
9876882445549497422370796274252261378595615081527071
7355225140609247146628770765759232800006362366178185
1820146612996897320450670193879122339028019679284857
6421517536050324280495374126017027574030305342271641
8639495786000064599523927669172889510347832507317813
6744237372764385742521775361860499615778451640562125
1200757112686925439127054804174130629085265480196487

98111143734157554754999170822366196507152797220508966
98520513649053554727844829720710781457451456846086111
15729565963175793886474842463316376564944811616192116
04628737029040409753672886130662513809093509849143099
15201368594034990362979313644033125901186964880120877
86661412089289781050701755926315932894759333535558533
96153573748647327788469290051483756269645692713830100
87192984228256114413262879693085543514821959294806700
80592226824590669598704980565202263218860004581338488
21338380107140119370503199986558912494606613140897999
46504502916697723288306019518497855653243552234794766
17679257654458205266051679944760722705502604025380699
30549886106509217465565888873017888384128959995965911
59322793045738084143789823577978122166391540010741211
33609571689636670057484589520547926161961736661796688
92473003181633077684291239817740029380690464820050599
30567210970470723585762765597152866857405809911535600
58690572447224208349859427076514578043398793416795766
81379636208339039508329344955805953637604854622311366
25167923643813542477484194804358933214598819421360577
92941674122615999836108550167244026493529026929476244
13425825638723031974350786160646176951012185410632200
30817166112414886764403388729757658443952202219610077
37434541485089168713374426783595270561315212526207388
69332183059356589306210304929209535538145495116012144
64198439793903718964398693487841589092209093907042755
78219059357094307672370243890053453120967003508961922
22429881608768643298293971488249604124465132808211244
89224188331444749268803634239182966671662822415678188
39675243796659674581169914928136901450227978013597699
31386539455472068457777174591385361784792669537103699
50896377227906128123654791570888690766768908193749344
06103684106738600541100262787008747059471065864514399
19698055970695055650501512339366658905717331336447644
21307065675731991903938811893825307521088285951614755
06809468839245740011135470274786982168621274321497666
51883003196033345622355344214173681170059576634936766

2232906918488032373451924349124656653297182384173969
1986587133313412071051707368341724123447237186724941
5108427049615549503795501528738224860538460676720392
8758686262616984382280560158757866830925122304015998
9753848922836009598075215949025807471715773537480669
0050015349853590369767229710437159210984693912049054
2624633960855082491823286584393402629382685637152596
2389474471783369996618020115478824991332365221595665
3404857011232582770818862150130715779346002743951689
2752435518239624083915012347392466865100222767351541
5339817443806293681884318702195394674578806838733045
0266993482047409308509529400887069518186325483548249
6626070665024995646819410647040712273110042154418554
9123163407340196180907498723670338997499343924399165
8806512836679056416957365216050282244885217573611331
7429473563577483084783309849295743057306045041284029
7114897365523337025293332859341544883137400581072624
2464775155356118977342742942596840194068133338141610
7490916404430780527772977009382768736682692808362834
6772591003284128128935787611886533037994391571099173
1308768414829814731674389241507081233304663866652688
5171522877349580865070902110271308011488070525108616
4181615255674817104786219400105612800134647100048960
0816181363398613143759537835734463470097380223970667
3331788471217421662379783395050849458405648043005911
2018417300502241962981137643255508625993667297921212
3968336262543307920197457555379857786033399361139731
4797358588046748266319055503171475107408265754553845
2600524391171639479854543854640477445274442320820115
8058940110410945383309204755983130127803405876733811
3451784704239620884935829339947629489274926307615195
9475808635230500390635269755089737196321432027975808
8695852977316219835944256689466634986556641828527840
5307103946038370553195307057814332061654837208763730
5421510498196398231622394615554016244819227488988991
5481816161014129111422332742480710512027849723849044
0264296807698034403160101365988085642296075406956955

7133508005356595188841456265319836545998149354703533
3965788049476127320900485531135908697471453536890157
1896984157754811777941076041901914565272888404051882
7950849008264536013242915228888937975199955998596828
8936089112710910309973067570621865972967346398430619
2842955042921054669160915418280351783236841552181865
5679634071112430643855154156248975176397718812620984
0873609829923836373030275059418770626263573784813688
6836829333969889277444686969662949968966027691600369
8605968904718417160001971841236265199793012787419139
2722798698022724595229401524322084048175398124008257
6371360670303344776541140427436996205412514504827002
1051517411890882549260977472146205536677900755200879
9892323465359252612737727695456712154449995014845268
5432597408757444502355950317588809467867062427950496
8122317381953193469955615100420921536083260321530691
8572722599298600395392547343180646477064441544096308
5983320989428117389795068274955248706628273280536479
2474107235461194594426537978749869795964322144950143
5206166327360833877895746269008896094162137344263820
0267657533910341892476105635988817008353964495585219
2240769813052521731950132374196453428976165813540733
1302630302854780224851925520783232374548326796328907
1256274752461229292964140942485029059474155759927512
4250850699663473537642940738391820741909162541609728
4459162561447272170590829385767671304543843359908261
6086284671963287516028608040353472085362193749494136
6551991508536892373512748507429481445196541693822917
9315309180378204728722389455877264858456583583688835
4176105647212556438640151856876957798827517422243471
1115526234369721170763345046653272476633423782119587
9117615783397228747084512798865992451832791641445982
8021443727018946266803579233337517480641049931852188
4550682118360856897563251181387504609424195574432639
7198431078927752472356672288395527355236066225987885
9853963287728445468369492367606484412273536829219132
4554704074421666820674612656707964301200553515990474

1708546163733620270386595227380642312621874062689090
3998167108615021958265006449778173573185052157715160
8476313142971006824824913916249289255612638476738621
8233345228302857013815543747650578465690588394825313
4199813962957362501919857105808466792364911059290880
5506833821771837094406269993826919706824470532005794
5408351861003661161560945082423986504199873162383835
7466645452188735902612193453160485833921266637725062
0851905462362373373281142581664384249889798596679541
7592003691300453703719627536068718301376481212741056
1482448795986351085005487349029844775297575349396655
3071505515441039093865374166652159183354997382566893
2645268620823596214023963409712242682687002533955826
9322438059983242690845304781419914989311071015716947
4955576352194587457686296301838990582420127878675579
6934723185030744610603483914598794977948711391354629
0254862822249743894474386826616436068318124885987232
6879107661623053800860292158338144328453224074623554
1879988817472381285359683828950062022524646357316108
2663603643148563553160504991541441308872154144331745
2537321065139611973306948781391309687331861366696193
9402572004993080999953482194276558781132352187781245
7183423976468069730697934060813830189077853792276215
4846640882312642981642123941745697135666153982555543
5294981722390145222359192574194314642182664776888667
7250217123294152289784461629105662854200714538834167
8148913456674450692912906319135618469153315855813507
6487541321859455745770156486652616746620858010700235
5681687581059596769989019854727064175312884874941390
7086642080806276944501319670197033551810041719242513
6442744852197815370357770961797803655471501006418180
5027540735836312100758486904203604530637430343887615
2382898355737285260279010965811358399008202510641500
6848860377851451703993626938322618882301899591271000
8790754166287483223944522850239184867913815692056863
5432170416335863703532435303860313824517889442520080
4853014804232640065209962960096641776937613082080687

0201308834720999166455805747029726506424859031007l8
4698953106901743228378475715367509849704348348205993
1223224751985485354554519808422814507464169325174171
1660329326677622781908348972275100308089752520503024
6649351264612784890874118303858779985696639063450520
0221239345266865799204423861484757242890105114328883
8145837001486782283330164072761140369413621157371699
8560584173544560803368889069252435338105397159315045
2592094012886826761957851131323839763615662128576482
6409723566892206504594883179858641010564381869889428
9927631481613114119783948514389640201616144702822217
5648273420064615301617015673045186184377052127062573
4225871506289598548861762052296168865652158377842471
4947986648770706732481799084249744141303077278122172
7570393898531076626569482761976332874465960455935018
0212323136168294691051653679961314965618814230797900
2819702075307204137744966745306251107845657924638003
8273856860224589540614393614994048484286969806483262
1084643807433365583986880899884715142957010145120549
6684289667486610186687651297313962832144102683416586
0359911438931807625644344609274750282537324722439635
8231441866189835337323691680869502922048143623720481
2347239217482268429106661178494353167259238504105377
9311049081572958008218904109965374052294046201984820
5947195654684300768220373283990614679207880313062108
0742887826745652227951039753018549245010806671435652
2034098520971712751588390486131129466673396409924349
2647162312234605681600440545721462865662561689286997
4683821633904331762863708095828508190969741762120740
5508511001815330208171813671768543487272089172606743
1810386875499096428742804106528574447831399487497892
4434353732018837509779534604400528119785269275442489
6325741629879428825506076395165108387117664673766387
5269750883793989032883574021122956354320874374141624
9491051576822711741771193223294539197409213659086000
0476202432818881161163644928715559908299814543584037
1709565305275679006103858220050257742242925697737322

9660938546733692944968154326056858234085073909150459
2315081322676781063632505958627455148922828565406945
2221056355802862241676405576952603221833562396373988
0801424611875505460955509412272002001667329078978000
7096384828312446295596545022120743326436571869713451
4468968889229510802001625680799512184131609833097021
7827686024487144336131445923206014734974814869132077
4245973643394920073088574820672241184792682405330459
8692754589707213154584154047723222208392080363241272
1763116289188164339403686931713558268000635173209745
0524378305263342982051585729293729065144082252093140
0383344286009437177696508292941118971208737078409355
2754759466760770073958299632583886106737450792618043
3067251405352116133767896036538579850322184508372345
7258939744824492032873386027992042012799506284586176
9293493474864504649196863533839144956932935349913992
2453664188100631030710950568773445423096458954361418
6308762414944474198666983477134053009577978720792914
6238962639088463929969063931016383363832131953937930
4280498487596279495977933648358480037841610186331741
4981336434273879803025779702183088650364181062215710
2928202158718144562664355191289239074644949104273596
9276652311469566584458181144763773752350128593126592
5760750556501984335677543209175069011358786716195493
6552029139094377173320155808072273331910017793165196
8154228088370576507598837597021754117738087137985338
9628968302629816280553011075577589785348238548129828
1050149628858871823221804888260182746164487902917284
1840002936746970665260086282843883088133563358024351
0577635285933515444380101024499421747818965531884708
7378155223644071454282954832131189537540818216074865
9797776227910230823364843617233660291719339926167882
8689800578911911569612861137304042229962444456452288
1179747153663579146150494381217969991171955339291546
7985888569082819110314718381129967865651453059679813
1908004221827090056930448074830083456425150236093637
5255638319351742703436292684023477641389903904336235

1851350263930095966444898776041743786038172342060790
7143711944475395228435516778885709093405909842628007
5550024872583065055751125935389631517182606584600128
9326829742161791308116878212721039217579698337007055
6004792659974323649525447559862348447404096171475662
8827532509173759154603250809601144327041198731596968
0079689360407597750180374094171469981885046628982630
5034062817302823197991255309183628077992511330655577
7258958054972193754768545034660721841155705309674742
4723286992072949541768648905189668750773621842979151
1575851590669815433469973569518765532300615728199574
0879843326878654838770370483886796963447104488103666
2552958628343438480462678122147642516601976422702036
2945087987909230689977626220233634117711820153990501
7307149399611554840371522826857799838510938400452826
3675761260563984391352940873945574277648499241414685
8242913857917756232954020168247866760021021513826664
1101300421780062344321791952368617769613880500659250
9070252900155749948752601372319426962837691739012328
0026029927806831063027349010110031406707790045792769
8279313684945163089509182904830296500190892833649883
7214396180793208742328768068649099204449507359860528
9572116995190404083389840835633066438541288900714754
3866567070850184695379003257164362715713573472251059
2958651109840684568971565825455725633355594576577747
6180201568378523640314978538098315662531858094257851
0581390461546524807271983290957729580553008835764393
5465714579135927674664325448384545582730913986167177
7995943499455031747714365979668345645423457823495302
2033793923089622734428667174667986556825373251453774
3009223721202753169385665977610782359801886307507560
4083677127418714589669834570275384870360304258428021
2778831071351132772777499204648303734048642790225836
7600773900836561591221966284269036366602985432353631
7994514703396394961386871671776305654609188565518443
4451689033462458468291241431753586574989124759556089
6040292155393977446428877279516223061246477930882654

2357480107178091527740165487104682655357030989813108
3104309486675394151163167665828140370086686554764383
3977042757125044749221422969080622787608544404858813
9957127475613019026375150738783118851441005371031418
3163474336751174923205317682583915034235450586022630
5868702779243229691617545953440457447399711712119207
7899158121806247094415506947273514012345502046290677
8337817621419189014690880707981810067348593967972669
3487696523832201910820873136965799813776062143345608
3913079919949161884261517662011516330133652402686506
0392172540343455286751808258010462905820836468110735
1833729751961456122525371356771188784251994023543962
2424775173733453533891507262882321321961008440631515
4799859952158888471159978375100106259724395444290564
7837242827112768861308729717667450387784470605049817
9263263011217932857796302134172905845069384231278574
4709828199036893333232252456587848649909717503202216
7285214998827754658306559015958435161855520548978267
3613579627957493305220111916108420802369961389004399
3283506634518164076838998792241323064158867250047986
2191487276091383655534567175784547455676041291332244
6095514831253536112871452107430436355478982136977200
9279301372617048908280393909998355740828980056130326
3280487840702588019085387730318818712730746136642050
9108749612419017691344997077457528363397832315611035
6482124709035953681610393841375688337715782793829413
5462391627057605602271779835321108182513441990267728
6104968703506073372089355308640081426599165087223597
2694624860483666589113039585075104744947269939264526
4605522665189249764530807213109352096231576084883113
8260364791668246800841422358877741046115569322707887
9326874361283795757685381840746778209846776273672446
9111891654655430484797104215239779289206763896635896
3835664180589442147453962262317532655797252681466311
7697801299371109824534052292709330039962951371443831
9513505731788056663961042122577872528013252838483308
2997288182507498851831471768420564258524847512955453

4338676312935464407773405000721996838014052269594709
1991189551884053700772316581952105304731149144913758
5879146562388719464798452652848026896471317271515010
4012602307121222879224688811363287433466723782055811
1794202721647922565370422448776329697012927691315004
2520225981080099979885590235689043290231478538738724
0209474483853941778367943233256130938177205017343097
9624532495103390555419373115567810960057360469040879
3506212987488240528497439396469465846777436237028014
1430030440166179011657976026979051136930909194130040
4399250566495624246838020340508948029000263762200578
6405521562274015538802800346895591754312762186131476
0525568998950949893238347287908253998423463975791200
4701170765647978008566264746119214932018152685676334
8584378214733117677538206424894580965820041347269571
6237038372077935659762005227802251822529632003253692
7653198445965838103755079283425278582591745643500626
0843441631129701955375474851856666610908213017688979
1760811867452945298815710766128602140319019653862395
6291513139154058784966307201942185802070548054644775
4769556787208960159717907361724140611419772765025290
1385992813531379709256036120684983888215192251813217
9980933359782131751070483806021942235771089945917470
5182470860381957578800281615561611665853316284726372
8589024423372208724629332702835317492841004434726304
1799008694169287372227838080288266378013920532928677
2928042365659125645148945492893200840060968418317839
0593490237620500022433912426191828329180108477069574
1558673641192801283496458206833115754626233916641813
9061988522457343683067739836101046423164641331355323
5227375266445338203004095510321928897610402878825716
5434018178388741015541294894921097507774609559771524
7869128811244829284257197468820488565490956157685128
6105935267802656143938335859217886735660059653009853
6546206425434637910629888573898318354369679219264775
7803797609915199776315697301911388169677703831361743
1644537854570073193605219700151735410076686960130543

7275881607049987789365017422246617438193269192862429
3010619842541634604853306131224444100538119042170529
5021020394928499328022918752088000318240833795363864
2237831826693545467637857036064039922133069977838159
1952595555888781915527331155001317087949016677272842
6303449497471356587672814394265492052630061908000591
0944956218545217579084841829846693841949558812074700
1060376835634410332054500161516572407201252986049588
9470797218342597016438475984865828098086905490367726
1462021539863629416377179304762721831365968096850029
1803114107970431138118641849141586975849866022454649
2765131028727586293867040021300059517816358644077874
8117806522568506530713507342458944602683462083458031
3392145354223387544486303860708727850277777750867762
9920222402710743245913400402892872090265865447356210
8425271922286631314180112202868765811959272712835132
4215088164501912089966902197373677201815546709893992
7354219911627316526245069499200284852366347161921013
5317944601819327396072124887398157503934618170208223
0279563933543167655527885029351632734594726579510401
1499734482377420658757303463602505161561392726898206
6048321085542322084072024003091375158254174604210574
5596153391984589002739715743753802242641556777875930
7934804245360144550080463441240544089726593996633007
9979725921292231094194707838420418307596625539432445
6096358954142256613673796928542371376474952633755697
7162602199226734913942723609178467619658720575970336
7639965915756363993425058788943766830142891641757385
0338881007860021862503880088175428204767842725956719
6048406057121007965616492691699693920855774499424977
5621141413333243237951509364096641340808447528614157
9712425085059258050810646205124572216884820197934411
5322991552482048598975130100755988479695867649524880
2794232312419807388735906666496491511396292448871057
0456434199781743710931692166476206909405665634791222
0876673531620143957944457621663868466016550053394793
1580877471076476445478396109992723489002851457979744

3446604405111760455861770084338760531480088536735702
1156503610454879928403753217591482018829939970865269
0645538159173954184083082629694214229603619265688038
5459991453155319530519391834550185884549349557882374
6509856082129842400795180417637267312953711599532987
1379480968186645546885144362583503666244085255859306
2441470020188212204642192591723459339896192579016317
7529238655519765624923716284656853893447270690882441
1028132584191077575505481596675431195699439784151633
2406889407043926608112156186190253032207139064676977
3268368150661123769280024896835575713723112884258276
8928782310956781189966969774094893487241466149739727
9494126610763694029582362830348242237176331997184323
9990398128493939858682724464605407169925940845181835
7188910943512641768469147790922528378375323956019474
5349090551016423380781159966344826207141587238135420
3240493175757533707935096946134828527465137714683420
5626236321971729619991708666383043815049988726067172
6765724838996673949347818640599731199005296600353299
4139326485263280930082210836770501971287076699002478
1300851305272966844060610164378230719136307049422049
3834196009535099939871351587925219739428061513565643
1340816156888490174782485783660680040392675996290863
0937903111506727671941876882462830751887950689739054
8979889399988362663176223805421655009734384360758921
4242046806810224873073877809478772352739016455743067
8984175586097805980115965586660818087835319300271259
7379133589578973340087634431188614869768378811215108
7712667757231018332511139198516665079680964485325566
3841758311669493885921614166745675555382378604424455
7126333966772146224467415869012956205456527681047363
6826097898649630056827907376919863156012916142569306
4781006989197020226538674162603049633997127366767957
4289555663140148811846154458200759316788356682670899
1945569858024090677654274445079856845354905475878108
2742772021979660038202099786595196994492933543169491
6491705544129448209296760925147990322588900495395499

5722993018062179262112407110162209578552704888605381
9284236085295701406576742213483133260263421770543730
9535701089078073022135498263628528878265586915685634
6625947313258519565913846892446244583440227704967066
5349938723547520909770648817090001106391255718740682
4704713672257092171222867211191725322046914596922156
1229752437455506882056173695494066627332526041370446
2908654461510442560076329179486145106922588794437481
5778912508859111341722330315674973819208639722065135
2015828008698246876537375323344666978377328101114526
1959958866650108216881597594956660290352758332149372
8643587459967647652582086055524473391211712979767298
1403798909728650697649203220992792404020662929795608
3333483709734371103831364074116484760050698520156744
7227783589996084255788427247353061170516020218645681
8014842377128587661659119901186881409516094024580682
6199051129611212754574119275346382293994753499107485
9561410795940104716129658891591656799255444422507796
9298705769705847914304011825454535611172427337139999
9432672127719323193391300068902626785230042044775981
3672847437901286717096517647825594600907601267703866
6595545097404684858912713323990972933798358653508097
2670948531616454530790871573529955075169242502752012
5482063398833903447828489486253263843703940445054343
6312024049568194850386290285693371915547792962898845
3250576597682076434462946898145244334205580449946585
8237505365957053622440381842993593544150681350709205
7744497999051667690538020962176636560683578158732037
9231026760528625950776545290579527210420413852461568
6057656282187475538902520578508007406933729429877374
6305792569937750402685661586616804076154221619388979
4638536338311259582441879709719549522407479595395194
2297685825139007695483654789933863568806182905679340
7651975802467421791025033221696437760044128207799049
0323308123176112379972572146928331218471752129580119
1664227393514437669837519984530645787366407773806996
5291870312319459500416451412022582697246981461488184

6289049872728361923812720419779161557131344799687575703393540193254612893602565440312289217822034095434438836935460408954458024790180847629703419796230216029451611247691650186005981164172743826673072895297840924401656957765725555729576130242533547909506761981919701424151665935744603959673618532551061007924434665329441818070121476460301248629639975333472474598343390086851604672402252161362633437520040663234254993084214185882609820943615743240845647108254431018830907726258570251121046551644620072039153746473932152920633797970985170747730380605362838981511969546064420171139295668853118832806264334317017170205170264852813184125163439247307171335549506051586713611010580504697835880611507229261275769395214080302982839704314172657091329508291125343176943080754686551329222427649904174047481341076368194957397746437968651830444656683983184841948244176059863821383782353137304522859498310101762249439405390186494271323242724894320227208505266704016110654929802726886299503201407835500681783266124429473379347087040369339474448836401463014307665488869199951906540631166476781940497301003751985705416725276470230859199227399842341493349983909786403439076922427293376689266049349441694715697155470599042174219258825753425929995659864267684514106123846878806544579177175040277628236063300794375840743669481319018301570912676662079893317001720205395490686409451497740977410943734521094285968471727023406450671007142874586355868678236996339079944880743760883533682031211373244515710404186429258794496377751457723511758660881749387826753877641303779519060504064763026806474269068794144744826935195393809701784967319464528914389222386328010121834314118098075047970243899193572725705270605470472814746474462082224540747141694065206969516760010559177013648971322784201700335664320555114622179331323222171193273734777157788658376969907911711317283537401935719863182444704761544818457702523441248455096812811666497090604322747183366809

81167836515700468018768170956889324184407362310602566274757479697637472090635484029491734727487308120195052709637583607462545479352954234171272681981339902641902576337361165697385508744130241104699836432221001267728682180949890762368689263384417518856262832129659941213751796989052092475199878946684512057837358450431489084456881613840158803969872920098540862969471322598632313171273219428914135494323359122937242309660154616499698557962763275554577984454403223040904934659346663308857369596023285750180268502120007127142076555657726407983020729194806512936021776113609391135939353081217736284318944173962135545360500246138787960676040805316930025093441825713421613614440119152289530653007779053972003396628941794659641777983559253422287774656400034393526477351783515933055719947159812125038017559547577745951520713908433610011537904901507890705670881844574114508830512299700209969611309712189372853308359428809904927007467960128096971829283941281989605213233610705214431760149950199589535897183369078283138634236857956796563419293049769815209974528878312145973684607567476636060039687917435421385034895326295820544748769241330366118328689112778317546029127912955676007418733225542908244759673043008045896634533781987536994863350537722445159010541656658870698989013753888024327585151411858454353652987491572718865356408268327225110147009098986271677224152877689862267703698396995978281655794088825371837645374022181884118744813382879471964454937570020723469625336015922182028081797260491282924207267129800242376002246485008798898857231588422146946147491027891554652123059424861027260471269813241493814990176735748960941182680504771730131645894682345409075051861661015410701795541759759828042293862715551077779967459122466163847477146011178964858677219186104417530731906039973098127182191067244244194814711527834303920653681221424655022693020656758812764754343855734742632815492770562660066546711848765207267335654731669662179781

7069833645835826830207766337778647489587312552194785
9905262932469843638246857387232842777898399257375979
0403828529778739768466380962417547759148309642991552
8145165803728244231957520927726582693230597524612450
5643557559140534014297675082443347466243985834376148
7938996414440251411300246886937195215783044869344384
0734559086690946448138235353204219253479168899801439
1786232376017552145946542744591797476063616652401862
0618884755864873805830893796002947147554901679495904
2815618428725158529398298363206691979956738947880019
9587980348283125992532975991050207350736266025424310
5132862940349554176399109476329448542904448102689152
5958952145732012266399103321680006863504862007357034
8589610617902776607181677068179366318045382379191259
5619843885760951802894321579652120565761109398991549
6884031377549393540856953449908009012703840172626180
1820466737959946829543697194232579220647470975895251
0700203774171942638934649675715427250404693871883572
7756344487694459508498638877616835505884970028685726
9280542129179585813191422838038713728275528683064543
3314868809215169723787601375527981104596698829177796
2410970709137037883930450645012459886785895788637091
0482165298115205833227808177991560826384334655419760
3150488841919860848406264900399587816932942706196229
7453127136532694441514636256973427524160873440269797
8790021463626962620120377533883677549691572082415463
9242413505757494323179537955463238524199308940605113
4109851878055884092735591167161670187118155350582366
0985691939227000646822883589944865752267148004631656
7521942019661830703452163928231825112778342832895414
2472172600810147875803021229475153440371272332167086
5724341810804779578414426518997788601140202712329482
3762338268920964111163011032217548508094335350434077
6283415930030005339134180446423526240038073951095137
0111585095347324822374380350893528935216587780976003
2571212835419531225508204126987603541890077136511086
0471819410808843111074253639171389881359753193233465

1579864775046845758698939196311882411441543035704000
8264012677795384110666509640790498752217172264156490
4349596267065632899045783760113685132624683460213484
5575646260777352188360216622168327326752466009077868
0934843387919815646612082484867805064429695846021029
3045372463348760259962242700993707358710745996846332
3761050293787408079067362688314322240026342541834946
4974943378764251078629404943559708895096801337624024
8190955811349132979136137997670206146980435014907067
3315063787624020662390702718327904802851482956061154
7406959972561949296683021021193050250622682638808965
4062984556197933891284197588113582007257435616857677
2366593657751853637407687010082486727325827055947358
7594991527183547659915630391378571672985440402751717
3882658738499177463098812776933411133364070227666786
1050716885045092417768179812507691095090914411583370
3031224085067230545812454372712593920137937129895049
7688964104658348507481945401185770235917246901296651
1482150074342299282812227997873756699200712962845547
2834859249657682030788467950647021947309801494977443
0994977910414662850090667009042196029866331103011001
1132782301805889845558966393460875372851907436002284
2303896215171619564180656270009999560134896928557393
2626572516364002040799847141233603888674683408729576
1469762659715075739705145216740388083854201531984940
9356208349418084482165184390718645485793541198865262
2403271917749385275782250382269359705400549454563248
6688166937021367054984886527552679637592542953115273
9433899621680155290688351370329024845825815071276360
0548356859654294106517041963038966990987130227585345
9041651175097486133027113604353170371718416809873141
8195850315648707721494788546280039262775184566504309
1260341422184230421013314028569150289035417750909654
2281408332982473259563282470188681318906977341224009
8245720477810212442578679419621797082454947151657388
0791213055942557982936591018592952145889198973641057
4890894887746130638684259775642325268634634396779838

3912119284815590463257268442980690380184893738786626
7760146675569302143175290992680576999578973176289166
0092351933832885941822434444837337458570622675049768
2736139292125659038855909391789629362867921956985672
2077114516768711517036384009901803018928779566852973
9177054160961453069937591596227833917112798101298430
0277489093587489453981173533815768461395084185638045
8329867284936286928780127547239866389241352820450578
0353885364847097327544925680094275596516029103457751
2451359419321526895090145445952416961136980647081102
3153333882766622987028410072488660528760375404302185
7842964223163531335034068040278444147520082689003786
3547841565363689567311688395043354563178010166155103
6438444886149827139560795729139057900864661298244152 2
8134778644226453349015863717448862610877742837337586
2772081968306963831805881107893895728376778301186999
7008943365606245718983815017674655124808934944770008
3734530619482002243872523604954757117394662913618252
4460576260094886690256581329311350674437793232025038
0205435018529949807781052599853044887550724931255065
1003885719082485478389932660286033739356873644557467
5696601191916014388077529876643945135794760714470988
8932776605307411631153013911049232821931105880973670
4743234405890882856400267975943534642742107841490615
4092962408338791081565789909498815712690236620043280
2152895896157752220606214579882840491823383657351254
3126936803798726868475523692007777213811796492400361
5902345770034941153406833557382487391313539191475815
8527625413375899842036188795513807317346224477123585
3672790530083551436917470851610091475448224060922455
7576103986096635678828945500333604333720846556725164
8946232727869309895094558630983011437878258832122990
6713291819397652202572455840081402413093213342095068
9694190778702026153575802182105123290813528319509157
2599255500671987968751463023457131529536331288669612
9887510461508593563383915070262224045056795116886946 0
5110946627672376072375052884555323504747748998015 82

8804575300504609169021596452383038262921063032258517
3215675280891358488354854871241265327424716575220157
9356043381984648837766055270265526784708356010243568
6088326611788097703136061377909360911308723407720961
8102390157853292035471271969105674794704414643312131
4395609675166261128944405036638819340286420485038072
8609879198298048731334990048455219406573465415553780
1466230578392118511432796651242618604478370665694246
5109697462111066255726717641719632000060679461544447
3830095878493631232523528518998571980916000681419191
8616838901518124804643471901284000707041173800540248
4159936710284924075593039634872403013753519911575118
0444162615222008808978968333245220004404297312612225
1617655694246030207535958932402085289249243527073965
6513885729183164947634488104460321439087648326551672
9876219997368370945854939343383325884620594015729844
6429786277862823329069044923276593292449912404613383
3969015247585601019682761237203430551089792317528677
7387820132668450988071802682039220280484921574342425
6804673069717712187345608571520356706217008389369827
5361917197371079407399906036839520612792285497138600
2256544673985627696756447769925630094957092701311919
2982495577576375085942369144906834763454604394437131
8679563025883826817928975790738675774381973520813126
8068528419443653155109713356812574828346979017890143
1827752618088331363997360282959901000051972437701525
5864819106123617676411458323693479550647375250340609
9278972307195082778497001196031316247555100872885217
3812918560876774785619606389351240925078580628668215
5371236246489693259203489219130671347171099770337362
8959597835095325235982712896491987110448156130108291
8581273929221101792694864281968328624692921084643895
9969288165507695822898733723861578329428488075621932
3095740630838089146283641381169292290832336756904067
8765357616396175426664593802436019747199540348365085
6887128855292429351867994655684451320866302074899531
6987888798390898954565467592320605819226177883766122

2533547338637096772959302399118055662280926495004498
7752773346252173326389389494914654038212432180736158
4536114046021555259753667202239198810502846810737704
3772329312159753622750515723876452943634838791352898
878790582155011104233584518310227611334335804789714
1497715251577898482989361644918289793177903264672287
4941867545326918364087982826661910595314252866309267
0701732405950706227386670579287338827184951879760685
3228061480881864659328799427797461245295409526712496
9044104613107390919211239694068351844696993047168661
7424050519222976752985296684867348368116257663525545
7522440318439224623325561652251410423592327320188833
6360890136050472633749620582370618484689529986526746
6363021958716136936786724835356109940373813996989004
4051538895903910328279457281511397587762747445318429
3053661363795169817274605208854646923204175911435001
3503950753484896156924950900207556055754384170290146
0781818699249239391194382411002570518213882300980354
5858781164004553031127717926661474377483736623746020
2073538403482309687677893623056167893806731785663137
1635387047158740172890527249125208271689304373982296
5883776322658705827681347353299478638582774381654779
3609856131567793418000201475804245593773783036039563
1349135763401350303484073992757865953439127551602075
3086523653852745037888597124866716500398774192653741
4501116049507100186493015358275730695530268422875848
3403210477930534852209549079257356397702757373204550
7991015818436798097731303516449897806847700924124658
9017179499863901523452423144640733146239704547452605
0430187421299438212570644915073546929709320638029037
7591188187390157910382794684926286064935577057927535
6999844748784544845580050634603101943277182722031250
8097750299519197688024658963981141533713506843610011
3109962982510923400768862681738142858204141106078957
0114950689430862796526028879535397611684580730435299
6562712209219685525228849169734490752782340903793436
6431756882408319294978354314731449399148449444634289

6571796553328504009467593996359499073386426656218748
1072632478724304062904946792025384859153980266086302
8206837106819258636756161674883003149343012810529959
9888061886299723689641652045680607795101009177465730
8145469192581327312967302559205871656588042033913153
9744159172871948757014262221471927891106602750761289
4427322429136699466592706572510419363102385981754907
9869943892438890089393463412138281362194718079811145
0300020433901579204393125539951922260908899971256309
2330271429125014403981870500420259608780870813586866
4901772444736952194467034906502238409693051825939968
0394210385832160164007694778919242121489543593744009
9937964372856130362894311674370894674473797167618208
6415931683561842400484542707621856066439597880214216
4316651900960728174606413768465552889186181960348825
8520748455566190896931890551159916269087256988217638
8463745955064737773914158066740766500977834149348421
6184403838831093760153938893631063154871345972529048
3703688341501686075158579902541153195356070770381497
1227446960968940953052600980817492700921933267421388
7448974859582609816278112393479338277056197406663789
7136621101115062064178327309094386421044341001568487
9462446967599935247049308499270573111248239759233598
9864528277041548277629085165037960815057835098230275
8742157795977900459206088800435487434735298281470367
6657845392604783416704594176917489468082537728362160
6685686911351945238338908918911109127985145874140765
0285259067209111152799259142128435543975758998522071
7590994624144692846326862634021582660249829211586948
0107657666054916190877864527586671942368344343143419
3812335002128977713142500052922497167310777100171685
1188822998892084729467521062242455347160799120832204
7262682159688664508167350961643933992447751146236967
0308230209426462536749134703407424587665240882619103
3625198046411371161227393497964475670145458128842701
0677196263293691784822862791205618498280909007323886
1546445788578285840970960477441126591618895177662916

6252716872171876251651363170897961764648430969865502
0383274990369013956616772453347773594284105079107689
3352387935549818784678123081259581976513263542563940
2346170289036070816642417750144015178987133940719540
6803684014377753823023274168651273314467179277067541
5259373709342131975970409314814919948898722867822674
6965193156630529145957136183974955247480194339279493
6563511497990843542657131020197144345952011778759458
0375078746251804995052213989814784595033178625784899
9775100918367587992192282605503802850781001785441580
7312470071381244813199018918287915172974806315354184
0272497319866760503046989214001220778491865416706288
9461437398736216971466574033954035465704207132586866
3336279283212156148810545073264464398983800241706849
2199129748009428508166943564929578343441236526169848
2303194094863341021660698828462310983387014189280782
0737483205430749541642896324813950945589933704491826
1866426541508190058270723588418982809980191510351770
6388164396642793470437736480269578536810367986653193
9037687765826608503687181035728366643365057717227866
6728793090809972219221888136713093480501052572641573
8134064117816210480678714302389166811566310728760536
9457537784780650585029952691170323330371671960112352
3444074752681289283558682868402632571605697450019444
4951701806618108190644105455246621997460551083766582
8858129814451526749038022695678115007030523761609761
6075164743583933410407798055772293477599134331367215
9786610550734227837116776183199485930275857261397058
6584709868667953396094629188678055717113687861007855
0074388185710641269205096699149289422749742545108502
9268848939544748228630674993942734032486837449145564
7358303787886059847569183700454233012866275857927095
1067896905333597342936142003288246714880458353144138
6307976265974048930871107688512323078607934245122017
1545580184169216407608957713287564519943138023272075
0051287556376203397533006053915208115880105429443178
5558933982463564583104692415828661178849010046341407

6094998214465414603928375102473505894407349196495403
7027224534572201244636870031384833984230918324024901
5795186593378599282947015985795919685983868598194490
5441298097484199553689636996354717206181813585525926
2329953991693759170924286136642185584676677206230544
4148858262537053208245437448229030278027065751038001
7108817832818651459865583258719554645852534197577697
8217729236120713336412402816349700143731500885527129
1758919608289738898020217608210487754500133361319171
3197620856414914813796406630406540354290848751446093
0454287978881572479698005656607995943883648202831603
3894667904451136553073142332760578710820792676043953
5354664225367818502682922884743863930395837400973858
7014265108634462047051758476054990457272575031508505
8585681974684292904170671661413318850438469807973560
1777104811768366410779396794362026848191177314824089
9769255512882631452780126438139137569639489110846479
9849976779062995454495157317870080407124802833371117
0264272301544239136020661853406307113082815207497537
1931468400971062250371959216386582428228940836492726
2824559605155252350591207841467076056758736733618941
1174276167915553017199436756847928532673226578045593
2978644256669444338925456899171225377298404289632547
2825553367989035077223103164222255936866224029328970
7900774962130957134962928964929387875975644766633100
1001208538049136577804869477100465384219708755533249
5654302279840491862319679062531769931473584528027778
2058867692232477965323935598599179926338256860793129
7153065955394878327773888071369814974640506625175143
1410646697548078882506210803693030149523602476687170
1778304594493982135466681201891555895109323779106639
7432171715286101290073335180113311141356684680079103
9962245315059620716207040285515702614335810300780277
8165023775172009678562401690788151274926741016063023
1957867333096674195332631791548690087721251730723578
9809252253022563226541902339914103803099343538458027
0680334442716519277258234253690592492756444995993189

3373020241561339217286882111688625658527099872906016
5657108807389487583616952170620956326824609039961837
8897872018226870237853568344181001214934617230676752
8635990112808891008964737792459795688250073985023807
7509591298155530669248953573727641323856684678177814
0576387723487604032512946764713594366594134553106314
0695856263463368973106516380743553431261006909674537
6501440519811834923531887194079939990955389289577987
5047727780327084660936154885596579969702595623280284
6174744164826320487241282989494781029882283774618561
8582323346671858688349581842191943883222041292566218
2136162779449004102238014024216582366043639132312295
4434656949827222067908232880702415137352416231227477
5730325612470426854433853224701279777859978351819954
3021487594778160761885270838202084834997471275528202
5796946996655381939593798289613778443007953018540036
9661661157628681207957971615356066202735762672471209
9348262911458725856610030202616546006848143871460837
2797925961459932392110170397337698734954390474516654
1954737744116188091670072852497347235348009245947369
1441023228343634384103720771001346205794664067075309
8240855503576886997007848143675510401545336522149188
8383202132791737495225085910409412646261587266461619
1769631733141663374449384885148595135867901870079581
7364758504070965634445300631086513401545686454609857
7937682284642303310173279927121446955531396650548172
7640197265790621953881325350963250947445989891087859
0169692258198604873251616314872253320868115250387783
9030921096397314087014632627748736199925160369035164
0181322863841015778913834548208695394911416541921224
4103725882353735283296622540405395995512541880469553
4701692796581808422169114677949577090318147199522420
0489058855937464416153490934682109527381906924934012
5854016212983888236067112990472785382510932212676008
6356788729911106064744385406313070269157932911471683
5748930860734176203484242255797432510010038643688160
5524668277328014816697873496332099634172377923788303

5166054871671792874001119144726245671247002497622458
2240273977070270235032377124169131491308448027264009
9449725957209235930230699273000522490736514197778118
2720305259060505366479310184870828397624359765410245
4719021296462440721878685429130719911560221343598109
2612780992449298835321421151688044305242231335172036
6875910920612181715721501323049150385243127601602678
0710277905987836302849099513323332565542428323312697
0826048284109116462935530797013471599928564268898897
0076744233464896500451148248944884381190520316202395
6125155080114293455938402258904452354916067705752641
7751973609044020448336788189156897027789366044998316
7550333460997434539066796812379833136398445862204918
2698598018506280836505817585632828672397021647037926
1157543687834566404659060788858593615726073376479473
6283941894928357605048336449401134986549120821004813
2550683756281970256869699549187621976397204502790044
6675639511937613156006454486485525074979942085002895
4444995335745046836622766872082483164135994803070601
6118223091561752595284900289952934287376173510267424
1881593715994890969792225772144013909127224617888387
4360805196751530314791141433573207365859049304277337
9844712919544464543044009597518309741823376186337811
5291280151786636009016476974544958957323146795549938
9337513949564362601454959264673734721602188531326544
6860088537207722134527510310595262530711102355378856
9164995911692208388877078051735438398567867015096378
0886864757576698253235404424542840088268747777032853
8261997625425819929342059121797780827787051185845230
8972985638768651127507276341003599414660122279489574
9559203609946037804848385525959910817123628274420401
7849802110317627877887503036302263616097666758010603
9555479978745699815797334233997424447588453139334536
6459175525813475504634426716109489081799689592267046
4402169176805100590718447352631235416442486477743878
7765173853478975014025204069329911353255614813604353
2968312928991295356152904027591317677341277704636530

8522132575486307933547882996383406999371515422494380
8824064733263123350461388215947951691759254509848097
9110892331137789539664074608345730097651170607524143
6288346609180038490635692685329640055165359978579130
6066404745571908486625041462763357204425087660332076
4120027683714720258395775725483081763522817065775941
5327083266255391096897305850456225936898498975622702
1582652652806200251844164898191969095582120789897267
1764613380083956648772201932046367188172395470493020
9279866104118469570486847004963864125953067376666038
9421761893875423752240187458159722844252290797737229
2551018088673990839885492144913863562538863789161591
8884240512998193651713692591695779199884949414977115
1943157558263070599485758635347496397568559703862678
0540072008974502527051939796981252968831191645904520
9756305283373094860238329627213225690073767533911168
2947198127705742624375221375825032008736375320464500
0573891793465923557708362204145285013907946406724667
3601827537985447814076921256085694481054166195675264
7507452390254517053109406626366824745757346070406527
5753577743201023913341138135775033225611439009760994
6952139817702841461088413608565918393299393135808195
2707069270927607721717779098763854634460240516904994
9774757748287300933978940641473694571985048299051348
4286707099150933044579663919535589147801094434509827
0173677240979049480592883846575269082140684093675224
8884246612560526950978010126057728228735993798499770
4573176117586941984674742904864012319967446222686363
3260076411070293488972360126214984596841603187424531
5843520548918590453569419644606333798849315311695475
8361115749766774070315448057897817050457319225881549
4311439025793450499895500372704236182645680415899700
1369770936461843182966606907313547636451203468088044
8544639479881054680984967013179549428668138827658444
6505851797712368142674585475571382290726634603164381
3750146542919453599343608206282790725318865751797342
4574661225027443019092242962776936653158716809444242

7862498407494665563526804502768435782113627340694356
1642515379232246664582371079321459044461110922107039
7604565101012869760593556579799723039393868396179918
9869179915938694708632424160101610309887378543956773
1482972348965947671427213412840043762106602005505662
0233395195755864512830241771428237157769328767978452
2757104237281724684454616224472703666186405467249250
1446712045654783572692144705387980542744365066549190
0369785230070372006141837257113091116810217228672125
8959485746004353295033114695796564900624462269539125
1125286803782637520834620109193920529940673531650175
8378263276410048994464890719508214784102083666996441
5554899731185444595312327881988435193646690756191744
3638748986263627650302720686092518809723608847938016
2529503922552102831859119520952160308797092230631824
3049051361251785266783098964840762618711211938564523
3258801563584566325681536597414006351665385278330279
9332761818731642928434366351615663255280877405476696
7000188148312992649497560617449945560484056516920660
6294419074711647405755619551834538740640464663446523
3320351037784476758856666090172875209822447115645O4
5567243110919573466626779195035111191248491406455543
5257725496566392678522191762385475381971083198322848
3946922294953737493172577554258096479498589102686359
4535761355146603751391496845122967455478424717793060
7499992487150391737113310598418240045774257591786671
2195051742896119673898889045793126077836316129782569
4113684507888434449786640449589578177977537633175038
6865692143284565907638770350854549228817745921346447
9036409671937884800156599085262761750977394016437406
4232153788341300502620171319759124864587923006850025
3381354891513199847109339798843654460482728915497107
6037317445126097171706215491523093558078456283921799
5093664990440960915699421493871124291466312054691322
3860633526874046418697649751346003184289825747512025
8445983103982014060948468707691618383086238789106146
8241972832005261503851106819905835176956323656969414

2362610152155381725036939198820292713685444010629436
7264512033344244808295285077865022358866847987292336
3345267582654613647644880992776331655021300744945298
9543396175266117490169121300581095542449122673852851
2234383695983978048397524645101205882674283063999608
9076557344363448014904882983571898164096451011928985
3987608822969226435420824456841236875233258977838524
3845422759205676092607958467279386639764013337529817
4103800839713669875017925188890050961627253029063851
2244815629473486671807921675743950233541879714637034
1359507068879603101501646447422392328672923717256538
4290147065906886729406948425427159052053409339014321
0813138545825970434858093148550633941870543754820191
7022688231751090132941361718770058091133457961642203
0125980220431073829668649070375044538684428440879094
5807138389620954492075996155366557130684404695271299
4878309713450788275724844899897349703962246511049877
7004179371263198665504248264504271322916123093246164
1353303916768945148358569270216603190656899930352729
6694254093298585047142854083867108892734431087363457
0026911192432496377298229400944020752024662066447383
9172044257548340145394080354627340999096406907207562
2907374684131198586578798855914252147080351784967358
9369335350075446723246131252582687046375634396463905
9479070392896125976486670569071815846029936042856641
6523782711376077456210903179808657173199934312817969
1294296204556952012433154460835957656507832446721125
8500776099689071429906214637225018370319985169220050
8139102053789839156229925006468735679795406627829502
2474592665617686886293211625656050084542945590329173
7200982085881735374687831820644702268612764976090634
3394156649026426219449599972627879887420686537748400
1240271202527473344361668130227204754587027046409864
3467573641494455604018046902564908532516727167095127
9005239342170274303288614132330961696475560201838621
5997872360962122757042109968737000712257994295186963
0041295418877962872409327846179804864790769989057773

3886055838249114228316374869287853814814314622428398
4962337855773508605110105092612932309949247418926751
3161881862075325740386848069649046982799912216113866
5663820326019135968402515649247906443983330031807895
2369616518405614103384347993761781821004743519090706
5480640320569117166432209921547124046169424710431522
2205104885366382704720835222072281792307724378621462
9860668300394431840318971195938758807198115048397750
8624928452053766099357035113846959784295954406485751
5050620849712281233606395414804523183037590392304193
1293580470253223963911512793759360151408705351460629
8995194074575535488686198226932469553821021947202124
8849044263655395305394353300405061335996832016669879
4772079525866914331899092118652260584960881496268365
4672349840481438109431985740368377466370936580637339
2945879466445168261148333459808948132301423449736300
0817700302901541674017957443616201842207605357765423
1628456427644093755689715537872869598722457407255862
8891627542274001393924890937205554561607842415395618
8399905582479740297074147881435727079149221043557745
4609349005737681596828196801977159064579660575494254
1814453299247981996657119645518856556568121513349098
0465803768476182660137371756837274130617046443808292
4391653713296382616530723106068233388999951025530732
7441209426994137920110104663208749715410587445826115
9959068772409428697875946269429880547786077645800522
9407570308406385160436059312555658553312311655463865
5410300726993447731181690786628011955322480995617127
8815563935190125607711288469473226363577830259656423
2697384744696747938171918349844120102842351921959479
4118886452077960231343063217174354069248862375881350
7950130195421256853896066931254279329444504604143997
5110593068307589866631208970917867381443106267328572
9135247335737334139964981622560564197417879844156021
3301230296004214891147848575264386251814296191368983
8222749415586395787408237809038341408792976445118888
0233720098864012236174176430506370381076927834118901

7294013294623856393085315322442227688185717288938851
6865007590484415709736636539394113814143300708417983
1172343974321289826930882817908454608352276168839236
7067018193779063875186889826789859261224877072553707
7468090917381219556815424678175824986359084577528221
0873370947100534470431626453265098754187048260408504
9432894496711511556404503759922183153150621894176807
9740040267805916373594923058021891056162459013019130
7342309220118774890541763475234787205095750830124580
0976089449020142283215170817863829715178754102327793
7785407268363951803722727399072461328116672620238536
2958044768744894558935561294166787711514445952150728
0112452897838988924398057926113395117163372724763220
3561341376655493609107607754288639415375091128355095
3891017569927348105802052447498860656941406880598157
1455349610211640412992071191573822396496879006157717
9595751167559084819296904355934449028936964861900866
1184836408445470922988082018696435010241658928067289
3461892128839980588675472338226628746748982760072139
6917908199418072364514382989430173553032251147891184
6799356943769256545351140882462644178781754058520462
7520911945517150715322322900317837863929972645795041
3622430738874742958842038653796300887828497491015367
1404487270605223983274654434529356528076960477221503
0240603299608201442874066463090243695582201751481902
1276959194854806500251596627667211562658270387072145
8114744697943698506765768823503506327908552602752457
8452139311897121959560222778010521705631239634446112
7002125867235737090090246740694925766696530479737142
7626570535957448790876199299797549154709567456866266
6933178160269499245637861334504012864826825645467814
4257343138486006682679767270827808689640327625941678
8405714594114804629164881285200102610561984527854316
7673430194180028739280946151948185468663109907951504
4881336501254612214135427912283913036947262589780027
3316800303122198079222723393023339699024932381363017
8121360113136831038867733900122525241037628839200114

6874739783019681516234484584417084711540563067122150
2524006009566845020996544235778554788006584453063354
5865197485172711046869237082103742383445162144635916
4268326469767485119161178393419514216263434572806939
3696716440345231117981331367472558103023300335647091
9313118415493271541872145395500896215199557764809532
2472817056846283525646275430576200001721499028964513
8438331171165858417764510618582508210472454698908552
6057154334736407766059802184462230230357649286983535
4772286691167255092926423495359112977808670676400914
1760472511018422954289469724271931395782511177137364
6612886109938596154463668555664063242409655115271244
8740813863395840329068301663750515937889447003244679
4442050062323805678075861405161794915675878485475379
8036133204588672013137468422277739338301015927873826
6403520793857372733187872211600697886849105688767373
5468207179344205169186912073390602304035953144024848
2755249884966863905615181291354913330689198050958319
5474524347688558773831041434453993397826437905600793
2577903268772389335970408929063798623591326882769120
4823532268533111309492766145457891114707181678298795
4873977966493933404999580445111442637878055006181201
9428565801151160496597514709360848244784071510165688
1915154763085370899177749500315467791994055430238455
5684329587122807033930365508052035902690222574326941
7733783302077850881645997749984355147030418993109278
8636285559549866645215146083224327029398653547117146
1120978644646443256180023520004864169280901251955587
9677770712014158540324292727965139982142625427852383
6182489744008376465719918339566033650949707852328897
4588149334773099486397383182019324468791919227598192
8705919005907540673571852348118683464347785054157828
2244907985840870058612052844716335877723304380827091
3739688950568455535232618221110237312713688142058398
3854747049000536093865761777730487830785597217538818
6172720052568368730086721769154412628775495032392221
1648722178615226209112425923774670550161255823546154

4208744584214964776025595727402918471645315888525882
7884286855889496508427039429899354427475578488409253
4243675964612654381092025562998205154148019374700480
3579667745874100884131811872582385785448488816682771
3303205091480499027063386694690993512487922710050927
7461414952914626105176485990692992381759902441618794
9702356571809514425854995765179296210567446002214133
6737548079860359695946856983379772672852875925060444
1846435778240362377667029872832370604527442555068151
8095558886835891475330457011692173081904136821961979
2444604264993438484527323746191137110824464335016948
0155025328418462873349226613456894413900247066335547
0788141605485679668656432669813031783621646465882083
0050364454812922884964461641012502375768282189455118
4580595732090939462064867750938024379128960180082794
0474817422063327914712842095137010561943271762928685
7909094932180555689790662862891547375486731986937384
6641958561889977082799001692464942519034737770398863
0149201118352837232791996507771558163406176196984072
2231867827012354034994384291685565051517438377353920
3225611599297568796557485637139984984188981872386344
7747735515013535079191081808269895360125247825470432
1409778469323294380147255125843760860998146021996986
3154847357629208099670223123077999976689261493709481
3131176126725029020251125017695385831757933248237475
8716019598930968995429457152380277892235685048764182
4139102326914916557444484512460830514578741750738717
6744173551101307645537897887522221852058613301075912
7201284158160990129182911856667157392998092917991491
6100046603332922577608675656663596534181854225605988
4318442721810942318109063106059677332784039505960669
7721027773614118272064342985384424658772847185178836
3375281932542583005499614614837350414191861619862910
6706387911951762050556410548148530782364372016539996
5209504238904116749764360290243419567622446851420894
5656803232777781828040235317172716161384745264342561
7030040388071688888421475795728132416306939771904869

3210177484934395220889779774606413216065912354287306
6349856495138429915119375415682688046403440740031916
4875242281289908496049108875248189766295443946375362
1983060853026629767762297282370884106403067983957045
5079864255624132956970690312606937272270497385849063
5481194665353366026431535545612762002245436501409287
0701038450249422996824698948193110638058489638349668
4231546895857394178100478027431024343617063937322667
1414047645055320685254527772713379959897191293351095
3352183762378144828212389833183922695403629441944477
9393386294782085653916027686547422160219745416250347
9486573308748696362146050796657211558568264712564019
8323687221801627370274085461233153148746531939362613
4387278498409826789986196912202669754590120334748717
1527203423784853201997739410893991832399353436083831
9448350407170160500197194079556929169441248268460290
5546024805566374925676902358565496305630549909473833
8682002253272751014765382015718968917643861760384669
1471270502090101667931435388981799261389495820568431
5254271733398458496778421955422849693775382141987539
8362804151385327095150706778317618492686749245187887
8256523880787559339874189923941064660842841595176800
2918996744604562347684174628436149052409961460913299
0782507767172994873799300497060952190291591878815655
0294869249948263886396823849560868500579992081259169
8095762500632522742977572235324464631299191602952782
8387808637437948487816007986691844944952033890911819
9684001436019370933054722190471786161995211092911279
1700197667202021639669184222413495352640541503434784
9373009481075000992303709153882015508200330120076040
3674004750282381217235331054698371010096575548811661
4281741971422411114653964940881068713531679967489742
7937226351358103899882545160553672510213641692900321
0731223430072399556914225246430126273566681782285901
6903399525588647085175428439793423274925339989258403
9232079835993252157435842523037436661389938632000800
6457475088070937998462747096657080943692993610700273

4538156848014398180022649796165498392424721495385516
6490613388699479448764070625660160551717891131051578
9812484067441540438634321808049603577636933696507502
4967546596535171500859975076400045595426370119626833
5042396940932473254073217465365771218978633545568241
7039103781824265672441578184384945382562034978117494
7104658950823214082047820539992217083096379247191435
7052689273788296301720459841639676597939924684512021
6731557594061085011084015014939584813243143264831706
3835229338983573286296250064539653232340901663455349
7614539777543545510180022729878166610572423124306235
0399126692725593983870446822440569021752720890597314
0317194993937576065170443081784358468902322640906702
5582563156527103991987874499600566965311694201789033
3193079128764045002452926077757355448308514991216046
2604079663570042929414152107851793951248929311310872
3403687549333211997169415582242253234526991651484270
8074964982432091087091302719220736052823988903337764
8244024821643674489283893271787246301295213777584065
6766503422548447952734389296263521706924829572233723
7260521214867559012437510688636168620684810753252551
9080870082393756679930005256400410568687321345774201
1004302127479640462677207960288680754533284461163963
6702961676361061209564091590392267597725612770823369
1017979324027600947790504939059499035509762328552456
9201492338038955511453694537989642439077531543866107
9617254935797164480344612666235380414555736764262514
4590571925802222930640330494317739911077459948051848
4341690301247105284001145301170159264176031004668798
4340067636613575415938107394902338459599785664900633
1001925807617965927489021730881865451249156100845649
2191733984184936400789242534005288512740978260728184
4993623344396777834164303286170746555744709588710661
2285979843832898882778608994982593444570625520846669
3362073645613299517534649909660709934312563597490296
5671846803518878764437192734322849675753487438060586
8393873208710712341196033089306023521935023796475301

5141593728621182295259067018575855904869810336131061
9537044107720860233000669435598229899720038942050712
4130963301247398988986501613446041636976412991855139
8564133480244010903820420980509818816370765602535422
8852064250474895868089917946686117193250332482302410
5980558476638045521378932305723502009715574760259377
2076776087468148213452253163020878882398535566840846
2018877633388938239400596938234755581196604405366017
0851554314092863357092159448116017532965834133347177
3027110597090517811589017086602990160512479450702430
1232306702621797011415106820022681399975072583213035
6166794912610054201286453229800672689000948209707585
4102198848852954596019747306361328429698553852265238
1613088966591450912488681312595353629605766031975042
9504118843939724705360578947986283171400396848076421
1909414275681202732454233195931119523950562906226110
0996439894838164644874586683075485778532874081993757
2748521974137180942967777411722239364135603321190933
4407556787838113045199845148628980006084838694206218
5271928018778042486680802995128970347329446317094600
3859512545338683557968905846517230067044888968406108
6304061213515520387421392844962022257546585820866986
4060498655425885908145530994843493842733842178645051
3985427397429095857008561462561834952700228141732536
7653979469127529747013170063835415965446342449683526
3505948534474472107805610781082964942647881002597931
8775639239043291785327634203752297565752743408295084
5479470152452608993138857831239117512692255667572885
1334043976962540393117493371399449529356801060379694
4595685975249877267348079073267618245233552121621496
8023449292542886551457337565576594555709239533428142
4629031727815403998341556419837718018982112476085595
5518999506207300714034520815503329814975070244267726
4360338737539731484313740709266544922952042319900734
5946393119965350568073329814865841109199443946272328
4536771128484736224606331360285910596352371938716345
9869636443906854053223193152413546932487576730463381

7030294479835226020518149444585049612032690923375271
6235513352623432072194300935881503393598974493352695
7874572783140396703969100170734149325310220632630169
2523701801202442268849290981955511719561208381550144
8583657166510269086648717323819014860992469913154608
2001992705047305688761418932981083102352648281081024
8545022087572212834413437948499972792025835434172044
2598469327409141782143949280179974565987369828742826
8267484421213715468232751128534165237031653070432582
8337211237137609695939937549536223222219746596193325
2907404248760251381952426973910175637197534300447961
7825043115331506758256273534347625253914251527570478
7843767885241871966346241992700802576108392749762263
6549001865320645499515580290839851327262732196728478
3025385221904792786893895387803686993188466031043633
5243732715469989881118644367011402842620261504738823
5899747281493343257065474637022451887289061255030273
7919026403965774176068989769834566464705204716359214
4830709958430647172750763371802071459744565251418850
5037163773818902969685284409258193174410055576089850
2923271016008982572238173945436985296547694901873840
4646564373071319590114076131742822038833136029708526
1451234907307147624534050245423763666685575740120604
0598869556301144154431416969860740322078860031327905
5391786967416355524025076765308563224273784714974603
7784064609346812991871902892759763070159840887781745
1926901476910300302349794585110001808866062162868011
1091511610409832730880899543375511871831737865587764
8735885449036687907416918253831330636233820582015488
5449827883821742437580549238159729640619631058215167
0919032631873093381385011309133277093034251122109721
5505610491387050426219805260247611607505971037943664
7715293534917186125066116367383304995648787793656979
1129319587827777406010753337243279000973240725607038
8800871137309659634457096041175089444791191621745516
9709648447620461741281545361284226015513015895258628
0695706392444354780239051686133854986192665676227603

58234985569192248906901164598660961567955415492157 58
12835409202539317080737814650788320352665134570706 55
36856992811242656708448017786461497404549211843122 76
21845224410884715075701908834950224375698850494542 41
05726624609458320959625728720309947765401073985580 69
96619801953450050651450105492401886108913417306075 92
38439461246569861605609094541772907364009391321956 76
83643332996199979654243480265694068369867061758741 31
67650506027135882574433707216458819522613397054520 52
34412809231730650519899545028585383522873787286982 91
72758078242098261060609015092521109089899643892978 62
94301411140067628774992078172379474620916899838961 89
48630773060274910267038839433524742434212367121770 24
61011602406111857168705082440491623894343504702508 13
64915515510432725074794102473747819226520593355136 25
53018121964249924882129926017740801037198842125474 47
21602969500277426850775217586691799380136258422417 39
85607669916543275706123045286728070763189470589544 06
18842131571380033984874086809414461845854230344951 44
85210539912489324554866593053355845782771442377433 85
33920824127763216596753832665064306388069556915620 26
22259469414290079987698344209146999897956832667090 41
41398068633325156525696878789922574327961396437026 54
31446393337990008198575307978817613374360948392830 22
33803279732038653646242498055114022469226478931592 94
49180403829836490348886386975644148556085605371772 28
73714822823686427811365720394749690693723991561003 68
28075141178473782562212775995916775814038532986888 51
36552323139384317422585356280701915440077630163476 47
78126132438024842186566985527400764105119010954528 98
38256348407045580809052214769722042873822020644511 16
55802021581372204663476633351756179905008977808856 00
93208145417644390323524193973154721892092020247427 04
05857913543792600976807636298756614782644338926121 21
34565944568010422276165241837640186734533914880790 01
36360352843217478040168314672618806793649304686587 36
46311483282698049202278762554540178509177497728294 67

5692935451666098641948145836444789096038304582853698
9560806656661903603100453047200242263938815720686746
9006192987308224849001681634892125543375444758038873
6158658859248597558742783859992954170991722511534025
4222296922343661777819296533134967477572209784301938
1844505985075080564948318967922058343414789051025761
7490997582001234724410285079412318437601471421378295
1102187358993917081686805598813354699137945378313771
1413549197124346581093112783326621759222758936993477
0498435251921073517573702005457345130587313221438462
0876027751840534893713237662907112055269490530038882
2804204808210851928761076772814548927944032272362841
2479431926441924307968137038294820058697420150132353
1987205199203538773142158822114859743182387909891993
2539341503787689959695804327251777689866125889395442
0139171306418862951066526896062414520844803403474113
3148800431386473171584468157548365316705126017466407
6072809596721327294224536104266792809469773583638349
8228114766916959695573277028279120997926086381701765
4210151115811092683287485950646262362992592496943781
8222916923432548536160415251420636925214774598094198
8926814776443905376111917463026074170414492241949800
1706238668169466079892475169718500094287444466241005
8638035586306506360957270979734930709648890760630190
7923346170197456656248712884986738267854626894512094
6229054827542330251732132827853165175869340541118457
1510483463283200810115252541311936795456126121610319
7427832103879548253917431409877065257006037671968383
0478692910032674834420668406673515088586091662976572
2779969260038682733496418758332118080916861851713459
1156850893149404481996107250233789677989918872582686
7053775743512465081379862813483506643132466270924523
6546983517407042651369882414330883552631945475425324
8458840939937846318673381336610310058114645515187730
5014130815666018061132268352963971146240104983148464
6043952006163735425846982605212545385915053646201416
5930615241377433037246441039759850015689210279093786

7860318517672404695991572340529003816760724247701697
8152148615194746270546142211256912722538159050203958
3051468585003602420054909455026826185095875426210973
9913220950295489582891742329813431540692575705880157
3654535178927415271289413143182694769025888547865991
9689883599904388542274846947311875197488771215019159
1908885813263769981215908089691823350590797148746133
3680397063500369555394225907345369332638969616238425
4104794629483793746381324612408253069026914650215171
9655242567884712722975266792028038412966157502184681
1088693939925268794395032672870087581886104930085975
0196928095204823757138945438844241621756626373518033
6432774017341259427455820267544027881214092868804120
3000227063180825931867664814904472579944670459423828
1114878611771247129940234353193476389778948446256041
7454780205295530815025429801570903923821472145748055
0269684443809603801644372619530044039908184265874879
5326360174773439634741333029275581820008464886098183
2917078621243703459404281501604147705818564413222318
8779392163988961273699240022516053512829372365573731
9319389834175384397083090358533308571836931136503700
8190565450432984220993814900454454004486182214055060
5287168313414262789644169533398029687951753666784755
0967256739077347181693399759001189891113962465206160
7188564911926642694016069590616104437801498666998284
3324652576088257101089919591111808350303650797428123
0709395198402548016944626205923636351071990148156774
4000123130725410256056015931684053281699733907071537
2088090132654154084084869464851337622748284991612747
4705322255057918337197829980217375914243447148663438
0845991013925549446596674237301191215691048473306980
5081712972731380662922967849316570700310830556788979
1232981303751031746783401120813530914660733675118762
1872232478968370382795012395444953967588537384466413
4720247001588520176131215481437213347926567224469179
5625863300306994583628910381264895233888076384381880
9211807397927683141230868809469171765526719713411290

2715814675294427625213295015430115336903892213841013
3788327539359202090519420939389794128938076595308702
7355077304221911430043256684624072289034021711512731
6890533946785187382278485158224704841277392682805651
7260376445844006828466737354482592496916214239033889
3181170113891315019222854107381814522925921851485752
5247618384807758382986666755676288831008601526358935
8635712742177728951544583129347098008471782896727341
2903707607159708978389934279262557381889563194050727
5208732599945654560858685535826313370849340641393491
3749179077218228531160094982703667927892358788431608
4918389533623615765768334222895927023939153868119581
5299606645208188618438969720018935678548949422165123
6101718473760460603829616442733117304025869985215372
1647753632066072061232929323962407800474280678081245
4299279509350514397602119291130378021703950829509906
9944393911955003407215741480177592299719328271630397
1339380034926791334165089113946424796613209472508981
9572375049113312916698427371148958662864933613397030
3624716301283210070841520262768996306818969526666944
0516751132937869050034120042841739429541445844475412
4677185087103019497637306464566186349406566159555902
2203881357136176823477846310988547480314160751831305
4403428112703507195990142030588149232148841293569445
2964317031319035779377391901656912577178069851218011
5203637847182328949534446551462678468202630396663019
9069226106232408625749366744026581397192480133160604
1280430940301178183953384569734777875549650522671989
2039806465088098765959904026835458290840488743590121
0248570719056955851822996481632457015703748335071625
6177878402602480795436431427740975204629600770595232
1593783926560584763183164539460869129139029707529324
5752831122063165307794559970935880191017954784682220
2981379496733900870993483523659252153174780914929635
9576803641645315741984082949704240390235340265922135
2598530803324317858577406056900395174966247831146154
7074710152410927099019069460555923968305614807726537

3998499618880204526136957223089073682258369725346732
9304349079623158386261263444240863141034997403655086
9216050046280979394259573707752242127755962792858417
7313693090364594363844939823392918548955367647356087
8989971977807038662092746318519850851926852624538258
7024957380501099381632387082022856447402189035037020
5158682457112552023153087809016194370538717681644829
3944911865897681228582282863058898610296052832544386
0452415988790437047737737136850213539382265394902342
1710889783710051749817597020686825702725971239990477
7313320890674211212008218398677787056562942469382004
2378134913066630287587520925647732731674636966474625
7061200346715640903478963148186821345980612729682565
4557674985919728673797538106742173858019486939329232
9785455610079000547253305167081854955082957331645303
8966728057387828058258833867439567213201612236427190
2399903185069780948472980377051720009130562696720528
4316649649046311031340090378152174924340671446624526
2478844479703267080920479660215253622977798413035291
8249166775473338610975771752089257006270953984192767
2468102127146446516361829038678557363550700118367241
5428954047748343284348374756430452744506808442943288
0032635348132660476017214226940034962856397025725628
3718828008954760238666306549747045349843766403859894
6458144805319509199259451498233613653904412316740712
9663261870424198840786560346889440980735651117190884
1621313370383284752715801402333740874605509019733708
3604720090201650957663760378637218832003863900863187
0025211594327839328479264705620690613940364883094552
0239437276001155644876784475408356160399848851377292
3034323530097939678336983009127779979497170462853141
0043534933382267484965817752353127196015906172828462
1371401030533439202703451301949107034461745657717921
9132414373649693969048965732825756691327645310834456
8715944990050920224047579942148514139245457253260782
7164713056995813723489622724101513581463393598573917
6240573446271578414318958068767448800803490104731595

8190724146417291159828969702593777767500673203833675
9992089814111989857577054569008806673510534703721329
3489055455497205353010557324696610538066099500702754
7522626763831652725279154833145361939167195510302515
1941717812391244827611145322188667771767341740645237
8993015037104211023223884776937314255347983888883741
3246016948494522990510710895587932162056322085156530
0740795055924944597435103491904411337915663666177246
1772342505771351604426630343126850879654103973473927
4529051405512410302983625893362358342678655320477807
5390909094601404380676438291147830296465182153861892
0140071149455186269299132473757299220611525247829166
5457569566383251147867252560975782740644380460707154
2461773873376807193067971913031072597471709514887374
7469503491520242571874386619420798287288310571028157
9364027146345327984734941390242082297386601437354856
0908029571346922653311826406796525443498999697808629
3470385536157001037969986461806802893285729725290142
8290197040501117093960401799400772790730059574064535
9239166638811445976418742692009378297052256405569829
9485145116253966586742236308393508148778211469714515
6264218397232451972563939261864816182977438532410643
4925254168626435227043137238046625985568603693929750
7794006992109272687025326458806166692255729635224420
4858671812045093427166631890519262450149084600688052
2350944434658274022582905129503049799614016607725564
4874614627097868897325672101929230098245476543800864
3740010975723666092413723186800774476268629452639172
4897026767456792734869542632962933077993830606951742
5895653753303319825137466974625871617196767115785959
1293894641900519594231162554265589036040426117249989
0930640509531999646107199705169382815884249161492363
9600111384532591716540276495476617295659031279164951
7360294213628187264592677678276896029664972247633742
3458853255432357319176321653972435014685762391656504
3387680021015694435823608634845735010253174638070912
0581568187968385839451042367958767056686080265948295

2325594202921026887529115436922289471789836330628408
9773039313699219114714816289438323424291014432554651
2377642992120827639297377159406413974052092305309640
7841631655435542402370608217818229921582821565401767
6697661208991448060059529906543762363612026305988951
0077235755651659219714516783674355129084511136295559
5308758670837780019199618165568310045570228278919161
3024036183516412231627090454293979674128235084330945
2202160626684268919718081496548307692214680927243772
1832739709295506752479298131156698823159331096989939
1225434479357745606611269046429656407405419288226478
5479797974455922998213530029315531271761722186319708
9822353116106940684881575528948162082091816672481312
8908104623769907013229353284450081408941518931109879
6540721461827574805585802435381615139187720044458506
5798479202545694411477966399229790532027713023499786
4403442182496163201891811804326478311151594681114816
4726454776176349787130866887569529762603031666841214
6343026244079468697828598741839503171394290013559087
7225577053738582054889531696018068632325879151070588
9327168594220715607638907859760776550843037319077176
6931455239135186223631142711608544580408475000247998
7582300587088265491462207872672680636374537598067416
4794484814973892387547284593438752792867664432725647
9556156983137246251045428885007334895130250436750092
4192903435436010856692082942618503883972921380359209
7852633481338499592725301325821086087156279941191224
2653301351854683014758836172716488218149835028778855
7539659788906106832228618619529827640257722566057247
4465593403252865816014386518537361611416307557297037
9994466511411405407942714753778111142551339728349611
5287533038864432816081165190097196055850403199345652
7982222342807299181713054322774875923028252948026550
5717408619925237062968149812420426377685988195395996
8475068012031703913507600700194920848368879638514240
5835386068968037754420706427165194190884481774909538
0525810448247391349962282964378356130937457114761509

3589594579265351844458244073484489100939557560229105
6062448456125600438212279003427907039318710752380856
3892113640393583540105820737115659807256303077282358
5173136193707794908570998474013366850020841298191122
3595969682225964545883214339357119483477892683458897
4007603069394540213572747663428486665759667909377976
4676816301584881677064603897065832759236308140929955
0394583781385143772011353236289264203778348412135950
8214072712089533733168788100003420840576180389037755
6602679235794264582504634285826405824664704136419947
4705466118335438791076331452420005378089995503474173
9824797089632306577880661508788209327159575654484261
6883205894028796721171980537283624065611890799913802
8832094343311246912681414321820670235390665254965636
1761513240156795689103548807955825825328000037407243
1650591530590694165524402644222655346570827097143842
8418562650322627493196295890247541653457612824093535
4062701440070911482095483336493865766285662502730215
4425979845382948191301499938168390326408029823488774
1471682495878481818005552066910156137025327368239846
4025891880120337650092064772911586849770591281425393
6463325614938138098819347295892191223730295843415750
5102177387600274257006713072132715585027183263165673
229741655699387896932382888666047534639873633285056
1625487738543568830057262497407546851550544779206649
5159632292580229326617229622139077254749522646217975
4608682099390454206571222320193765629382978088618030
6195284931320849673872332603947486973079385678500759
4093947867425498820242041547066719824068073770803660
7512546417371045935695061956103385552732209810939122
7546654430715285653966082482186099201307482570769885
3489469154319716569722476616161766706519887344489662
4586286251522253842901560874089954342176951754103353
3634037347503819645696162086050486414145849222445318
5569392104609170949935123097068124336744992981962743
8739313209704016695375731733639240662067336606643546
0951047803390531981705875771238273780252473499031335

0046117788777274760720342573148159707116345409091842
3175198274427763779402582949213724616142441229924935
8946143400880938914467003203465218983727289742397437
9715319830729612428502696561588466394484055677766860
5819505314833709768685171863066764959789067888706831
5052510880215627001700423292565257755357147488071703
3038406465942131322256028818375188180027781993016740
9986441062719982527455695968061642301490583712942032
8352608681346582068661120881999425608719239598657559
9068847944798957284718971157601126635233211871997466
5840035802181340436535578951022409632306446703623964
9691027401415384026186040025210407010782659915044971
8939463765389742339594447809412596446793881059540177
0617103493179060102858481794286927768025160634698186
4876158623370152516023637181787999034379595511508651
4540258797264232005725044441943317120137468254190708
5310721520851269398466280177111026424456380791588249
5667434928755342215336739800701841459689064069363816
4153469373358041348621682362812485519628207372517116
4471004843392677129047095449821771245997353794989592
5018192667009919231981513968720392407813173328822740
6183489936891596258827794341812597822374682335655696
6161707037739424555386021304334936400753195339849108
6996333561330849820046220523794212611273864697734352
9840614120293357476985428379850731423029608600645839
0203266832971066474859280280706222894119846852371184
6288178252337801817573627386563385765251145897691425
1738657976813345533125961089216796759921620047777736
6017396981570051893055632093476805752082392470930750
5648653540147096925863324475352400235795142080886099
3523690497927164952973307312176270227758280584330992
7781993804392172681117659365167746348586811523913603
8172399380436191370870417832458368707326879275915124
2132793069249368067256716409379581154374100844743180
7984798184562295337831592094370585987930674314958421
2809177146937159988393383659676333285508452144871734
9872982802287224597211100337067885195708636469326747

1590612320181128619220703781668982540615318307653843
9839566907198048513952994033311865280881232170777351
3270097093437728645069055248320188640544591213111790
4412799490178046065119346213835182007919417118694779
0208647877508811872244776343972876005334911972357654
0686820193680867718864154368080099068512360463269609
9408509903969483621715707186981568085994319275507836
1763106352286244358297577723971617124439000785272578
7442545538826516658015450494415164016924498904268345
7345626173402057140117938253155396691786508532861602
2779776182844821286729462937149110911635104493201078
0072370744963492569121890493135736176727370757948988
1987081480329456080270142457402799289156950749771877
6325901218558321168332456383246354264136630474420695
3390687461804874269573082528928549775570446553072537
2653544909165505854490500908073608133007772880591762
1134078127022071577217991846919996476476550010097766
0172796469665522445266993376993129003168352117905283
2728749544751298134025062913810954966857283877952905
0442889661515542282376017449137351386273975389837428
7836457085777879177722829418378625092628139451414028
3818107124993708570349600679768563369686476720810794
0134332689922920321079715285432230936878389824978469
9775589777584162128896661183097189395618055898887347
8822385862983606148281970638853766387023699719056887
9142001500589327815766795541265615324077751183701666
2680233363450787841366979508958640285849250404849337
9617118032804557079693217083088711762591995943564413
9714952826220987117939305641225493818743814308084727
5530838917269346294899678637549361154624333630992676
9946870192230881483903481460666095445247038837612083
1651420100763183375893493999620082480099860462830624
3249049275733571918058260532493551274110735860302253
0576982049946002145986397578505832139101615820893810
2536640432838562215605048187456596545074877137045502
5814588687870941584142363695825390500441026067279434
5106150247722806223509663315736125142667943414074665

2063534546583922261146323557839399370970262019162213
4738856245398302016554714620847629726525746426872338
6362315313553556816636051903571895244664237586619804
1144802521990130259589349990701834644312706805990043
2561387112104506890557506790767625630503842626225137
2906522565494265482732134998484160539500471824985656
0184731449074269196457452658113570265107162387733468
8337258997786436725917411557016552924842063152187379
1595246448319525159806339460374218673502419803903429
1728820585693526786658158201195876069825239649292348
9146794430847865749745083669623296079018930796244587
4843725174922985957167812557484233167303580613732387
2079009336336925742457891953867586676113412582471603
0708943168749649125716727634443030554358794396727983
9160550213514833324054482959124012187831340304592944
1621292750499845016709464849534409152471837782557600
2939794250148934214633867588345896387026112729779481
8453034442314713124407849821484673586273265362862550
1039736030179463212553543423298346504287829926979525
0762117660416274969836359499603243495791210300046219
2014621172228858441330174290915199745617412214254815
1012748826300946974835123007217022380251636265424573
1668291547444218100277761276072978923682586116206667
0221065842569816638217984120387841026448773556619962
9191877992434897623591813120793421262858085245471431
8097767637213645840433599276725551587668729268189864
6645773718156713727127327069174460768370230278792424
3821403830439878125470062137965560074412906471083531
3042028929121521463452256753566839681362781899954496
7628545917295133246231268627120730703168674086601595
7099166315864652238874562344964515990749792099459034
3139214940679848302722545724305143530454907855046857
5655985031418630848348650105751244928756718289912764
5724607259255296433846454729452076411844381790283425
8464092984568121756169389359659757404292955715291507
6829707802023978916969252199569167435795107281101358
4487803476120458651798666327827480150636490227523192

05745215355728802481958656569366630546997062951144222
0612555045686914803274486028172682810274248864026160
1893418136540881702751903632798220382703820898351245
5779035361653446984428513345483864352548108312849983
2894181729161208390713906848687892138101036796504435
2512042221465614364722014989420021438344201240970146
6236343858272692961086252555251189212907148360456679 5
5951686533053105451019783216574176268285593469187149
2472425479123271529093087674353283875056316259536391
9968592580225709413045416467857239737626687111622820
9347413738395309745385622839540124847808174876622182
7151570135695466014514527686148821945134110598019590
8637689785102904745476657438215181134510250246512821
3078825731151825135732082734220695504162056314244492
9775785418104799619444293639466973981359366911184515
7777992643506435444778907866885667636555675101330203
9206410994467486972380018901857712373933385189115176
1529179033570357835979595544489785949895424664715432
7706708272651894378379480640334221433772404786786940
6306620720565027737002585237439345579494371757594822
2394848213090905392311698664096483735068307724944504
6879290633705387963710117044090375203852640296257030
0441750898478007785474006909831594109288735871895339
4735828847351202477751441528641433301523760202571654
6639313515169580288664731538340423443705796376486698
0429174989729433098371512156861929654197679087543590
0658851911949101426075596946536884456188099014507149
2835593653328658780734465634881032988679671646781357
3843886039356754812673163834632608152226774336103425
6279904458071024372908605563941984409751355348193689
8944385856295365499862588604857134920887479532609109
7703034650129404022860283613342235054786831175310480
4378870682882815000408270466690072446076608132834861
6471638169678446051551761653711887635152359278589755
5315812735391683028019688722025518660905434784145591
6346645871317346752404232051378466514548485210827433
7280013486724506825679412532619761659887652396688873

9252407998493528141270966542050697694791090919691843
1017573132250706433908790786086732900306398353516515
9800074053082689602417037023267069764470297602604698
1576139529220964311841219908124432974534969714035525
0428613898420598735576265543534721231099641807557394
3049847583589778675340827775255945346907202949290138
6136911719381317360164647662549743551192372319033756
6403995542447783944298351828774246412327916622487261
1661531195565153183793037448307105649190863769246430
3925333281823210265162984382479889408161531526098178
3582054204757574239190790661089061726333635275358836
7285253235669995686270183081216740528018000255368992
9336938678811674407772991654278804467841356200936297
5545691518076713312963755314816579909072931041379628
4759941779072489099883994658129887050988367268841726
2456006546811448063717429508458798293125338228957064
2635558951957479210047487800990674713490817711659702
7874004848558461844362188045597553613978187711001600
1209738065852060227467398432198019506909533162304589
8291761625860113602182893530594673628712855570420487
4035807380175224160105364492072700313587362744654707
7793526648446408067183202372794201434472347416804982
1435321906618542992546902783902394693675658550471520
1141750008863744375280986335535666305314474278010855
9946161239902956286531638308517479794311770261662110
7506672591136795657312610790687425271321020893420430
6864426562190889801026878816775863339198793606817796
8001520738645811186433222300276066097923728491881652
2192627682090188547803894435603204243307256873460617
3702324526804366175896199744461169110304860590553199
7056360086339357467682549327302610292972652219997013
7847456683369278196032684120538725039677338741009430
2331065949859757562894408874945023244710451411575928
3854319043656099794417786945650590867372794050181035
8913300780225183670471294785839325894520337371154096
5251726143245821365109492864963727906756677570290788
1082452198737279403819995024928527363407436172234914

6013133741882615777539449981758244493738170620495051
6722942428537471667761788064434756742240965920803803
4162784725694268290229252125238485853373479136715924
6949973608350100908415999161378048415807766179309191
5947188569099610232560063679650977588295435738186298
5180328592847704137243664895146145050192069446910055
4517531628065330522640076777089331089867475923631142
4921196473798385954255778544792305750680638612690209
0132506307939519222133860918595065125949662111567524
7703172613236327036342371720466923617527296437171794
8451046238560425822273820574669416399792178115941355
9646469298483067092014257797423800935295307642131187
9630325003384984642860363407249831285559398096274868
2443195781859038887550090259136755043758914720260583
6213776664200909010705930533871909593483383302999128
1661449330234883242862780945037445941996227719259121
2973961871592020847315534698082291794557411069295622
7707464682922506407698844047590191734774146649892635
2361347170216506656059047328036496824398368514816737
2969698615723107288050308655420546534599832179968137
6938521879386471374152993484829918808957757606697549
6074257348997204999445924747765506558813098857791532
2125348254266184290083358153300425533240532142124836
0436128192217295783844480015460029895050118238964630
9785607091078801026493793765650947187932258338844608
8955461529462664001960838911309128568907495244109929
5591850870862987749246293509843054163189206518901022
1083687385076860449558367008844971994147118076441320
2319502904398729819740588179575559249432469415479940
5009578634649107893595827727866007415601127758927156
9746691856720880461585956897392923464608651825973480
9876278032735783122824239714918079349952189648914987
4891195446601846485979393343279816952173481047300746
4831253068075519706380791066795589876645300045068524
3345844501419438076057321597242893329409535776902892
8132730950373344137444842404965267264912307457602800
3390297467260071906965178930568460857048624192192189

5529501206499382979045221890711544996993791443408977
7511702614082584014441535739125752687821120477956764
3328875396972624473187903295250147526864259374561435
4879964023555025562412865641888711495639054779427685
8725041127991197852527863555532605490775416372818787
9421667567485785614162304336382407650378801560198809
5772380487988087768681885315507155213567545824862107
4536528904229917221564124329654859604047603401089607
9300019701881603401870668767349301168067625375834101
9449395995975179929452901381967248764294353034940384
0454890206908114517035801938191590053992538549904281
1644320959810429725600989398297814280181586790336160
1288340891824265607696572754763199672245330775663565
1498892403328017852909671591215592207966059200516647
1511308401574719857331559508149313274438609677728968
4562866947981365031356609445293868587621647238419472
6675264827978036461664616546009384320096386651058793
8358408749636141970634373244074406175963950405430230
8420453116892618690547745512308142928671314547872493
6722627140194805807208956072954026603573000165918462
2869442970735359279779135141545723636680561351033735
9416057986935094459307645925364350129491834074656822
1068024329544721198860420889704487725978522293551441
4359586661710840576191296480249689572963361908106473
4292915801249233415950727908608747347297779283102211
7036007855457695093136478690591301928975081368206937
4766244097973906085819570369479562173390922423568787
1954488870765540417538648949470012505326595490659115
8579312704816593456876477106138345165728635656730814
4810722413625202977151230115216366436175554153777313
5312607367384511610608415135837496499914467183123847
8172661278322910119023056934267889747688453059278879
0534903105077614255429962938754841751963997912067148
7710566996732225389139322340156445850150383589716631
7796965033760872633416256651521685707737530402247944
0398754569309976118475950363021128888392344959267003
7904222571028065129805454695532114953758276649718670

6631970791692260507908921887407617766347269293430000
28544451729601647189015641206994309986169366851898 33
40105486629594899465690353605541570293708695036832 25
81323917400113357891228386438186621427639112016762 85
09170158013239407971658678751993359226067740973101 51
17422689023963220013016501117008964085660915964199 60
20206039985951675993361646087251610537383060830138 54
70114691786335380234107417724977129694982389334746 09
78146663308890623941931970797684886220394489665218 55
30855371967667509635988051779747261793032691283182 67
48620578092652565290342801908827059014054663747001 38
33025889361601715015488868935910569440449721720211 39
24564847498885836600119303093340504248190948084678 98
76003827669218171908676966018297820355992071897780 42
92361836630667945726051889605382347287593282301195 15
86555180178283133565830986137316525935030805823962 55
90828657292138654721957793122433591473237524037869 08
25360216462948132048845739870513672463824650599873 11
46037622534977822103010145429875113822448213673526 73
72107046667779742422508385846488729490960052186147 74
77151996602465522606962127062085416878723951599782 68
61207922328864955994712943920009967608462614452966 88
18191118094488991955881471379931481529329582516076 01
07116624951556593720289303451396577612021582500179 27
16582114749398982521641313983316292159116787863114 90
35037080467291134675664573677603167044711872286977 77
49372049841155253422836893646994865992816389029333 32
95967050233310631864547159461989211746960440747930 11
21141664974922769663879341360554362801356521885916 91
91601265017034734921985779121559965545078845902800 24
19792836146886166968729406732657944855141210293316 47
67805043203023892916209496994094626811268847761941 92
98887283020453254835559564561049750752610524425580 93
86273420635892776828442281295814647347367073622559 72
82943726377487768399286376323453936634450584640322 20
27497341502487576599102151954483914148945282238326 81
48398051997076559268898315447157695118315551061873 00

6768568040571230783243356862781303169868434085520815
4148525567720882895194035878583720245447816863460841
1970360599469204205022389505099905474622524732812866
7550756059266570238196798544369738430363475513530101
4223131859665983550172909122663560683962724430840465
2362865284232381423499086766091580811529985673188270
0013109382157619132965931457997811998234263059141198
3389412562602152147455618597839408169145555130001751
3102083825563173616215459035129106180015379921299481
1409282642152270031879995958336372414326471243780519
7524452586298660913976536312050348551968701426164273
7394425538704104093401178690284831732444669643839274
1321758809800049498861502455683932136997226039513159
9909527837014548012759527925657066876033943012129119
7850593657439413065417846820219003381050360714952969
6827193408249954082421616396553900931007144638876203
1349134077562755620587774807984394581194951882007132
0525940334421340012436887461101510266689010672516105
8423245264855139823924004357219566830736578634636284
7139477187605142398707775353711988544676572934635846
9884781033914667847854708731515042908742459239461326
1089291950371011825492318420721043794200626722449362
6435095955033758381719510544311687327960567532982916
1312575435682524564650081228052263681461159753211991
7066036435974480828856219322673936865606254293822490
0619956077533306524648443072734872088797672926968147
9748532637460821151712272174344612137262433632401184
3291889863711234758307389741059168181881668955953718
2399583109158955865155107939110515371592823919727238
5189345950241777905515151572649757087242876794409731
4530204095907691738750096132648370745595341535133134
4900038752803101337856441911474235022452362008865534
2053548595620977763485917378780621225544022592527384
8307765179996373823009096197593801747525783079615632
3362664637308389573846711167092750644157476328242109
6818667021420776326837552607761110898894496467327753
4390149108961636184443514027645812228960506038155546

5315732310359157359227589113492569009356071479776887
0073195410228342717487574907091871630476238872340969
6302534078247465097250027224145260335082791705095244
089375733312319820302154163507867782655093217137817
1231371611421231812440806330981536074376395026542557
4723877774795160370760254863489482815303352021941346
4669637514327152808029100281629344182641082759195249
5181736937113651514537697574630355039688155770393898
3487154988401323876915333088608396152038792659834262
7243177492768626963541310684656244843493116508966874
8446934710340534822954844042494454802901500871209116
7917658606527872481257977534747880316689003510874585
6817454970704973599571133983455436889482161005371526
1453005639912424453996258031328022857815556753370517
9614335935483126071990025851110866754290799917035370
0606436838865760343284213467493632847981345995095244
5941366688588636535801656439672455889752138057636159
0214842158335086871769121384576004437560818745484730
5378791606854309384081019497206105993813537730874309
3608025437461836043691487411272220989011476728779478
9539517337789411265620467480077129380534584076556161
6367140328136450673896217852090789222464391679860438
0401984876059535757297848080536465417964743416320801
8815226299727917536184190105725391023526100666128579
3861284828721151144331217481644040056183601867286327
9325396928232426014000725989950970512646811985138070
2881169297470365138827816180181143146872793323314543
1510250480273769735906929697188889345453153536951518
9875968892467578877943475069709594838918709199985678
2139078432637031657987840807613470059542315330631356
2919723476311123701263743716988364569939180204023601
7065484741044827168499151197376191570800687543188967
2294915351919561931247877010488364900343071220216932
7164487378907486971688905566978934955748268599453881
5934237990106692455380866035693093474851627199700083
7266625946091636911918668340876039816053457187304488
3787844599456410661946492827902987144851382966227706

9771018679726274836234505524986843346217582535199489
3583892640008216042578581501014868869194413913566869
6883144598621199233497225849337410553246237223178411
5404225791978377626057249768611808885686579798004500
7436559961684910508497230735345599811561675513166841
9573344751329229644291365157996632553834794507661032
8968486987628172465341810383001694345217588085496074
7849007567303970474979941243512618421371502775616682
7369026168815088921349350732182693228066051823157630
5416072303671389874837793340501584198071058315310913
4517143956718801250305988695063116544061961806470807
3612376753683950228243987528605105113518141919737956
9469363868127453165281063503834682835738440551422934
8007993668210804493606862164600676634483031922495893
1559540458169497241368057496365679330791963671525698
0718618747311974596214344478745387556767923021361485
9728630338941823745049891769740071428842397668062831
7290119085740325033854522235403375989440089793296037
4120546035438418200763258809362869789359558085094975
6506799535033458370680057239672681329462140754013128
2863688230745377375496644876706623705867452856052628
7684371031208597830707742988942333196659933706724709
6895252498544582098143949212079393436556802569455922
6089370437968088305921735499028081872394235008316672
5862989891886521707048580918439451741645600902259738
4114577622103057688690926001961997563649835723155371
1648007804190340734995830844145512294281776021521049
3581318239662767871573714998284140803688197142273883
0208745015739858523972510668956996234962835492727345
0826891202714752570427050611184594647182853205454272
3345159408698624119906327074618266795601170127525835
0864738849204602785712850234784675339031318786296649
5520706455076895843839186767906678569477562130037760
8024518082734525214320551606331504036994154860613067
5953365971680023592160894781404681853314710609546833
6916218480655870713410551936174510371015992161361995
0970865013778053389759398836693787395482278968333613

8162870025109398023076921640166695991773931224839082
6447912809985780640532434091371862188532304910289011
3716108308680874867238115859742181565370929674465518
4929287500406515807192028280529922444162735432238256
4952517424671577482689612161852563998936569445055835
1032818566413199762669743911116778094871793696573690
3761431922383476655463730888517770884496832380581866
1049089021478787690731157126141234536496390084269944
7080255146246808750232039897736546071330432841705373
4413903893234842503181688302691387388458847942035579
8916511772879206893386316827004321470084246757852941
6604696403753841191237643274386841834063376908656334
2469199412844660048651217780432768985518402277518098
7901106967645819095323132112878909014976869469812633
5988541954556717571267667883715576312670276152597776
9925616437210301084993130354371898992470093957528242
9872969224107111180096726415730726186239271371578280
6114143002017111426482584889272072257212203277130099
8294999206516489224551883661421595958912823752483606
2172044495902485756809969155862344381651671463163310
0487392293226221943195365465767781209844862542039608
3011277674528313406762294368841354220939496071825983
2600135575309969096916373496655974503738055417879293
8203007569843076695439031187270674422084002598679309
2414366290522730548733203749931817038992564161674496
4935692665215348150771057177730318828452217536385509
0902876005207644455852172346692659704934706045532756
2960679060663231026724420595594961738355220588146676
6722615253909909223734700140656139750063356944875096
8771520848323518979421232259292010069507200894275409
8085726809450590641548218430923520598808493156199837
1348630863160817753099343135682611939113131975531178
7891831559922775024846159670664664052848809211928532
5699549573270133662891020185295971510209873181846394
4280052695190904330022355676019778705309066426484472
3116541768542221191674862350821405552416336285619372
9156026563037485196337983124937125125846685449441793

9893950361305635853814653497429452024282804730393196
3919665686795097812119450371518494839553461857076075
3296976206042126445050572723736239756574400005188506
4382554728183969647559539802630009732975909676681125
0192035086291396430856380268780542422691206708466695
6677938084288797590663333851918316580733312886410779
6998027738812163216461819722979938232792250343923207
1557065563693152921882935546190236507169076916658534
1141620278688145206333435182444586927795980466279701
2743488199998367141615930040385490934771232162308539
2496688136772265723665048007428664506672319728135654
4204314442205442795569289248165106790636054314227622
9820510603946839073598742744190252742980580057378424
6772140507586893275782389649330295393167543655826368
2933422786679969086201880895596168898194833665683085
0488836176460312166876308659491983675260971893798508
6429615427801315738842262690972075493012383476933259
9468506448850358448438659834930060179435497954684711
8805815318348388546700900036239016567509098519743826
0891762970517516264463210468224004535486063995436349
9876179329543924509327532211671851255361710720256583
3354352696800851148963267718413948971015796822371232
3777998170453134977953440975537822404146878310311873
2043028771398412667428430830099637565151994256408235
3249959052809366080177081548029774680469873018442817
2248913336690764257349831495301549179718603868488641
6413668203152784150730160382876090478550704679705102
9723762400497326796540443757434614926309232993953102
1385733403946546431380393284754601668289457039852028
9897056199050058566331681025424796122497994161923808
0817186956176833555930820343291711186119406614245930
2341685892832540228112237514487838380437533386005245
9837715219808574847409115965605101261021357390877587
4485223070727867510269340595258400443876916608889983
7692971157675464446866868225821631593841533402184543
0200445567878616535387790068479648161154614774256766
4346667861657143213956058051068524929302034863111876

9790675737306130865804980539962498308851629744832342
0918709324865566316859988821807634263361803803309370
8068565521163083570426598784939627742574272212686526
0875537703432427972131624718656218150212721001533814
3527640446743840817921398524717656064532716663053443
6065878850200216200832224339754743884656908084822976
1095989093615129181432469419017754297201682602926069
3628677593944130399807004856247227074616381708027260
9324491040716169431127216932459613466992507264935833
2736922363720775270567953185986989393297451923406580
3202897787347160231747227460805571629448861637690049
8802876817834462006220327101913512335770364900808455
5442557682671091481559673655385825943822195135801563
7711971679367020622245686348546905987780346589745658
4478524088192688309227770546049463556158414097453554
2394776734275305008393605426533858342924088679140936
8144232858790092230529152799915905009065192474414170
9100477084998890303385633941944104864392190202256628
7739846266971296286873757221618817127821235633490471
5499470758868040594632498061324505338954523862725524
5644008130368634687268413792193795785077498902782431
5838273235033109670899330157066419524016404802598619
5829466611465708789699312666194868191195168899785713
8260889720165022115150000459167591831413795284292401
6263075906928102591504769068722721155458369984724422
2110127830084214948985427598928757962681408284246202
7344811563776158858566660628707597984538985888539735
9150678466677440735588962117649026388773046312288150
0824148640352512607007505904268551545873081994483614
2517500969198812147177694290558463647557815938880662
3954201893749084032275102063489232830934102823789408
8989531725707278146698392546761505090113385627347482
2826535359010089528178689238513054833708012163757061
8687637859548021488100672071140270517244681807607601
2994222575783785152623729804437548974460375910029399
1487767534772424112309983814691187957254447449604200
9487586321049675622617109806570375457977845287555067

0681333011940370283684631830050668075306247064752243
6185743826977839833542129935793422513520349235503103
2263242244961884867197753558031652147106199584104062
2119083596690789796338081213639892843731074315823093
5296944233518933301343708243264385327899082614976500
2466223360134797174646420807983695465133294537755766
2272524462050785169967201989999340101333507207580771
0403826558364817000188252048738626947279776868198412
4497130143674446177383603784505886773104155213802955
1148294138196811165301623559891014775948075941177157
0116508692185411471004985307661544503263624343420505
2785612868371178869862422845371122780475373192140489
5699108618357580954983844456601308474652765120152251
6404600863882540892101024615835361898855615679545559
4341539377450291006612155870640166529816814549150487
0910453636023369626790074204524510787476711108581603
8833504907301984545965754311516272250132882641327327
2004586475040035975388411508587123667673305367165176
6723592347410822055662073897145896736124428601318448
0990099661941855126591403124482682150505096821172238
6212311526523007131658654273609247852137350152630873
6450420970868697527750651953873468375231671703179386
4707833381254703056742160675957181244817241549006779
5539503394974344065110140266883233981384072995257794
2686050980385710038847224848237875870249233921727160
6723237837596682806558779053366211090973343419979784
4334852653243507225336039871946098706430167889540908
4338318327380095549056808509279132189616199663626200
9622696371104592305859857933213945718121849704692397
4684711940903162865482672781602334664124586755315372
2069865707588845615920039276376761595539143811410641
0712803664182738485702131664766755240750094911376831
8919687594594576779755050543591395884768627920441607
3899438660597048755736032076189399290784732570121893
8635124345040256111531106159816041062934721259033810
3396221076910135920641844272757256362449921884192969
2845048865571680814766789660936346427193755563009580

2657166740606967045070055976452397530935669267213475
6563223105217097972678820117567338442025858585630995
8277223767012426195014021412415170170625748239199375
7315398056518447478149788507156871414235362443327302
1226810854708277975682257009325701405883696366404360
2801865735583944918479033626350155432400879444552885
8414945115640942709224065884058157027469906282629123
5403802457997574932701823914476262867197280067415106
4615784260870892100884337577739685600881081369285756
5080184359309963392248748227346979480784064776775774
3508106481311385412026681801462169606348638693518228
3330213466133693279168595030799246911478332513082076
6910151785797169586601857927299671940055236690498805
7652288340540382957232949566661061816310978922639940
7301157002457628374773576308183798161638119079736031
5872121983138196203449769816206423985075228327733732
5833243728821605978861098735519137785588140372986508
3817511632667454759408345296967092628169980844889010
4436399294171335504917218530444136127553727404380196
6167062333829738024049062982586095139350274958987505
2807206731837864730248913057905481562736693420315336
5934803545221237593231989288843748037324107862086056
2246217556691406649656032562701524693947009851471160
0994089128351947568274832967227217668299349444534067
9073655254171748865839010393192874133368705592913800
2772731477548697755411984013889604528855416461552959
6730625328830411855501188829841586911577081853736377
5060713350543268713538259132850979585614230274360395
7649606785980108937955654308759455387321323500760300
7730107917896408379918870861963208679051158891374126
2048385411578488220395923594332704524591411473507805
4550403358190338470648101424868919725638116863193321
6497629814849539646083874849543969458515257951809116
2665452367335174501636083335297396558922109211569119
7831672918442579257575500403074904666422709066432242
8513245440807605030997258302179025792182169082377943
8834808656545165803224926641396247339605115247595839

3563660744382993667731189712438143450714426349066833
6746267965144875760421649004657804566810897623735246
6304486513946482319660407267126392740537148060020288
1411947914174210964506313055942410056398778030768315
4549550715730058201479251595904602466115178393678560
1612532406275242043711925061379648508919819095877705
8099249084208560010388602394315570640926229658546194
0599098696476507408909037102041073773228330001943441
3131898982911460768793479475637926076608714832586479
3093194260650376503122078960597372119426458074333932
2946456217152992869877572374421745243510593701522448
6976130482977756125599017514547595430957297867274007
2177410442876241952705751308188966699039025134180461
8791396391751999610688039579457814086994355792939886
1400711417249894605120231528627326923240980270344039
3572135190345840288779882338142266098933184966747889
2669887427451477269584954658077358949663204319684227
2939054023420699365934655175946440193569680887573651
3565527378275584560031781257854725797390168904462967
0212624861217876626477548370653350285918705939390177
8046471759424323624082047979583793498520455967990254
0515940494547744176083957797368673769058556479182794
8871367856297647036197192772736875817422061934212158
3922147169397052507247831658368690921206606029778144
2291451540611950586526514128818711064918796979431604
7996512884740365408399967083236314679272132449362698
5707407047130584348263293188454405105866194037558732
5025731566527539290215236220741530144657593072275088
8326706370582148843196156531041890179125548457874634
0538187505614044231461197845130970565453691268255794
4487804258829500769682321898626238111107236874499214
2469150404040031332186682305915435954967189013559031
4696913736224077614431031226486037647919392677174935
4075341625710255887112283009804553940333704873282358
8397650909521672656474725802054809063989279095949756
5422906184606794762581291478097037557429248089816541
7448550967270889203351426936069221576656346844171425

1418841300734634211770913388497281568269755648308409
1332633671724772500819038332573981873743660364293416
8154160852874645872556535076211507308850688034569682
3291535406069272424136126383496243025222734822076674
3282446251953438168663298333351231892641537539026898
0209230690246918808661653425971370319964656365517841
8522520774003775501336093586141637044909219989129808
1723596202953847731659451593254575127621826588269650
2308327487795212641617601548390384586984583193711720
9368199083288932591601683152378984175095542145657354
2447703079868907388542365511940355299976861512185036
9118581180619655257629645824743954946388495230522939
0431920834700408595906247758692673900383856751959635
7188381477340819136711013050785355589760197241318688
7351769388896092097081199608639323761143676107035675
6751954509856677260928307830333849478963026102904 98
2778516773496566301792404581418889016861103714921870
7553685442128423002943618462997844709902389209087028
2586775997932900240687404128958781800155126238506698
3520761015258102922206918505898797807610317527440586
0231492711148538125250221206590865901782025878010422
4742644030508149340088935536994215541242794024288330
1635560084548061142519736000794683676743928913680375
4522585250356433749230406118813478986495017912039521
3927258935188503227479728304555377830648563922152548
1817233208066725276659643162054180594480709373376273
8519177931869947302459081165317084647784897935244335
7203806146987009232700653873512378118935558807382761
4313871047732133065852429217695200986522155832270762
4047867472018538357834377022350425081424571284375657
1479689310363418089373299722019185677528387336013394
4177263815622028490159604474228410691468157779660193
5784853607566769164334614029548215017627807940265327
5401698834354341953169825972572703237672327104388502
1457551160781078490367273413756850547512833318281681
8068792348394235790283046532570909969893392931956859
4221510427531613778589133026238399572492462565209238

0982163100220099385178387862912847546474997380846098
8491571783874044619813744839068734059128423831399226
8355061730953776007733441551971134477801136367129304
9539990019837057605614721077458334412357522587981973
2015437084932129176882040450744949109305178697056711
7701141965616954874695699554579466490245988058182052
2678280554400630551716201091706454966577280912312827
3037117934488881927575250994256805720457156003088351
2548354531821670031202098881756656024815643339856121
0335218031222038724546176492771976075616398999255588
4482471253975444199575287553381607523828907495794340
8478409819050499547223491767808998973555428169119866
0291050548190046634137155573699308168787927857366696
4628266282678449837678112704991616800588086998350764
2972740292518890639492218318736923256039915873045117
7448334267912516853853293533405574069546616393992231
2566237332891308244660515960756818641377290495515788
5328899061541234378136169948046390831109473541619189
4425279561023915503630620246485820967519520566863096
0889827424160266486843101853579201093414899591329401
2970080979209338728524506836770635227193756121070552
9957700568521654739563093401459950709068494399952610
9950370114911485835658403569443924401813758729506682
0507129554209918508765071118127238305255094270970180
8367332460479643916371196248167128781688086672286327
5435721997602795855021249960429342072378556610335213
4288354032472088407536227477076722657032216145985308
4807014327113323279558962703353311330191389309874263
3572074810144260604122194998739520982043140940011739
6015001163746493739333742500935967983549598081681749
4264116175890029744537902464510011573256811536605504
9218229071820849066730683734741354087686468803628105
3463660636396370508400603225944423680684070109041479
1408074372432537542045294549232054885409447277733086
7728957284483519930624062333601489465701029537276931
9239800791374246128988062094042510718216870035418271
7824693588528969368387123395081807112652032316069421

3943504416422118045865972847019874705877049174523614
0621664718268422765043098722130253292080624410135651
9019242976831941842247953230488722276199998130593434
3540963408734149400983611379929010404902165198001696
5345105093201186528895001349855481579764869199753554
0019962392183037155704956611140749069890614688031627
4565043092929685649204939142663256004909546724624506
0386702779065977825588642987245795155182788954929379
1361864335494956074796062993943328989560710185007412
9740279027598329853445366547373825443050649218755849
0547253663137453707765892998875453181707580339014469
7612612080121923648496013191453758738420887032773631
0454441031752042168170202492424685433938481832397201
1733672714375914035732497421907662839518726209487736
4228903555507099568064481385391213949440769981765712
2176758776977413747693008038530924147025926730972434
2514738979433308220709198444296349368273455599927565
4308492145005895949858388972194908048002110310107746
9457503027986720456029668290243300490195865656152338
5066010860852470512592510723689039144260448041910688
5691711560554676557754131337846734357281913557921521
2524416342672879351457987262360684768794244924543295
9375932256803824124308040441517032330354680593025545
9808401814193883913099131378039516586688564053442504
5976863068464703852930422812537128881240638371919682
5131550640458658212202703344525150024555697185204494
2712661755709779076522016314099243456249658234632741
9996696591909630315077216229597379413304490691235466
2899976787063964427543970530720101567866765146256209
6649852069770710283319925355222101790715425549106689
0989015714352252320882493548326405825443322899338818
1433246077620211927635355605540181246640118064490748
6567924930457530305406358998581036430506691788360446
3376007252105930721019229224189532238291589093412822
8750510980146316239573671434576541180542227401565044
1113110505863376808693783303841959694538613373289247
1602532229369731322493799729280059926755672479757909

53496341417002038663419614534418285905495280534099977
76308793433997597891056860413390126065078612560823220
43804545448631332510663723910133243023185466526509229
39118054325741690275954071908417628867735299986425446
05851448070324558508800237492198538899825763569046694
02036938827446681840139984177397803802254291492591195
74927731783795745010729161896999328831375419807672264
91800130643899054963744840691458284642612420496167787
49190028707996979188534126248108795055420743526441779
47194040344796510910272448179190320555977587738427227
38131058143603281196234824968770668087190298872341772
89527936418706406948463980108956431535233422900314779
81016684313694791961472060973085801361507055166513333
91443324485869469012049247732557182305818857471699144
63603187793301384759759861120961707162675773157297661
33468570481409796915486001251242060298788553533764778
46511231292899852004061054006583425034516346856303099
40509789836252773934123544691012936261701989971395667
10393977453100963054901054483657021671991872203463557
56363824411161709378889385493935561250090479363717115
42240806103438118860330575561248673326845606941752779
34409497025997743501467019950271079144783210978457115
56218883274107109765306341681297933459346742756707224
41346943145162167047337673582685317196054281712858887
08301593160488414921903243630580503307219660182809000
94043271791799057699323544388105032406749186069157884
06289873447376709234225782244945434828081265677173221
25804023869289361125565308204506530663405614901036996
86585620638108841621125072437468721892420926302653334
76485100648824018753110775238799051914751301701155551
25936509502776638656047599326777269553478063270493003
94893489795980911778925653744326543937422785062716553
26171004913012764765858878130048084071184414269029006
25420067897649616996620372407499901838392496230801667
97133423606997570874906266593733463493311747238915667
34441286767762850073286742893474298435416914069489881
44641854134452472851022266000796138096075270104077227

5968926615167444366066579171195189270912069311570060
7846131048600959027972951465467723383197530039299820
2306775086147937831034092325166793058858094449717802
0124060753205842690453320997327935806562077891445512
4440482552693035330351335901481444516470171740967805
4134220952991809060291260715683392766168926744561155
3209280009335523419347624811687507833375052144864123
0046935912724551684094934356435920208400866397288745
2644764168122243197900574037675204113145935690829486
3650288665138667187098401912652087193821461049629177
1605611241326229847291819735019232646934760673591792
0473460195021468502042227254990605003905271739830882
3934696132954605823559616885961443855177325568257200
4086406671572614287421586562936534666765030539943726
4337771175524863334654686617471029474257074711452408
4069357459330581066479105872577087039412159583497208
0143473202166732002959217783114654794156763235809015
9344983934360311394701960236806436471015526548330332
2903249484887401748716255541784323459835831317368567
4119482164882832198303929378208900666864165635632999
0025891242536746598742757842445313505681163336650379
8136161033428769139977909395653758746991782052951266
0654358748024953105462079082869238253173487093488509
2202989974921677075930466511581133383047832465845377
9995964224511055205282259855135799543314078046873828
8309181716886504746473420060161594458175927648783175
1006154057153340802777001733934869159725448358499570
3116908905023479004182696112774389101113682498365112
4352217329412770603813026341816523575148350077986826
1068935781083578681115816638025428639454882447431521
7144653188212265603371686438855279490837229672715058
5998390200737043520451962130066826186389711245753979
8316748358028660902437533659703795530174286017098228
5254346628226050297822819229679749507068469470141114
7182712377179454395424754558231763707200209395248030
5502486152919425838074644561247566303293621119406438
5351261733638375785199093033778956350709984872647563

2881577185877829735370548456218406166516380521163355
3595615713655488304023083904849905346402275363505322
1312628579847488712087974239197129806515715622514535
7376229649699578516189472586019304801887884140770642
7789821115503893404172990784288779139603190909475642
7746282346450496185589571867270893050509917508259060
1623035608541002615065995829409418822317117668561034
3109409015195560359419762952715191446546570114627356
4760341664073355301078406612170688048776765829603445
9262386475855747734122855909945512619726150335698046
7429465446524107479989467997846489274548144629270461
0242772494517285427202517075137258797350890288356512
3730751402162131170264048251331975042822098951384552
8136268841773267008825254317243579889892752698716039
8846333076402741020725428607539146320563341900517801
6954816417949317348878122444725186119410088351313755
5490494181672864244172188026388165627539857333600411
0595994336004451093825257880276648144257909554842576
3256667686186591270514838980441597550202022308442316
4817145782010230876136894006217159186369664756680115
0589395069179486179388110691357391119291194764571638
4243985067676092701560138762547387755513082161491786
3311075676996932639836360199843056398867930350363110
1462125926182324329202305048739735551038806183963033
8392022445021877806341801390292508001654765599060390
8806917718524407509635151958193085485349436375269314
2834726012863215569558934759037521733395698605355818
1334040468345871203940744923635469708013539670296892
0564153270576178507436941042162002813859740994458039
4843717122378085916106254729128941850143373320113941
9237769927284987921679570485720848262117895374509959
0160573191451033015921950477998616399735136343018127
0766361962564218296135571474141968251977845129239951
8094802761577905150529962456467685941080031965524777
7844310184875368377693048120859493475149575363519083
6645103834811783830402008724954329435980183990961161
4425208081024615483437721590074746546970789568255917

8102153564440601396893159844515933212019827378778074
6052798480631885339359255326405680475871392736171274
3904944420640017604659600972674199501118019442199470
3076386808201891521069608033731147927545046918070866
3306830471771276993888434975203615083864030923677079
6659624074665303205885579543535898594777542084658544
6621311741732136381937611174526719845377107365956594
9816596887595352725655305225867778743619699554527008
8885043127590941433023642951983092817046363671057700
4082355626345787874785234482399846943780739800388210
3551971467793674394843979504226155553063937295775976
1213908282947761609069866159952443474072082548727474
1637905412032043763264976757883944911594852617550814
8236435201449006844903755254950452871127759024402682
8966023991381226668728716943914242335690716721974852
9854906502866938531390027800622281054659491949657677
3408717385262225853195847657410136500168835483562369
9235440122359693600605122970484808670655982833625461
6249888405808056838747680592416721489925466769707042
7975947074439924019217135876892945577148244404837002
8276093844436672655795920533328636382118343620277464
6087176456018236920499752142614111949509141365939939
5988849687325390545616898630962962774556931596271112
9138906875337142858168326558229153167341288902773433
5549344788683553410612823002184662365260252030829905
5735996294121284036158487698284476721665060508430933
2357791634125986725241074116285556088741764834982071
4209069639040582853918262162289982686959759493805904
8857536815235174514964661426965879562019976643810050
6150418006870765847045347714700596330723357790794376
7064211961192058242544441864130889629668960333915001
3243279609922778353395891846625759931945266902421463
6598684615865059340714840086040303385526382246381589
1581183633596643738185621040582013281656985403167355
6381630196805645873480396751605716449040168382782016
0310032606803266839604685589812913403117536801291255
7689000970360499259145265139775772983468530585536936

3518247572333780440075047551435090756127219522846296
0672210621607461237715153711868850400371478628178842
6461390580536475028946907239289094722636256621257205
6919773693290313934135875697822879124283350725027285
9563234780250407896120197892164132387436929916913977
4347271497800996496729789539148727048958122750145899
0446238905869642949272303541293353238761892115645887
6442971363897816413221384394558034626557913144029141
2501168851998922870799882033327458850878739620195842
8491699988096256663978461402160950597299728709612433
0457625312926815643291803738394819151464952919885361
9766896498777534700409893333797271594905193918030312
4409381216360642720597499374300957961622047067461174
0857341097442874902407222407192008491185818151812427
6338523114088091933869905247375517969791533483698607
7884734179237590002069645477898046544209616558245456
5757260109829279462120160358645909800121461108129748
6526766493775485550163800936391440387470440680741730
7111491203955955647637863687252125866419965181552726
8261024910471618972792199637288140577295437189483001
2920612558250088095864823435031158427250447144179924
0885831604436354263131199883815034474732739773265725
8291837424868253221336201914847369762675550760047847
4750713026331527914424648458310542617927325595978995
0216364980568016721702398636422151384913678946966518
9599636981895289292091091581455804158302963877917869
3541218300409986888887076505606757845234883714489299
5803139722692500263442393372937783612199894600460805
1929181573650714060521324366571174865186510958665531
7669933181738303448325237239280960676905236851464558
2723843589209066695738354627801124291041420564745807
1394447904816658809815878347299839103102752287469474
0469677382116151097247127560918182160321327115448287
9902209158099544671791023985775776007593706623699315
2851061780016222800130689503482824380598897428078097
8633732375367387515639962500202688917156087205681980
3813215927133464986079783246988263250521724677323215

8505276772769080739518020633239202228935130743426597
8605937025106926387895048939556321921166611355155598
1326905757540944016636894260092675520406533365539514
5959443033647298697252246130287398349730483019618694
5556575297910677872775472113472308106665122026618370
2365900835311812752978241047417681200547328540882448
3885466837414233650591259942286879229483507726271457
5470462006165094003489129260399554319578326832004035
4268718280682549652538315835325773079887414298463873
9305884324111675854532875489997195502300338352132642
3565271107017507937488068307856033254146019433209677
0637493574153953300374788399099007025314629659804152
6455897799394876475410724850931927603294897917174136
2137841981035068496164039387135610981878533506494822
5067534562645152529774032989275375616918174853755507
3371637048051131082092768493599453069558121008228531
4541817055339623762767685236462658936773733428035587
8578128082111574306197915537124356435476888116318086
8339377582789315224641995493001697844790900079766476
1987833614645661921975754528302389984112801986210384
9883015774370873841082808014473728766681903237096742
8941970934024336445816131807477228213377537599246894
9488568872590487141814602376469599508013866043470594
3517498600905231831220139459184889075304017368699612
5439466721399672314030349362286270110183021106675111
1569744130936944850884308639209469638005567006340478
7656103708240980486788426585055996477627529334517217
9481954550738493811330423859464446390168372344019907
1880860774745846502332455205724897116515037354612483
9533550370716633546955833592208900331481109310503562
5241575155460739324444620243895162945071839767616987
0974697327731850083632859062863381325773471767970860
0828636578477101424365570873713729405753606851996199
0142326153519121878183240382660104099322768038702518
2826899050139287494337547628268055926443806446358529
1569837975102408599405715559620169061180606385304794
6278101163688371115018556420832409881625698054524196

1108050107591342574231162743886126499208689264393552
1215084790616735964953417920335729931922987009457311
9991169784226885366510539372307341483362776594610820
2750720135484799053771977521102080214881391072844348
3895833745239607913126446165738853182117046599366534
3126495903472419700891057207310514031003142001607836
8342775492638478125557268114790797901786907065870634
7495144162525321346591354161159377354271127487844264
0103209138695354514175104568359401016226775468370908
6779176383299513414680468895693528680453620097557985
8801075441759285242964102754439417498319758454369167
1545375831879858306467153427646260166170736520150241
2509413289171747243577279364230528420491538431367186
8862378670068866990269549824223482653556886677643797
5758217353681724178526139621292352814651019033040202
9598086319943328122202989858917413312941254825530968
6872331162921846782131002620265685696863338986031149
0682515184065358262028492036911080130045106582899768
8939862230200298730202666823959834337214834359411418
6800944102423948059712951621528595803182583624588407
3891924717130756271362697428833359520054337402297168
9775651438500239796312208322968868544151807687575048
5099198641600385192906490187818432826073803657941537
5088922473330912890232978391570165470989902590963377
5625832771152197699012720654276736343144359633866983
7899069142731429877121028098135403899051819659025752
8717110177255359810918897180596906653462252559961087
1060290385068261037365951903659809459038756802348958
1209837818456632847510122625581176153911397278786965
6636476038783309584586952129741360212303926230727583
1620171532709809176029470213889754474404764535418138
4402323951927105008365411261449874776295766461315292
7304082624646701708792176731621559023521033971585954
7058024228382702797149401860222887247744951501920484
0639089778470639368376384247027691843714011326399534
9055391609284364993786270814923084851585691045365720
3421411183827241925996098440307151328839084613953670

71412105272205061025340510194029407497595745271749 29
5390793858606386322716975883091315775480834273084500
3458209437567851176238291813322850072395652673288180
9023821928341494144956554284260221379058861020041883
3919731786325472260696786349814689795481129245649195
6275748589910851167660235201086703572062410419111398
9650805631017762544678994028211648920629930993950416
2691936328525056590712236826429134597500011438126624
4639619402922612493139664600821783860242226340290988
2607071413101340225182292518114507453249611798278098
0909040598668887394654345337415292835273206845203742
2867061801875774419308457568459008304866895218185054
6205836400727652064823160244792294576503502716102402
3604827609189292591418654431079730615857216897581301
4599779416671685835670145627974813776287791201997077
3376009154885054854373491910724448878268507976727424
7498877503716950996456850662105235981331559735770965
5906404999570137621979292143842319021934015133733714
6388569975602575260969199204167969823087835133893409
7212741361796713331802161065533514784012271805000560
5899625441087429177105963861488871216534202742021940
0108982349163214334109664552364564157442547616280614
9948622628197947120995332656928835757076874231482565
4762139665761587018860883087352063421381805508095387
1062643310979218340123910155873234497899286404340085
6643324403552063429457083508674597822201907204349182
0981652741547556192053287163770669883912653893258830
0907859330973252798030071390325461116679061262209148
4958642463137460474292851212258409058847153194384311
3310747680446329529101441178853360841472418307882287
9553889265428666448434674012601752783005323779504717
3946198949841265861788389973276677309259772363725112
4093693571530993445334363159572110004778061319562566
4941902661002920527566702498156483747966409720938614
2874282180671772944466864229698980601045005527182047
4193533036594764842861974188173599121091811051783171
7355723362048767977349797951642582972286108934350157

9983963113356714420777512245221594458881235393183178
9842776790776195747512520272576345924105999269154185
9509460537709471536644233681603453774944782038031479
9452485419024158225473078010510922138304388873009741
5958976243928516827241735402495335256497883617447651
9814621487379737335020138996317498404803141747331125
3576810877282054402753015794992122482281883158599032
1764218085761179589830507631045793941516754013599164
5960889661120356360724099260713876870353530836023137
1618275879494370788026235451349994710057516165840831
4018141609641484855569557304840323932205248542084091
7721499157966705053940949709130942603584424107356659
6751505941297650572681495317756547067231503130463608
4548358457214462467208837762651946049230729108575517
1808704011926298599674373996670398429978562924491578
3679456050193823228919978420229143846192877103398117
9532791964008706484999273641610192982836441987022
8318235369601337295269640031432055042715716563003478
0171924642065185460756811038794804264588691923654859
3033626064402769482209740683542342439780194853171920
2606336030218984998773957051431924279415742837146691
7177565362215383803955612588336253255619898881383941
3151905940783614415697879733902202666436676056612603
4177238527338171700746543287622673577991734420640145
9575985605811985204360990748786201063309505039894971
3531747581834943611833358525756392124646558514617733
1430099874708293493663050146531674574921491274225822
0888494609209423211433462825171607831824274822368063
1197587626810722779638741191448120760796135398449987
8324587780855847079140358040322793321570138959365817
7353967847577538591986059077025714985199792918862071
7554066504414367406195975690246107524513634966072493
5824938152862368659264139236327584459542351653026603
3702306645558408623065624456971108791978300610297648
8461105742426529547417648662520787040049090179046710
3598496470060348647617110294936726514970098727032847
9905999347892818513060236900749309573793718138695168

2139546812959146498623414918326207550263876824895095
6748676320264693455175510292818249839119646790918239
3524187155525228632683189420876997759678736117498348
5889930089824631185447842241011310191145821330652805
8112412300535896490363692652436919364069404865160756
3283689485719246133771989589253365265257048202672064
7698022098371415108748082727121455265654004946322613
7117556522557855785438620484397274512811246989303953
8513275572087385861363328451549809991216221760819422
9832953752884308497481526598950959603170767549866453
7413763046783260728838516515898281905983662442409841
2397675433819956413887733902556191040434070925405873
3122719515004390733257007402291089271063985702642339
4507230166256217803265052508088792039039830239056304
0930830181301726145707308395001842861952901257381244
2180643661159699702227693367937704896765160022948925
5184171690301299072120129650133350627007142276635497
4111199921981966469870956664006653242100394714517812
9100001780324540645368945014739497490056690622425714
6068056925494622647946704886636289350462532097847012
8681090305962783791319601090907816037257598889091566
8049431931958905969736237831810429437253396100728725
7463297767480226244825115785530275005860141541908753
7221131528876724434954889393712681182357650797573755
9186226095475879390068550537922635207130175199884858
1413739120823909552910494808863207734526534495606973
7731565388547835754306823098580903306345184634352421
1935900991772519327329122989298239984803314307134208
8986768649183176648276455164850978318312757196668594
0965467399168666738031142877256054767215666764458975
6821784995803697938800350918275358548373510238035096
6032255255659914155544417369194496215692433112650812
4794987752339716009896404320451632415661243250145503
4316605675360644354019814710729774780115502323050776
5864292355729797955055139760232195070145877926441473
9212118715575931178810856734946743677579086970048686
0076104855396740093966826692529948537691346709983406

5831062322136420749971036766488090636658182890886783
6544765605239961168746643503885454965793933667829994
2212390575467578961132002146388875142770428485161410
3790853626728543299282619009124004269300184230897419
4723371882770765364599634437675073059724489094684373
3502536860175083172039515236001787907322728885243637
0333004444092781290593453686631414701046593418834746
8928262998823630130601376692698821779885172124541457
3378488230382467191665951105174632431279031560874148
8607081554831102132540133568685405588343101887088938
7613937325023408807965938201480483031644811231762015
4024345025897217767005259876857529110799488761703346
8123201993231132192874341846612599870181746561179146
1186892683702520165299119898887494882924206169649654
3089442346341753064626206632041270524790465222259474
8526298821801665103773915209569257176760513915129079
0833063089131384670767807136082989918994490539984327
4940243889710601762751648654324350417468217404772053
5790729788190300647621795656051593785317469975436785
0429962280685938360583506521637181437581203594638980
1353857890087538637799944252751397164285764558538815
0959986542599611112126352521835373754089383829940714
7671947955656533381033560920916513587960431756452149
0021087374521940701660790742117146389209287184760160
9024923191104226715102906017895674642383409519835911
4240864264571107074853007624980220673638377984459884
1477515071622932192031026050055150907697894319437834
8211223131719769687308328746838329398680193191653702
6638200348246498888280099530802191763804197594627304
3423705049816862663146331381992449951350409336852132
6486221662614304563801554167029975670181079914598371
4301340032034976529521643857783420248049746048135655
6278767001411676453276570915946987857471095170775617
5897195470146914052898762386344666075216918405152920
3706434167143445810148812459041088366769369630161221
4043030796233418792780707414554309612195098803307323
2712251430746743792949084700011181578721760472562843

6874440299990349072352336477956148260727543047507338
3579416952085411858141421163366331884361393046086404
4381203050087374740743035198125879556512101543796185
4018176835163955314297889210979335064421892206382792
6017080859661513409231014455095980500497093334182603
4628222661365245786243689338287481808083116632140886
0189627933379691796702389260039510884923222262487914
6995246944822132220716228187633754117440717644082563
5977749100498441131586645655216934794699385345895276
4802986158402264099994210004334206449394164465158608
2274972790568046591058023199814041816664689710703815
8991782599052443794164767665313637038164955688078417
1970666908881871118929635540970894493508380867208740
8738589167828057846463873013356329005608175565705186
8983518288538558189418761846431885541883532205586551
4919608401350510913042964386737267017692094625684048
2169559422438162836317605490729939838290187707137864
8219596279582737284384930210765170111412097127189513
6778113363452251194325640609290920398920303114286931
1029961628971574165135312265097656638725415021881894
5769606338265402520174627484331378659366835358927288
9441372227122323730318997627121875635903055240593344
0680406716588549910892233951031852280400316307779313
9388788124263739945766173505780454864709713365612269
1054526803233517094657822351326347119775415664800121
6476189153839445438270741357110988027382524358192945
2706387243028983887862373972699481019099564763398726
7797438188240786469637206134575020404386514048145408
4863722948089187495333684538332918569261160013609052
6980748507788080971992207905493846449129981160445105
1248201576813423036975845979313525076499172671897835
9204624413558353968020439011298808208487927059939845
2081514555277160455545091663910861465981010943648552
9958341589405011322175918918227407885854505737075417
1980793576571347642256640078552027123596498418147809
1852475405178298598358719450900926456203221456793600
3200980365891403659248029705970423893401417849408983

4058894208281375410845327194765940157849180879884127
8671288346973044453463300113407842446974676100521632
5231469607461797235227518889113661084728250447333876
9880899788249617457143265389593198919380945373562006
9779566007329207375987877395334012242628246381176046
6552954932776015165444339879779657596421303648495380
2973362973405409536566027156662095624204018972541002
6903088730688596758323634848608003136749337880469881
0817924348705558586126044335111341550683472102803886
3079884248647959934426910709807053082895065139289872
4560947408991150499159326607612639813504186421268398
7924382810639019024427167350764624024576817524129773
4377047211534086168041782949967650685806251274752995
0655953249881866111872216992147209556065475521976145
5049109990697568423675215339297355971225275715087665
6596645027191717820529388510936944721032791299729978
9495953721796541482204684847107971331529242256581065
9649076885751212831515757008915688390775159233949705
5571554396120342875806175188670398308678334081013481
6839437033921934197424316334687716755401028790595185
5469702441074836909988531592235768360501858756785573
6374585771484106340133489759087774905833455397705795
3521359016826647738272508556557813548876359883200278
5770631642240468395161657169656331177116454122497180
8621652553084508026356189132604359629200964032384354
6372129515947537029341355782056091034650314827936141
0656603454420828537160023113668131909196410287304920
8500417437038337446281046462094277696378558093427578
7598784183334039966019354202671488261281948862553950
4381515336088819835281749473542529613052058898947452
9789819276536230214649271640863202923565929941917524
5477614408430602237931856760648303943416218753627041
4749761396384298639152870831145938517668485369224524
7913397018567966189810070204722123318045419230799439
2215089918339722129466485426691824780579878782653881
3387791747999298627164543339304246091128474141007611
0542008971258536672836314860898634346456934102417486

7567648864999320167607691395117456163032737449804460
9078090640304676349444315588698973721502306022408768
9628089967772008295400972862196936979990856286378189
2068734312434251912571665860848853313223426184326055
8367351735755946247449424091891352020742489171844022
3182667020736467686101862473648492758014735888129615
7116453077773099187906102988862871493020467922752516
7103708071639439123743167928668219344462247657260407
0545998596828789594818122960996644984189543550512697
4622222840558216017815638489324156294294102354724474
4065298275956508523080398810417675310953945082956686
6700980596803972387830710888730991670839909866670302
1614657172247840852262333842572081681007339653460 32
1498432069726639309186514925480137010383870547849580
5692390809071470146803194411882916774100108676071463
6703460970165877479386198655725149160321261997199738
0349016484226754491259673931239799007483105538506866
1830482906443355681392530449017556754977224586553701
3114885452145575276500340012894742742237558340321677
4265860294150285405959573417873490709801590858265302
2046578069213686344182383358550580440690789048769469
5230168242268953030195038490457409477237858413080942
4481263867625452617907185667849491594475752589043298
5971556253916870664050033869114702527528774632307639
4773662205021243171119766975540707333112675955811430
7664350837766138393741882119872814024301959257723399
2497745653599173737048234552569017468386181605906850
2523687172292558204547178143199185807494916821191010
6141017546675307620289154632134291872260156914532339
2446783536092923925956317992477364265588541429930289
4571429764367323222629236024015550305643202837051864
4027032070094133089307407897145934113546630626365872
8571889770055691796392094089540494967577669166831282
6151980538685795163887456933961269736698722204498574
2652078573393450055218249597364838727810394612054451
5637979612030291659476574699341543271014074745772892
6544229966008021914307516320121147122336288689110031

4198269762081161023720046209913211643260706919886802
8640972266780902380740359354214499157461979683557148
1367714201028436827004103443187994214361381197705387
0570251577675008745353928774720196545049062159447237
7056510619675999908569487775939149115942015050991367
7419640531912235392749755102752262125932903159292020
6322743156316398835598947694912780282598450835836799
8620353352020685460559216786552835764981566953231585
8857238729888822191559448037870908916485672990721373
8605360437121439616910385695176160284757070741220885
5744548038615549299960111090089529305615092834665028
8039831552918890865902817664933855036021130100426140
4612185620272908635851705705207750060330829518090619
3350336573369268872311459864004662237348473629802877
9881021471019245854937487774531159628979254055017807
4749196477840674655279039319556581389669254392861168
1270286078016492471757947690040713838418710229217335
1898940764080897143188308922163936596875379870142040
0378491301275010036189355286464804238014072668778949
9470242525139568329366720126727746887603228486942873
0134997355463449841082903990246143112485288482555246
8148762739942714989089896406588465382777488201549894
0055948650851084658197861933024860833800725503537057
5267261626271208957483857078107167903963214061147985
7589273165206651387414183990141524080694271641531248
4146575073671621014372856615067280484820945901214115
3970570484622153904550532054514086490834816933675066
2852070850447616870476424706292519842182340567119317
5977385071213843566161200541291487091099968133185503
4567552502739480560945533332426165004974273699236895
9557120323458164445061839809446368120108418926213314
6656721599470819817686659148832682318546016554172883
4534167044930916637484656897676342312018983264383910
3418758413624196745799464920222197983459305656369275
6849359777671093103041411307312539564248638501455500
7579436042665449474702259689851026633743830181532607
0463610412035069829100774024752336575842434925980678

1961067661254989366947457932038348011891804623993440
2048605474005397291988706489083532738462542597815237
7016539340906639616141813699362622724220637338198430
6775264803874177190613456070869512882942134188943261
4115598374198430965061807992482485995574739758659791
7835001625124791176820566112456878979546722894411612
0724622182150361118719603867594046340815340520931954
8994528013639239204558207050232815917711079086385994
3266252683370835162218627906963513461000189272878972
2396733421122488552537949623348050174564571416968863
6010053871749288214974692896253474032490659110794774
6995501662902714298465088391795743901191544231663338
7279050548931573371400843033387711793984550288105152
2538785585885276786724654682252601394142126380025151
1052536202028508833681167117913145351827458269079362
1433828736367147855025406183150742638171351310767393
5765006518722579662135584845251998140046504966442936
9446264325353422704810873584386515316574783693494381
7561843938910192099339207935917302351336134333617409
3788943324363676621020575206404986003394762611773065
9790071733843508611904667283091919140548761824903540
9603611171758738428295310712978874130067815729007187
2028525347373683052683882088519006528889920671141417
5614821804859030161269936302200424573036545063083444
5212718140481106462655021833491808728134317000593894
5464777807178007554115944795663687523130280968563849
7664674164239794038097802400682239304397514877618551
0146807492444313049368424027979663806970107218594446
9466756952631588382852626134002780565139541647267978
4720187392873431743195634271468612868703186802680513
0778331133364970514243458619433993760383134891953616
5221985717340060262681642333152627532561526998660446
7428210001630787133567564176057061036539724403434996
4075523914459700042488278070090182478520476973060681
8272868950111230402025965464639168826534406245138943
8008685826309926370738304783630389808601099489941257
5125614015344638442370874909562441301959987563891046

5209667545877660086590395215269307249475934637655249
9957398136870468238357822213502275156277174392239955
4134549014307806588887145132813370761485025768523236
3829331474280596688096462099842247620743942690027942
9172375897478932798562424729659085321594720533236949
0434027966266307402731316432230471242896578160810904
6022568044881972470679934948937439150755051735578827
3674663011336512806280676387389443510734047785428449
4581032402153026889267092892734321622288665308079172
5525366482531922248604671904011881497966918972383904
8992144990637834224725829744875713871639376603835319
5822125838995005317567009552936485078884042900036232
4607985108094470411877669656985270022423654214840823
0742496591289909650888536308725432732151415989181628
7567811307051625685105581512671359344832178026783508
9604725800542617103328951883638910324473716748320591
7873365096282974559694346240925565281665664281336902
5930758740440023467313737677792486726102625840368808
1693860941830435421605123289943113775339106511731742
5791903877442755577466603040662009904063042605149202
9870431846013273895090998152703064336944690410044571
2022354511710113287564039593702423317102983934900820
7273903649597967324607011744165743432549961178069176
4675964746879791515572781516247306058334526364851289
8167784698088189911321003939555111869683602326765781
9460839277758877356094075598291775428086114543301395
0045524655124291004911372885966068671895355711890373
3300649089756833516500494824375020133685157284996369
6746425914953603739411549609823443143510932022180970
9359780329549759598895081104350136062164200304054253
5251820091558762332175442175880859419299401661600036
3439101534009403986138161418529659189582746862217600
4007540224052349144874115414450603504256362329696036
5972082364925594214765207713745747951220023253307577
2735440666725460638556600200246857044600372754039232
9608743253281392448927596263699974608198030761215869
4436812543464760058234517098658868757896434602270548

0070837900413305141721926594157615687911501913402974
8585051714860817315609739898187117889639975438593851
4812712285659202786935286076096100145004686282143308
1002880034237990803160388504060829762941823082783808
6035227249810236770590604646347730952402490251187179
8642433919025304589573203908585078719522550177703765
2162664218528198174050734002666372515280934052081167
1011269698677937225985693349519432693201259024230765
1827771352718844725327780205511448358644478230115471
1844183522932511493257269886174912260328402072777884
3300201824351288952626434850401801176692189400301384
6230392559573128981537243816953077315894785564602548
9012335984452603058421107836641770498438042272775618
1463614970822052978940468419642105195952976344279449
3800876237527458736540436860323925681203968153978062
0318441175173406354964644946886431290056599239710398
0260552719134441217649315767012502328215868291339917
0943472186019906149947270419372232448113657773647843
3420225999696279855298823483581345198218142567592434
9886313315764355498522016187470094484862457290141545
5918948887077304374958672079248383857434010982500628
9616607997109441836998747844395676792923888624160244
3690271546527600224939349036905471674482965770830739
2421001528327233796093569239903388244656012980079191
7643031420219423739939643744425088139872031104733044
6839944062988196979371975773532541936499970332980309
5057301944905176813411652445359329905152911986147095
7035374526557874245185688896013513044654670270758809
9460903301835695366013279171879444954101560343692286
4802222470447675869609032209684225636134056348368297
1743439491345035015456271211307069128196826386733221
3184044441497703738450944461754830545368993606820580
3889877247411952389292421637467845624985427985031449
3299533158554300276671540262962651669580914607881017
4714306991744199865847329040165535665857626308050241
4955884775334898523646722389341636565324794364510059
0252258632136464125849984679616184355234035232472111

0521226636091573602713021329448208976614103780709193
6558026221817849571220758511904228780008745928677362
7633230096904378031370895252076667175727182998614393
6555118371669223725419466798082166668111039566043933
7503728075545148480681660436746789432640453711566586
3750531512081271327549205306822200052569298501430885
8791838338588827226166775683455460042038732166503756
3085408359999738344203187925351510988383385390032909
6587405487398852972968379972293660129231230716020550
9733930936050345903955144350530779986167924716144327
0747624508513019789738699270993325789524645547506763
6682646452715255225433388053548362739162623925296676
6458754894673447577273356013838273729005393896656592
2305985710484827743980497205838211155382009892096613
6946893177199111474717037337482698105962706129131399
6060882187721485255788982496057151197409955071399286
6920154565838343101426030808586884932719229841589509
2643571831409247104705184512875869988410928735902874
3120393437627985164110324412262926311001109691495544
5030945335769214098033156765480642125772776756252536
6210180850636818295792871608398234021472035362598206
3645520085231280580032671686683448151110463737048499 7
3483990721027211903580088432422211643344450800225977
9528179717226997323743864517946984457648063948949183
3438525180428786932632752902447890475937940428598452
7499222779721000238911215489383823913828729899317311
9476173906115044782792876911023764755025225717321948
1814737063013088417889819598162999541083390244410692
7067375959569971195359309384961102865740765063676944
9089301855864987037289727234334572249278915326092232
4770228772629642491769808030278236217239379885400503
6257155488753610089011456864982824376781505124828205
5049206761472527146521896630049688579599767752259397
4060305110289858039626218819712821705192632230895174
6815864772494006634762523998541731960261610369241957
1597760197169490239932872743974658804365659364968801
6852863977515522475999764941859502680405006409698435

1130737971104411979180057464654930780215212529810087
3140604694735659064689241814839126360000736247105564
8198258938088745764536277429937681358765419179735722
9612700089296847136964936836789635251823038913103992
6337585965257961649644990890955243550865890255302785
9907755325901273060023553112413722883395464048657778
3316157682986151786509241374742372088701308805439522
5927885302394309216595649098407706095942612962824796
7788111335332629528747975409878835566787900429195451
5767441486784044823639223350956600727547939140169710
7231858244127989233882002377940639757536572516250133
5163672644359159774750611925713016230090937345100474
5276180163807096773700943768059667142294135896008247
5538324597480393206079604490501769207058512367261984
5895683093796806254340250957462165951887975505779654
9195504949286712332513375567387160573563800289429902
4885121880124056867923618924755604824874955328263873
1464641642059885385147743343317259129731711974000426
4987222438106142110327499241363713375474324062966725
1815657913864370256202430396479048900504429852624465
7566236218820854094942368505732727377622836552938642
1319461785260626049990625479688474585304413059373947
2779307753507819355762734410692155894072757362859694
4966388909215851327061017106149797620538570852812095
7527632949857667771947593521524216768778681734370556
7423740243650963517997153302057143111464013586402829
0245151732610767169202252500633762431074161787476243
1101802901331809722311238240044665270255791343338648
2338478240836415091426303214665547366175962561696659
4331206659851276704614504355650567632319272380345140
2535421280961853640658606595686500900542984050046093
5485306206267704765845643230355579621397040128414507
1551329589154551669286582783894033915299223882332902
5388857260584924330742050480774965818966106091810585
4541979324802037965568280399926145969205463805877136
4903148774404809112742816548241917457211102449742316
1561692475479084730751663266098195237956764638767882

5343150882208179566747714706801635105964756831889832
4971204609208556997133757414465046934783022432710038
4214768793221424857135656462834010324241282632765420
8160894480701691549541907889908583899738707067015416
6653384195835071697145193419203744574382051040777729
7327360839324163745628589224133765386367495504954305
6637708434508365177004646646381532867444822629049601
8468605036883440776084482397002567762132357213772691
3923930959252379422056676983704392607890348267373475
4528332857659917761012569555350192620551799398021571
0312414311453023069858987010303589421288523151506444
1420652849534936220244215603285094454454628741407401
8450857333734350776305942612250192525532512991863421
4765821403830797952738737610527302639241822426415421
5090646009883184415256430726001468614601161949130240
3669382475017141894225920208067077454915759538454237
8138860870217866424786028682455382570607007852827332
2265105633445664908743615822952264506909608316956172
6052652534915020704138021903400570178831183123741998
1786872388251010597475120234906541684015733501431783
7335248193861982871799710861170481956079258642819561
9770249670042110009538004738803920047245467873090629
2796860054268202283888668402908313352076886505277918
6562901289213124031511478404650007571261779711587696
0036259178899584553203528776418478397863166507073750
9669088361316739147668310680483001761136059412558390
2618497547666962172853403585921903452376715116431337
2600671055941435933213580593431965154632317833809081
8578233195716802322563645435465739653891585126961726
8356652954529933665361650739802987340183886461244163
6517466666989389248273782645426314272038650117553097
0761558733454310267608916815162421264870580775063592
7882007357177805690888160798433456597061009242403609
8417826254172021527883071915797667428851450587738133
7614484000839126439568917135693227613352816047973256
1164800243647813394194939199814446345033897730483079
0172218978761141526758491378276713640481452224170097

6380245927541667269859014203411158804515183793647076
9448992165195823326382816833363255130234263516944400
8445773426489193207412771550956432261038689103857009
5852192162848418489882732686554704236675275075298431
2290873054198395044089420214166780821968098279767077
4928984971242388093354144951082942562973278266923004
1011618064786854216339301274558922324247867491607615
7695411688302134542171596584609084801971964948722854
2292491332269577189910652192682352097342852936279886
0921167917076294858474961499783598343087200700477102
1856897441261723109103558622624994683902497802482310
6107738908043031729059847704522430321003304957569659
5575909808971877355132748296339886457187784691064035
5644896125273514486823105300277818843106768143634883
6868815197935919480586451837858659731027120780587817
6828347642204580417485465272557925932127542209355067
0915217460741863450104795444847280432287590427853279
8925864532242985233863325720785494344100713049181600
7509571981783809560002874758275571459591214237982410
3442011990429800083484667984779173667633916755981233
0736044998178330002714620794715396260742401905177826
9682879307334273726355455968205132147577968851655215
7856382150610037574421068786981759087972310547187859
7945093416353173097134275573684804654936846085893279
5193878054835351838457955127888971075385264812591819
7952271446731488978306681441294809043876475417203288
3679315394873192784282061408378211112385518592573720
2642344646616985206338453406008552668716999882546853
1836845011643354224246766319745613460084963085605745
3735903033205858460474211719831580072893001356115756
2071742314893304796447468064963812931642923523581139
0296699494680014506883857295049880031742947556236767
4376499424361295901887816363422319493407258497317389
7184738742749355098506472696968441265206780502194204
2876107362888938588503873245685564388165788462809886
6182032035782303338009930591300723334132345096025973
7460520043570998600298145509578462832001513573592546

0273515967564416365301122647127864032448240077379969
1764665060238733966563567934039835656807221965404885
1193224882054279809129711007700450022775461206617166
9155913980976565982271696317371323802330818946438128
1348664524959954457359960273403749531981034137354585
9961495498360917612628539530787384570759463293037148
8225193038171751154383500267089958265452638110372525
4887743926013540602522145491981699579873716453513255
0990520879677994407822530807758169956027111277585448
6844027605293945142888002909538028485411012261577841
4915814077499984149629224019889130831785966691538822
9009946947450247844902571367356972639792830403286063
4546819859014808677414089210890401057657503110419221
6149418743145878476136714773918305353143832266954538
3299223940456133606017821411886509292207929496640912
1600359051153880564921627054464191236518908206532775
8919973892229390120026823222236977367233003938217236
7465305265074391406830947473212603208840098990148026
7801994826858553514806570539140057693454713673203875
7724513069759606056795390037265846113845113230645833
7250580531679344725994305521750085317786363398194721
7743849839416646214485505188770661689027887474197775
0727859461678481964887923839242970123021952643848769
1711692941913676453989753022131894427468986445119523
3611358086995256573849951322723448589323113867978311
9517843877135064823078704829980344715507014188205310
4142668229481600816095024682359788933239467695015594
7575022359260204247226384941003113670440974536586103
0801205930892752761072852639425752928436218637764253
5427818993064800665696367275161697181990722601937571
1689259479744761248762888217986501367475075006383234
7988396497740048841235756668657161421583110847360913
9345000273200513079812815702225616906552683303083665
6381413470070819422166484822104159343491908204056408
5952240388003780734926165030023171799931482592911800
3774744659501565939981386238692869062652382061233623
6745964072098350770110829907902806903410917509635731

4561823190444770495486618716069228030350137359522412
3316964183487990807480804086899822172755131619587809
6775216539898309620348940936838565394211961230810211
0347105174241643465517192077927713852950602675186424
3969265536723344784100681455951149036782838817570535
3800389460027690705631270232301414130663168017467973
3509725414626095995788159410727806596654228530160830
9809482798087779954151330634185197787230301266392253
9995594139496211004195460782520674425080328818050339
3892187565244451699554137647784167163730755847972338
6593926352240182260803169276708468269071288406191974
9117656286999690849707308233756477976874846675305269
1929850792803668182143767960730508738080830144642975
9825417007864397304961083418619696619596322018403591
6356341184358185982051413631915309125174406624049390
9245135885190762706889366270990559464689376680069204
6828363046250164021027437917854480248512861821612512
1145700035734674069253679036890950259239891548171622
5418252452080600390600406155405892907732068730958006
1749971920364712097884192446667092044497498748240886
5905266935889487752575164013543674237920145307220235
3576834544681206865951393272635592769965995737774411
0379091071568358658465622085862107139549354539122567
3280629527519007549409048963943888064254557072622115
9363124394916459725649101842575408220047228888466345
1280304483190178400740116764773956154364713952355819
9976935901084177521973362030832625761659684119366145
8533114207532119529276697170420670518598424997628346
0412316390812279089005602391472762547230446561373891
9348291198454930609446294509615911536755252862610591
2712122041446841774978186305011129740039411193508189
0835733329055114407430444467585330390819867745858706
4667531058733204448664381954737084809840190145711080
1511114446629507460652330517345945257725758930786370
0719576792849542202391372656825993183849635737174554
0503873578054083223542866825098340742461917212410659
2840528111662009232829603017213638492851047735852983

9208709892631698435885742206374457995610541443705248
8223358026756746099254402277684009359318177750785767
3345320731185308379736957382024604745009604524055600
6415683554046864181064155915986925744890303471460863
6868420714152951953996888639944162985012621982654789
9506312921479605647184999313392444952972883378335522
5306656081139111557599979071382892418373574090519324
1180832753210575834434078628766429488113359530078115
1425957827964092837812763167468852532329802856792473
2045320938542101580714740180947946116048627767867343
7757511437592333049254994572062768423364394693270173
3610844401875653569316078803127015677432921109546037
4669864630589643299619579908391638851073583655397358
6858039475629404022863520963472170394504703852571085
3133624475454200105259671217835787463335941659232356
2570393312801881979934876980850885387379015678885924
9599338041045070956681978068909791304753127014469119
9081713805793823536727157978743995647891549064076938
1923678366723218190582136399034973143981196742174048
6606919650658688315134834018768134679042643955738590
0654837580715281812895107414409604501704396548535905
3827804348135830772445160037860973737431472179402649
5307729429552473216428585864193133904622555731428767
9022533447878688563797709322070475438244713707210817
2807262161922034516766385698542146002937106613178446
7434949460342745909707794025711988737531399123813260
0095632063682362858307898741532274462127591793546312
1492158563100688909580778060593728283740660451733814
7569406689687907446437289032404571746893162279915260
7670087495794636552981080060563205359323461491322115
0818691711550065566655477454978755929060742276192490
1312867775801342108401628830872921226157765095221015
0804466374403297822650479584839492908809137834911052
7018915866659781539522433630203819430779722100749295
2919014175775251699297714793500137189964488911520647
3629667612182183930848992602460041899154669973852196
7567293096434216988980634192953311166015202689067552

63925108101725929474115970572467020836237914457657307631055046947966134150629498747616641845707823855574
37047474057201870953392723123250003365765441821501626602351718472672153312107509640158510189813774914265
45299866692087088903694910230493046303417508983514679902569728765115044510267583560834962773334541439538
79619686112286271837770262649999543789756618638452242447394924921505485101221670755524051021038733002845
93613184584427867338231426176973563642708421202831884367381928347131950871731221910120316721141109395899
92288467480174165676099378781968770763447597018787011536350704268062034322219624818967910905627992687206
31573443595078978529230696719511043330556678384953840961252779583891054879684848620867971749300845214435
94253462001124108426656675868978082776276840134698294192958020330574004749139789710591226422104073255789
13140477467095216337310954671071478824347469732253620897184341401689515209329372895796179900974531322807
63181828994331896956895304723703995389058396574835501508194701003364946075415680939094827544998118100311
51431124371620602850821167716052901503038399817787498619635004890805220896906827949155038157223974665114
42040712132800560653624460196857385732581380945079434740660360543591168103854745541390100521085682696417
43645926975730111231427615691640643829304144191258097001501476260450843029947397770443406025584831551837
09862104371824449093244999094123969680727355749909747543929025579847970934821903280850591023318505659588
56934103697521787966167710423049423523510863007281287132147932780402066461426230078561408403259834892557
12085118985382238513620972879195187746506418610105011000152392140198811550103331906715391496612736381353
49062018988011860626488814169435292751302012074448506939497156569637005281044364579654008558044162484257
18544837208664333866575252285810948289217257839158191476913646032684476202255833788430706626820136562567
06042916609673993739633372559817540236901883535300799

9015939672492877457231001781338885062942677684523610
6426208547207080605367376268476687684621043656625525
4577155820968489551256042709483869900453706023638867
1367910424911491996301475646726002794069393629208526
8041593916556942831013701721500246134125553880321201
7480246619940571602598114205384973309909585864771311
2190057785168213546567692543695868395539592269791198
1510567862427873863559696351596525780100887775161394
8594765302893365917624022970657836985360711004953475
6757228407933874696396982052754885413863809128046556
7957867380247796245580749357238874918172010300891988
9932379533927562492951430639175417565236205655375337
4784035475143499180169624212773057517531727140899284
1799710543799766469304839985765697038891618026889482
8664983964738224030523683857879176549873616284716015
2275110555356422709303412906341214050374706538761104
4057631277677687955828396936068797492992473055757014
5071286487760372167136663996479516812181508956359322
1450808534862645244138042319376365335527483533321558
3412888866477801396224946024358430223059175741552754
4778471665151580601596831434699386022411670339610331
4344414215237812127052900415296832835814274572054807
6341739976854032114278702709946582145669614204935860
0517832030749599849994536775963901544332983729598770
2158798404530424172368853956543113249128001668861432
1335901814598815345115649693087226879981544016379036
2584744940276762231405838302463232783555897049122876
3755160993522863875948264709234548966040439552829693
4963273296194539263412540443583064912727969941442577
1537866021215962838480080776486006844211951284281111
8606563381627587966850467909393030243819414713450444
6109962381417080458893859796343824476120094314750139
1451102903534584642339866533775034032887511784456217
0190700826875371234894254845267952905967289911416216
8717207252789541303662531312161687184002908491401088
2474192903310039585332809030568981619595841464035008
8183835447766161764083433565762829160365278550533429

2017344423999912982156065639233096831232606113498474
5904753481757247935228998935009434950753963734828911
5471101729844079071163848822988417921854283174985756
0164435622264612259464028308647766387359459884245047
0990867877167500913930038211751981118425649944996192
5019393804725337399459333773125252463064043429925100
6362772644042522129335363983888712558650282148393519
5378291219232513295504794270774798175730690981398175
8364267491567563803402416350300899745884664455951052
6377303887533487344021772585481657032600335620490417
7355790973475984394759958454297654636741210753515070
1385121261710170943881638681800325344560780113893154
5723258763168759141418393365682229624660091462055145
9783379115646479266635436278233025485821978207099747
3109160351106470097487400073152228766473962912778621
8446835550020204307191420072846279018318639787025702
7722687823910369724454864110588891669111059220294449
3294362703541330988052688008793461709563048460288276
6011907089407300282006643598669430991288386695237929
8666281788984272697048886044737676094202615377177917
9096775127871974710440809155990679190772341720808085
9904286004545456751422771384737823411005311824430632
3887152844086268756605069723478477362196202376584411
0337215904381189469829313069286115985645313989313994
8999833404400924228379125175552174621891287605139468
9884726771874665047852706643625748319169084915371258
8054145403632674786953964910372400574613022029319950
3101987750602880237975002552157499644642453349885915
9093695439580845280450049366398305637822541056262416
8321730113237466365081832155139049801939199625148203
4852336035202979892243773111500916585701032600353564
4447517742469891589357347056587514976256326803969581
6969490397599461063976343230542272130876246685734670
4606223493784199198380130993928023652274191986054264
2497117922820503705375874271366672714855309460807796
2908093583854654683498403635552168457034303500634102
3502853487766353047125068844087232667590565579334784

5911332126078901928698099359636775783128957269704288
3799355130392695124058919984490604631927762990564603
9476875652776188987807508202115485364253791970754729
0711263442813605992811917357099215255551980276056037
1809051890207185773505552327139139625015942725393023
7186445017661783595005367424528353346296600400846807
2728533180835272486343316020649687387392161609559277
7074720918638361912857557193948445722793390984130659
4059996512638479997333289827244713523630011731945879
7298546955749664148606783193641212157264534076580706
6896025318540147824874797280631201916673807223776392
0872324754210132174021931705246883119566130365670703
0521912378617793925690762722384770505239270622228374
9423814306103440377982382108877414396313901510708203
1275456607954643371353459928062871969469725559248762
3405608599760254233805356029198699095607613736827707
0442866704641224740569967492098598383612829365067744
9802452216780957009937928110107393230867893546477556
5148775074794665008756926950491325561264280600598839
4995155162576408277981605727557443961201814749780045
1321782129786374375109974767337631313444066932216989
7906481415224596057960363749293853905804580982603556
8195289395221669574156422043036643722991469676043864
4194213013675900169324226934913049246702707782481845
5231134611034534892733150600012302853833423036382471
5502555136874563216693665604441464245556981823192741
1945082796887046941767029660195074502498516529806186
8054638134752514384332789919960927108581220089357662
2256979935986992214992546471768210127651995959782470
0501422174754586194296039239459092882841814687748419
1361418987382812648355343241016964934655262953455634
6170833510950168069402286750567763445714741321777516
7306207782187706492244447520828000983425570457784919
1916884176771786886303321268954919764573940753700988
3714004870254292603296387792875437706956043733999010
0294852638150032628997285511301036985819326744885052
2284219180545540822774752760746538989050643747981698

3471777490761037600628289061945763967812992880200275
9672944986154770652047321954189029670858378060569502
6885991520228616831779318138111337008466482823164294
0194035016674892993924087057564794251319995861278330
9735265384335938416752475309684246977819186243437711
1559925282827313295136978782140742545163126864523275
7808217439154214688943590977318156580220579640441942
1225370147991892788537903377743287340955117413783520
1919791527965001393868848556937474821612927167295727
8563847386324693858405292467490402241343895188323801
0793146018342916216331965737001597942739004500063841
6513145176559085970026470032213022198522497413915779
5298792909634972898511760181137446922094253103131383
4496135599318178835441647145038785547166597698246724
7974403116606061989122504156909044764662457128363820
6166742756470277596897462784181051476701358042590385
7530620365729378401649166948271359285692735430676917
8867000492202732316264040702550279562093496216227338
6194868110608449358956017870858831338441728876389093
1537407240007280253256276428402648656501968697974430
4259225849580474179227925340055252474495023408392656
1723909309423006093663032348020210867886808965918168
4792736833014327146956844570493654212738523641976274
8941760376029752016153593894487622023573913546834272
5946282950905765143194209595516072612741353598331918
4123578419642134288725668738970843831141046585600376
8822032463086565154107992964690465777065237950534596
0246149402061560544843064378729945258226263609197006
3423456958121081018804294851368286739852132534519851
9868065279201617538965618411825224252968934634983238
7862657383224882146718221239216145216332527567001704
2893990524654825877851241251856125788689455531665497
5464304753503190355902321438128581792753398401246082
3890717054605835960586771990218346528305718682775107
6250665370945298883021196273029318588927084757014856
8998529665057338471038620599638943209413359577964476
9922141537865511246485379439254073621927524684823828

4997312571864576555150869582415134979815701743782733
6637993430650906064980929838630333539425021822566381
2732097435466224458876434994073553886358770672063368
1111322942298365405268821561270245962885723548264218
3145461433191254583311812559791473648401291468622198
6737581897719518823278520809332782805285034388138019
5284554650513932469026911560267684358544393507628566
7261266508394535898309320803700107893243658291550801
3223812988714648091356440292471252441002452525445080
2324615782206358687167110556932854380162468346196749
2375522775131051001267760576438799157199485970651760
2138914640631735022338464345839483543502798190272879
7303202850846884019875949870378146179668646287546670
3998963042483342254904924467013293924725832362315311
9971239894462176588427193382546662161038214006990230
2774264438571417574587943978979594815804972905977772
6218748279191542139015671040424987960383839873080715
5042530393201138172623366914341884766267550325863449
2672941635544061641605812600689785048902465409536738
4485080441994812315226437778358928010770528723579813
1917642254407902629775223299431624405682282404897958
5862042095903016753077009841255041439517370577205755
5508175512601790181200735133417723724762208120008604
4079512395142659896434037642450608295996661560889038
5710684640291412717376571513488794464268910769410895
3101190999299956309309050352227772332629147014017864
4514635311873837849554388250085692730878394745287492
0176886447311783104110199160063149881892999061015277
8168708421621381839557079184051198067597769599875313
7757726887891088645916544689831334742354792980519109
2151468307163238553510387271875446767082952974905345
9537652593192516594514793368506381679734786636883270
7895439596677298327096680062790539599982945777316823
8326073880180654102514617216288678835870661909367729
7966422255933690824586710321214530157614065638488320
4620465511573100331062717763663272535510511401137294
7974242341799659537348942142140023658440811338831976

1752550589009245453137756058842247628652387606272469
9030212670470780945124147162949557027040189986666320
1798423005507008440753327962569991771876542652570331
2549513970864944719145272944883050946018415295562514
7404095257980099014633837977690212939408531024885615
6735060633863492368448950752823340100752025828306207
1137190594267815521410921860570542096103071329372555
3682257947355874625677765164533109298228760283792259
3025131851658133770605209210865756174301233428908469
9223497351511631421745254267139789248050025172232090
8212457411077611635359168604652376411852083104556005
1390958949873097070872311142547023121673320381085480
9201787391487934488837268545689214877830390016547741
7622812605807283554153313690079313963000637697020076
2535050726123344151011142807093681940223698999130824
7424654012701940113222999932048332874671355383494579
6358368992886232904397225844938171077259058039497162
5950663691604242881282548386971596653055474254354559
7343320165017471694261408641380380466595322388060995
9689304939813989144177810804401776804126311873070380
3284078136515237865950551008740358384973781723210016
6230527219947879907436057423140992833458661530302659
1088028489438826271928605926885462526118115065543143
9186047386383201495201419924016510173976740922604325
4842945659258581776899771652026749864198907493364258
8243030082299140884230370334920003210947642357493708
2515388359612855402857151199968412130951329760106062
2384467853304303605283324594771517521109132184692968
9013599203990675174666377175408931626352691592231667
5852838151330957335182944234019485759992887571589611
3735250073352994468645177277810729355506620011166278
6406845834742122015354618427456277813956310035038009
0185222039972627590546827269914375360065865512634531
6534223994033256987619903270018293229045380216469805
3155309882953376189673095344571303771285992545818022
7261374655690582259578692098980461167400939173233575
4451424181559427904164840501217527511162224841376487

9395289487689110620834678757632368819950650817234936
8185004920139539693115045084063183316979565001151633
0083782711074977286046415193311497771862005817211835
7176588916463557018448873306567412167110459918528506
1221968011073225482951877407666997960230384720072533
2760059467869526790514319525735477141111573062837948
7172387990101107371970337951113879024422857661195134
7093824055168672986987094588552809896555090500583947
9776816362135995896454669367741167952365593301962543
1714598281637637734830415853528871062820092867345131
7867057905586242287769770380335867189664400760452105
0778010902637401436327800462862893243121698489569692
6812699655700961162978104880833322640115844498657886
9198915511649877595008201165471079495476162725359744
3140698950143479155214870180524406888053182445054861
5105575082458334830601530515271410340134615871762049
3273768228117936382237726367695089960600576457607434
9083808672495330340119364736422164031877350174262838
3091816033713053081947005481456663342292943943791296
1361179742997959789822201838204339375151390081879567
5780849881967116995779814800468611110202998559769628
4193886876123274515246277330804465733695463654938404
0081977760970663913237654253918686820356685427661932
6843902885919967881472483502319505887747564159106418
9912406912530941631256195410954353088146423434083316
0970495044930981167353983129373553934118732008867086
7106762928026623131366609838364307561568243371003247
6128660874213918935675213059506263362049826465500820
6650187746333184048109653726939935499250846093222363
8918187900587249238610783215777979026003556222664391 7
2544446062893294594542958310015673005507543724742621
1846516371207702459968277475890212707746082328108777
4656437622050892211762862594923337322306799176150246
4359913563816206074085843974251331593898633831027241
1438507532080538973380115912508795623407291394530386
2707068178014681947724028939617221644175848630204516
4887958376109298506760537167764010412278781795500182

3319726057461761188377946845473203989183811701977866
2208080181016483471431403292545031424952200821114330
7446640136242253192598750915751217391324329653494012
0953928653470846315882150495516801442870649484831563
8437272630481694795792035566844577863829722889535344
1185206100695450417704454744925970866988636099344700
6199388647273449927912722231658528362329253648259342
1073552499528548442312732204674710780642436699584238
5286374322732442018283439734000322418590192380305900
5872229289610551499388306141350064936910473902129154
3977494536051080648720801311904902231107072307706242
8339195289372209114877839087904495963152229896827082
2053048965601639594558607553522222159573834959609286
4920413661120498768165209416326912589404845282229036
0702772755091042347607151026084703720499533073566165
2016080315883563879622431208900709419217345047787877
4094071468706792259425905227518180949282295331821489
0404208439333772858902536650842632772581439486019593
7648754924471152085966166588308595533616071705852042
4797750579059521204948991346273733393517973537490955
4018050208624252947155610087991541470696535457299922
4070932580384255389746776351480895187698836463589425
4942842207312036451005027160780398336131700022763357
3220580504720999012877768935337598574166458520076392
1687804857367539294950338409822939797065831425555392
8295919229698068777227966397293907779082178517324761
0873556418967084941823230292691324944913403757697788
0000852120689948851950118704243081970477677656477051
6660073820649884857171048322723457119259180652711670
4896922909858075153627517095505284290392243650048248
8074481318574364656698445218053666467548379873567916
4220132961903517086414797337157754106151617424044495
7995803155611157910308731347209903530189499946582192
9921964771056882282986141014201946390542328584438171
2408363322656243251238385947763567301206764410850147
5398144634285931049496869363826400464162596469514903
1961110485447759191706584392706760240031145212717647

0833320094175688738759477706324099809206846305347743
3241945220021763000466228202380827780419779493389391
8898522440855068666909872615099934327559421951361489
4603275485400282474638185574303867224848145712041289
4022001415268847709624612221149992288764391919108994
0007645004267363603359564464427081897852741770774513
3958409576044621143276559892571212464070497606068894
7521778888467536577313088848317041308470830281177925
9466701208771841286594199018778750963200281102375514
3635612304865546153298828299904617451774858147760123
1334341731387710905577093670657365502031757900430672
2930314450241991977428096762212425199286327402583704
0075297281743548041063934506373267506843468818388748
3323541121663418804241233034034909675779416537790841
2586879828861032352788573321553381519883058804585315
3469043130898096369406641813704154859314966671159813
0899440825457153552300652508228496172872396746508251
9004532455823574868772006674794971216360282085235430
2783865361171124532148648794241321331700852315433727
7460768066376699618895122880491089111765955157364984
7388606952471668475237514464152133654892467276122585
3936148416514385818691738416754348278131766314291169
3785564618171660966340291272053653025444763830653355
0451146415224708651212131290099900196815169592152430
3910229496964390635521990651394321630365345397471515
7350144591560970031479537382507222864326791180222854
5445051006686838264972907481325848087102088749505142
6964293739258136771841690654521561087615737802053527
9580044684913613691746825371728035367843503618901245
7777583386467700487187551541811503714129454911427269
6877208861952903110006521480604793894304261211250474
6362225675393476819222219200635168766825821506827988
0160735706110805557861686704947486404202700061440097
7944187614785497645823956249805444955125709106402708
3239081446009251177787652063803937135711764476329221
6214126564837394710745132290507372055423326208635230
1211192300993282164753643692379063250335682531355433

0347896291533044923115380919957555329449870528019033
5116740752763665598147220612180438573003072978792173
5685000125623318067425988724010996896981385239730619
1955928336948611903239492535944158365958161383912185
4119515199265507043722245110633671266896256727586677
3882387907913364509386511722013962859447865442962183
2667845170020278188419240093649036627235744372874485
6331087287895845848235250820156274220792392203392045
0828194622615291844607061975822852133877969632367864
0133130410995564537477406457775768490351379167325321
8265004401500646241636404631178277966053575667603716
1374420266916172189142991639230484973825428942219985
4547868948256705457712083060696640151754701139843828
9931933635228885729811548286913596032428511636219222
0766798190063884048427152608659071553704971859335226
0571181041504579347353963276399887232560636896084103
1544136421487826119285418384995743044276868385145914
9189187424006619028314479859226445963347995310286302
7801831150089300076576280887077688855667113610618616
8996249963879937029113619500621550959100266394298058
4231778964965066275652978344151497338825523633644652
0362201321628032135004936195975070269172783237013837
1576432302888101329633287393824573874624509689508223
8330844176192408476051027246860191447430391510803777
4819238710529112795905517498822939075512744080364169
3282921255378008849192870285467542546669735739705365
3624540072223989562013067604811339156349727760567144
9640906404511480948248685117962164044280689719576297
5356223618168885002728569433652880013184412121411238
9838519527851194814679016652840688382186958683066129
5903977459905614870361228980984113820061585914247128
6229860041718906453010082032794088580385760890512269
8760086424606482694850486186296517221848751835565288
1466312752376870674667526917244167297354569567331667
7184928439431385995773850485061097313805782920944994
4463239430606875958119003860261904921098398719699336
4746331140662945111471520556948040279873582430918597

3826397713404114160166237752693577223614775634790555
2752166482414609981486281328766311875241074174329874
3627538504577774254205757662739315698404385677291438
3783590182351736877088048034374286323665907452288539
5228357786813472195376605008468619896263300503309360
4099782229148151475257718379528138489156804919218123
8360412271358296116497247108026125459205952486511422
7833911564975466627448671852187561816294711964713806
6877715360854178694183861466074853653955258090169662
3778000065583681884719769824445873229844530886990299
3378877952657197280899159793941934367522718663437829
0793682444032062463960186699049723190315024362504081
3055353383065316109237895273339143697925936907286974
0426234042475870337917002214925834352410157186453983
4784545175892241236136735291362601712155441084930332
1644230075969711058854695869571172632037985137192940
1448711950153715879163321253830793896944127468922739
8610118372085142869319715028646909873284817207387381
5201591163794512301010196666203644541295629190355481
0519125343871316001524124785504522454804170858009744
1643608403759638018838607489535266269503532816480968
1679448801761593992935630643145711148351667475654627
7594216725378229613379520048290422881659995670507607
3487042908590849968490952949104632685176365224631701
3487998937688779842092948512962785230159883015336126
8342991776614639254947705193020500310556054936776339
1632839538955788699776971431354461013249618901917057
0120182066702116577665414605136853434511733028437410
9752675183557592471851518890916898948657604164533212
4170280811484675907730132725479460941550972867896718
7961801220443350296879621964404832788637996040983882
5362938230395839698373949610171235821842177439742703
6469112581075159452663554646751436788687978229022955
9947715764306712652597185561511035747661042096417841
2476830158403397936011211878112008231745037140757004
0927108374354010891993459498375670612740971769921395
4110952101250839813654956204515024366843113398973858

7361130651247452423215425715091709831141400860264890
5393707127744124406690768316708542405730036117869052
4323205442682356865003270303065015077480473870024026
7292414480020515306773270191114548987396349292489206
2897129047830449268538002831487535810599705614838069
2737300968661098886378902955732473773218392968237265
9640999994766795576053071821618693945992778164452536
9696581224500935645899443219172791686763985578925399
6966098824872270586902492001784902712148635395532805
9468288469933578566896852991438034105283369383898081
0654163125074949460944740888646908365216571068529029
3790114001017104187582043926198243372615611135685841
7307628652063100312749714469978191038819926090166461
7975406910297256584746904507192524494658335276417463
9951788616905673265931633412458545178258082449007188
5017851428871766720831159955592926726211782561190445
9506934896175722832507114752044127677657505086893670
9803366686786798668095854516356450456700098282130467
1244237580255335849496783455027631075616185676110242
0062290862217868011243459156477566136273107961732523
4681409070011050097945863457244190266200577367179046
1232744208537976097262686877009464227258685007163645
9572360638163384749434997520654319048826875825350515
6910742713458684179456955827177098027528168620203131
9544359749164686549228763080466621431318534173986503
5226451902580545172423793197133547989443130184302401
1898085682842876356115113592545051759006609029317534
1972370373166267631045526770572841082661953953967680
2350046763923223819889539040992691678591566892197646
205371386957697999088841317689151137434627023755761
3560235953212929513393306880410662258095597530152275
9071143072612099804061120495954470252902245829810326
0460366610790784614562477180721220945378224379608920
8478369881533998578438358476233111145529449931635895
6545170743494100864563756823652452255289027971719998
6729963816213783471325388298076612871462553478352971
4630139379788432298529584543951967680386477081199962

1581708094439895803825070711358270798334780985661030
0809248363340106643785171705008672856057256658249006
3038166592502760359401420724325335032907153406437209
9104955441721929617285189318787667540913587710997530
6839422832961708480658343497111814009227825302615934
8139475517360355840894266447493309584619986206812924
8429990069049530956019916735927003422770580777725794
2989192483507500026253575382687483236342276724808711
4413930325764451626363014157737299135855264761831061
5475055054350039788791534532770215960445663570306770
6610019201793214017147969716738469733339707056058598
9228309253129526494279536183676079287994017708601760
8475303934791124788612396945329823362750327417646243
2178205058631210032808102535309052281121335769067348
2789377192908366864035028279947062486247688670440208
5953853472413704692825937227528959641559749916775787
2687009614379339091219386991136301973189710945603730
1611109766624424400181780650555724623398592568655386
1168261270433407009518008868971398949213194807654561
6095446512264314966934969743616983961687411240926925
0879164951012251867524836356160571234863846892796456
6676498484647671650465612669908654854037052810502823
2541582319648245828614979004180345759695865716578935
9912026924047544686256265670714416277114320705726232
0457058642548648538642871832592358827210050178192591
0321862102525429061064196493219738482292467214508802
7677363100251061465898752818456725920500790060992633
1793502930263391497547899805591598387407228201211160
3184744607313092642236057201406831874074143684735669
3028138596844963678163554669045752531854826599116179
1886475069922132683880788802178150798752627095916528
2807676731436874076260554027715833284666790562252441
5031604568648941811259999795035307072839954188004062
1837690405286046378220683553744365548569477836150626
3598993634787027909030997462772184241100176482159012
7056711820876687822757646769942854113054242846979677
1293655637190811243475249918894104423899887655814909

8163133828344398678830561422069948656437056345681695
1020971434238126537052902311489174162697598468906754
9381513688231255317855349374611505045803566781944318
4768513482917267953046515495980756411689982379368626
5452254476823193821655598816897565449898472233608040
2362152126378985700273204797098350733847558808852656
0046011136366410695357973449086862143285777692977138
8386687549173683553591485305782457191099830298313951
3757052525620958095401789769541394615172026176496607
9521063305486458118903323277535560804209288079548131
0443083142541175646937964493700880517884390646505986
9952993456228849781367900568424690668982348037672283
9141414639383441970505255527456615243070311689399586
4095218468006890113619130090893426827828837570563951
9533012518018235004929310697272580570319664319775643
4141864919570951944115202257960157942174332998712495
3984816432158840168231801567682205889033443405769062
3837262060541947083026988680883194001517792506775717
5174593722384717722050820930704159311736220200038813
0607888400911177396641888367333204465296464459344197
6859642812624451256257788632315383190956542867923083
4512402761688835965251028829254701745885087785467432
3354131435813989005234227038800607714317834252668029
9665251595805267396825662975785411273245999963482719
3710570217727679088300584907363364013164799368378780
9427547616082779063539802635477089489921777344518972
9100846164905691264458400492070830852606455660554188
7941017689160277283372571529263391248090560002823379
2017752640689351180449779199732378020380548451534642
1441121574092611717531775253425231265652789565479949
5249996613418668561137172657535761612467563936363465
8529021988359358313921924913934186424541359344281660
3840579430340585830595161258412086641797040450055709
0151031427979014579958567197456136453537244757325971
7624162216656098154776510792433284597303502214180191
0437784872406174681283719961462839166425348030966724
0241178483790511869883383917902679307649564913279665

7819771695645747593331331342626077489713671970058790
5164110575608680390392686582634870634540551576313986
1676381077414451225941285507544942159529485739898630
5684715535148771193322794310386006628760697072692238
8392110104220541823141878387002847488883890566750633
1222092051480787061361084283744060089044614667973715
8262720291116842293247482478917896858777805977609418
1864431634002885026453644551355067121340118869078555
7499410205012020584369359438338431421187984966957966
7123182969419117181580494352579524060183758509979343
7113088026402154288164434430672028630306124498537156
7180909678367412752020111345413499839171117253538517
0214243067321000314413728871055440789582470237649047
5232030970596062076120274233173017656903200367749269
4227330322757762751700794150691302352338229522938042
3742991955311001375700873574048960493014910013105148
2856386998429294173647557855294153337949392024402317
1942716029023127159436936461304778015704697510260615
4335602353227271325523781649405525365188946498390345
1788574439635435801343497602714738438551198478108928
6682294577257535978429545434990952690776186980501260
9732425756673965166418609433503841496183873703509380
7038010169536630461609240729436221133735537225631799
2452086818271670641961004506900017835172681539217865
8474812406988299439446928475392120476967040089175169
8004473501340113780055210563049882543493206799641734
1838113208226066190871983360217148256562393710727708
0044260302578643754691414064673799881333062749804340
4884443393728585142092907140693132785150534696812734
5232046363566630091702633597632388614244380188240491
0085101582522932565567279300996617151675711037022790
9005772432264519348539581533676201804307906634693859
5228276348528073739266954153406812865943469956911804
7243766089383156219243865641031313405891150807219328
6739168832381494477607199208103547538842345389673265
4849684408063510671575237120307503288768536916612388
6017735354040091088039589275121002549716693707978718

6642922014014558824562342414467031323035012810324420
1632163057653771387099052725968494087829811861525888
4925272186032895221824026282983232700817634556129971
4577465837854724296746182466438490297865300776315937
9196442547839488282680655017633162340101463270947259
7201882333538813353025456539046348310517300849704261
5373676406203301379064787387371214645255533658355828
2531298690853960026366890725505682714100041735228984
8217175999402680746491418887030638147115329646789559
3180865122302663699720471131747863490527766941427342
2723279580870002598280525801378538783200311808721509
8462322707431627761529412804042731738766997695548191
5380842773570932813737605617669370728806121195580689
0081599398264876451200332178564869843209063760256299
9928889730761247741228383269616110751648915228250644
5462683064172271803383431737658197124639514478783200
0935133331865522233895566025164708101900224467477939
8750108774616269889409502813107485697512863709006577
1914143770296754856231438325159503852517313664426550
2859810576418370704824060732078711770552954530969818
3519729294115418251958353083363474955997898763199425
1041743808773856423077173325405197636112390635219894
9174390525500247755923939446107776141318676550924068
9222302864431715623756205532535947588834188411031832
6056616085707801241213297274916600004899274741402158
3437012481574191298877094711693311104113046464573198
7871069570113548766016847239555580887290897174691220
3692511897246805917112574573943911456518061060805637
8653089578477353911887809522439697800184582364439544
8244656556278922902372298460832554705494068221878803
4731789918341605402685996748721800813207788553382430
5278895252897097708026085078817526844197474750300894
4242700277303247293819695799127666762690253597622952
4623326878831389484003936817044687218510139571324767
5408223230437822817504981212218492381106072004410173
7240292002257194762803149945154783344703033464531637
3131527491626927728719658075797656249029624121342739

4749494996058045828871922218243736016280861644682942
1784486643308194165490980503506193935373441844498848
1855950496502632226450862252087098394179516137292615
6316664116379086259966195848326953820640610268251704
0494898838579192421686509829170475839529326011944310
5729997009488200064628250641429808857846119843850931
7434993315875405684618443087248169038284969445491491
2112838991744269803554425667205923759401508528758412
0562381867054311890166818097178919022122935188742790
9219551551808886890313447708457777145492835784529617
2487465733320431666064130222407809409924086148619661
6461791573178124113520858151699182455410544125676216
4334410701735120323683692247023947522319864680946657
6700844147762711278182037067394773027252712992031143
8451352019946347089810581466817328707877561024420480
7475308420410443016348726332678834504450244842310917
7551673670760528410517521452293493248028426483884846
9002099445286264641842034722165709568643477981110366
2204260528403203412153418624039613679265397630572391
7678082394197387937396993118579045446878731500279916
4768258851740784400056127034280313124159400625600806
0316896796077526780558515844477375732040986866150895
6832617959871934719108242562218826890023341053971891
7603630564713421885935991661004043119568998668547095
2963076078686347518088766821399004676927370948474866
3262578526322996526490290475572655642628171067361227
7932681885116213824241463851419800618482560202162079
6477460122499966967228685245060285138006176482634185
9670837636160177178787584787211352425712748345276492
3176264625743670922662113077335552056833860543070541
5874024492900357561116555571699857931310006819332853
8001828777472325091411581985057695450951287072005765
2266342717714995753519942375852010832755725591013419
8306611780920840327015963024197359149660681090192953
8788649629120915479131840927623462313114441025278015
8536443513026311959243846145507833436871321090511487
2712309587577971222071836071360236141627933630276200

6615130143175584243644725271284482860833476749411206
7999001847319319046961301417860432255267100830950296
6151612339140072324812740694337484813194585701844194
8519546090713962540695926556536231923829498572212861
2645094639194954110726922190617568177212932823950981
6329423697312472408434620676415165837242952236930174
3268413874102094132231590431123090085591780898098639
8114724234312597727307258749654507988460850364940355
6360642136024636725029758258821423970906963894751585
2196010056708757617434222006881840186782896407971342
1198798942420054262322639109160808328221720623832181
5660095637661311507035253943137043847640671257073659
8637047057472995577056332924928706676715784263974841
6481898746442062732629180918634569514082111262111830
7784231880550483902301823855961986897258663753836854
8518808900275672391487967574758714477044963905966647
6388405551398320851105180869467334421469643889365198
7429294500796357933677630658354791344094374984874917
8110525952934946088696039617923756363527056828663233
5693875478284960491495594355811233762962794911089627
2845630669590312923738949873906464548234526530112459
7093691663681984294019757039611050180939637076776957
4613416736594918684071597994097749212951447536435570
5104049071822680477534689219219223966559889264013833
8542564942877508240456558180339798752807500932132516
5955626489684326472095084572694267619620324652425361
1380812855410809861388993231752761000593682748189193
0572687927052706681650947384411241352252246216405896
6977377662669472230604792593765890405451089087271966
9296026156460569223470830776597249422251734490049588
1032587117349851293340748288962822868451406587059582
2088546356622237969268457727080062428624708391000127
1322746693291077595784116055523253942755096006076083
7053344808069613574798661003456942905048742886541585
2057182474934302926645020129015228508576137455210972
7391108825404901954679022537325594370109330222334353
3679247915548650890561031509201053297700331149099331

9144158254039376710385615125897084131515152832179811
6250944078384463599269824851477998262383671542281850
6966671626620176160970567094861125093594209257672502
7370408358338931609750567830354428400003970020286423
3335323800308367275769471670632072715632881451354026
6545280537056163375657565561560456839735821127233493
0266571373761578088814841695460695189245089776145827
7305644711560367234013737512391133422213509952010356
1776430770908044730482689991779764687600344803644414
8641346346899950784555102030298886333848328181072699
2000889828571936841389153981116763523701359960477679
4327352194624941833983034177275287197321653523974783
6661598830001870135480072549967794815641222507319820
7773749493693405159261512147252409133124328535226609
5099178862420621470761814443651682059066937026972848
4430350602521914049775512614504465725697123159579764
2958217968313207204976676286570471311714659716899414
1419415585527921329165534108035864539423604397634616
9533528984633901443709771031718852633529786009866930
8669843526393418436970318857439062865471068519080023
8247922659706695796912922827797760081119896191705187
6577047154804076234201469014747012300723672006374794
1465248848800218697254454637056860047232674220196980
8221422778472139305509963930666588351205734209536232
7068335190537432350964241692802434924637529131968240
0703838920994682079708205085530265060841729064678943
2924890422666237149391487185204533360509281927220026
6783545090067219293448357187400486452584361950094552
0255338530150193608275455081483677629431734666870183
9896632736870667173889783817058514355616626575305893
4928379956883715661905117469340198535875250672746157
7175427929766541124306616885792146630482653762362917
8636344787320604811685644513319633775974521089142064
2621077337168716290579109047378376399934031761013295
8656601786861508413202599439184688026600919412070156
7367052752104460254246476622255379685622112998222213
6619496927805123461358938800937870064458889553009309

6780802205671222911732449621946663360850150959491680
9119443188318875572728807260492048422572695023477727
3625668174264307924072732430554923883140548639459295
2167346120593473310812267074435854430639051354861245
8469235277295559595081633912403448808461557483165110
2805659691260473822009624298786189595228730193832928
5962799349844728509583416182154386610869136544064259
0891158969812989744271470860637344088950811207925632
4343912160486709803726929822324855824995708943110863
7654840373752138369775403725350930868402408318659218
9475434254656819933192028697364168763807805594272649
0940543186086883587062399303809324893776063958808816
9278821723284010080077874468552849300953561681336987
8218813198796091570780060658751204136810551501389914
0721605454073209842445711240786902752955637176499692
2732097345339032749384224697177560429121963794004392
9139399369638343123810572712316016546238186080341693
7792323608064841668516700884580070799053990191389869
2886788685012546791682529572979509077814007758372919
5952592307789852900953749693144648856042018307248366
8164534888900616418487213140176197920673842538852829
6023984285540130893392381411757012080830155418887191
0117122035439604125881368880489293940966294766794056
2265648193922536371686301060079822869890537184707195
5248636144245298398605497226673813992712323269791352
6781947508154247515825957071821517477933308380538542
2525935868790011201769715069484687232397756960219053
0277291344179895332845717225695951393984500808185505
0284617312093463323897674396686784116298253664412222
3505420632363899786130116046064276425247211995387977
6525470989913875733370958754446068818103330791592335
1694002680509969009201681369502875893937714949331121
5724159021236225154972698785804236244127487899865593
1459382697569314305549817950653144186683273288195130
2751995672175382370571522522917889973898390630173991
7299478596473288245148057315766239727468127206928354
6859972049158200654932142637378536537776577631054256

4915637728001898109944412679472769211509560728023628
2848806019820301770557360355034313476907412760284240
7027856245018676049368092179666630506285791925690321
6092155038298157384365049568333881991420375632076289
4376846024766103943520229448101014041168409635222224
4652669840429640253001700640703723317193522076178313
5003574256845231020154486184741129462404963288983640
5515453802300065762431476684153339805267798198723759
5528463455943508759753081225407831152856459138266985
4061989075985920268537279230084147530346162184574848
8155548751802807500870114860205663005110795262621816
5099970462460938200305624610355311040342287433668786
9698965895714605263601339474029550277428860048235899
7616032607498570471221845586671322714100697884695817
1262714993858982858592146253686919607865356133568450
0757664674333108632471158007102508635704220042755102
7969222298288679505160390327842273887381118303297732
9682416042683236925334343948056151302774756556422435
3863128341393926559729662026384969453375872939469025
9628738874014880709463380659799698316511192908802511
9256028581730499108412164179968432523640204120266603
3906135644140131389322120287306294453261319831333565
4125205821195292932149405364882300337137813069933752
6267525734272054718259851934397122847549742322825404
0766261730855917977487126298870022667410474706114686
9802870273514819887933069078040518521698287223191317
5583775553832930641934332767058711857276687145649647
5004679237066771990706913870008711630394429748229895
1815094107419158382849830008905156337499709234458126
8411890557203901380937449267263259914733455221666514
0583847415931793774701388310929207297257480726892748
6362545010226180365457994969418363051852033485830613
8860891537475768145068904830504743231041756725444567
8677540565353246244263845011254015881133762879227914
4576999324549020710057193890243323158180677444472667
2317664560768234838627921390923872430415110888337404
8260207564476757768523134578578434649845829336240746

4801415979464521435092855444146566236726007057107442
9471480899305462524615053954667294574547752230988593
8349227321181401218199372897274783639130242229542283
2265812729939788697581742933644005462333979847923661
9185224022625456201012629622742562381753005177632863
7553954276860419576758487868293565801648475164681085
0672530738693501865443186067821174807067102386061328
9732571244782397944323926851468558707175932678693358
7685471603448961677611716341299667336450589792110754
6018301236333606911449641883879341218421949353787499
6652379751539176305915964610347152121465642747239185
3965021316325911230816528350294868195657135887476772
3962901916931991677686458730587552887475970901591888
5841340231494453516371582046119392953865006309019687
8114872520024632500585955097326566975940880924885276
6267444246672639849454940963335885449443855070827786
6043263766955889909423392133453243745869236955358408
4415785410486788230887659149925858776134937893933817
2395661267326458605886099833130238817706692933483404
8048944814411828434656375280816656029472988769848951
9112897279950597630327730517769376782997599259573447
5281486283944418936299209900241358060024233268552032
5343646493589775270690343598809038877648352811417289
8522061576657718974599693126904994274595857748900715
0177102800505102185087463337480281826723866376110593
6721215117891006644603828092457244431417385808314417
3658768637249757501724180505564815645888269442413625
6973792922559459020506132103923172279468628689561996
7419234007390131687938180418212831728509935919804705
9989085315255060611129029607455276541605554323462610
1670881285152031851635233054684501630978768688625160
9553921820193843043220857880847407848236511516852457
9205376362858300472488949906232202771501038935424792
0672721803903231808545493523002157032240483035751683
4614195045183770461312945022064049232543372082330956
6895789820800186133500506875378551738982296086457926
1346012293364278941293472373772411834710567199474988

1325566302417485511254461353073138250801775959002057
5525688293535410332940582476334595489119148002630594
1122856290209919829708922675202451182220011255715792
4073365057461275883823812394802411040449071889793901
7921349177985750287517112124497071036628890526745050
6250939260199653995467076685614565930046721731049675
6990455624866338937724735088968724553086783819397974
1498390808700712484440614848824896131026973307385912
4987903741870626001051421583904088761237036537135202
9294878765054888518285342824841003839855366231337645
7670923704477054434480282490886232209658664498591943
3631192058316205453101445784822337399509556817273414
5085035650280593064829271852974721285323881753240007
7625654889901114846834623218046070717582378163621157
3793933365635143612655788243001838705978682745348622
4103312709824869444225463205182429773306319378288052
4736277279010236575158387770445736380232431010591330
2522397460325442284253047639249936874122594125253442
7457191435790426198559803235541075551717960222573100
7940379296322417785536177805088049496315873832128675
5542970688659515555216828508491818467151197608790723
5697860364620410919720649300412315018883777045831139
7629187009238052681820978751665089521199378270494551
3992044479894337884842312224876272598929978954825746
6434857557926460375862242308540160622023123958379396
7906186082020738486233385006386037167953946412827981
7191131196147821670008308089507854458132669548181663
8625695092523164821917822134624141759133543797902834
1423778353888920595668461979810523192825766529383288
0911966876104818588964544244022054274896911206933353
4552351730720139527954521612369056345572702560858266
0648538391808817916070760661386419533907534631042316
3143461455149546839215254571765230176279071346702989
6411934810468624873109129058828693056154855010986571
6994317948320400204615028922428325398720403509193836
4542056806579819349237779739923616433923284412229124
1613102368123841230940085955594392123890261014800241

1843673257218566928685211711743895773556713694403655
3418826630321967371220477846139672424239343820655757
1014652108390411058025499057430927093659718121325372
9453276129255716033811599327295120079560078112053314
6361127222089311501471925179535242820970463835362813
2373305039579921425714309960203390688005783518739818
8931411813030607371891553698731169760498551140763775
1095130499113983932418267750678612930933434194460047
1092978714189848380110311027418843783292048283906220
1145924541976522126891453178721214513312677987845560
4456159595249754529857535222154681714898414511138425
0814105501308548962342436284979104194400095354198478
2940988043650418337571950530816933625465486685067659
4860210844032088836979178574562358881459948339271283
2581756647814893076301831884503175187702543381512172
1747380864741836917805118542713928305925515203696154
4265579274334095174087902928468948476751281828993693
6963800520966258096309522447841141623347063886751542
0150413175353850008132877162808104700188346896555554
4325765680376567350202672652996826688446781066574956
3509998596178894324430751583105497717634532380639058
2047731811620890740012401492008019238695449202136793
0242580246465176983105671485019785695873997555678750
2966173286707949082756613216383025793652727117235764
9371977298818387650160046960198076260717297338719586
4987099099687247174947611141014782590188011824536102
1522042204981976181635300339885999787260560375797598
6803073314752247564007571518792201644858905540902200
6921571006776359249857239955276320144412168491543577
8005526595484999851247978090199909088142554531420111
7411693915483314820573828041454362718046468662629137
3582349664828752208969958362900365962129406854850887
9007970609715361504930307097144793566401490311205530
8082409502443841987582416470860586489907394068754413
1101744503519464872331767462602490066115788861097162
6885032390425642166579301971423307771445355387862843
2514336021250189909452614454005268521377177876674943

2239192593559065144012279953236573878596336671886798
2100511224848175402924750136695005047596708441801263
7345880106622530400185066981810971904908813022918728
7636601125981634073025813256626207879429393773088340
5816982294048034393246806755200085304321405383703681
1967449736426643299033781535255022524682542776448272
6049064908269772138869837559244207237728578067019484
0754825024590313667177787679458280971200747091410334
5397923040702124704789597241630631679390426246365331
5382468027846299984929558297067770367361780912744865
1096131947837554214734726097785221422069774417333930
2375889154517446218414658881740633219251066399862636
9368800156129053977489178692275227186600000648688606
9438433416956897619104793155040737906994805414239314
0342772120564176101918465198316227552484709378875008
8883031786357307282918767303563271353874927186278478
3688027045784857087073472818436550765235117062567559
4045614751081746288865138894890196169187358664401789
1139650331248286282127732341495662570381436670394964
9642201742696525412503138667154925779242582471839304
0622119315529395247295659610883491077311846506610853
7822372276507233007298643368167640663482712659661133
5519405462150242795988260339343777851368966589207774
7703809123061987976754485388849348861517902298951335
5151317816944793898448055101604475387293460208105642
3999956531329438181181794916820664342298614174888862
4688910910401578383578904647033850624225450915261785
8172096015495904190198081724986333236532399620352998
3358128818444003123166844128209215710365003177932767
1655485926788764291194438852182338965094495108299292
6077565435181399883335003165944187490157382515153491
1730252921091190759492242574830715970710339463947729
0177111179548387532454308219195710971163136558511133
6167799267450022545674617270261926582280969301676526
3536797856218150440685524250206327743120789245928073
0831609544932281990578760422714125469899220372255809
4193132907397952984690749668849730282861276834244509

4025419733427838637133216298271886802877048859039816
5426651312217466506883903438805406628761324151473649
8011322568352453964459177880877465457020953889057559
8696204337257885828876643040148602881347856677752231
6770085281567031351708699368722839597977614270430443
3875367404610340962659814007959537338037660820668071
0192988098346461125352172626474614809062450843560918
3266233286147657178873670704326449642375583111810861
3532063045273135501338932584698039650758945995278709
8686773661665652729409662709536189923808343523250645
3540653636670081238084314212916084893926064757504243
6077337007697801048622686697327206131308771242883807
9513682325960493546269875478506976087090621394446709
8705043539971706828557773245561833859672584152958438
8095410036379769074832639223226967631086303998106796
8923416082594660541196578250833791005176430719600669
1422677808031086358949549126660196821689848727980997
3651882678777310532850591912303608390383406391113657
5293847387327396475869511903909612921446578009656278
2308548525764983192372138419858188049149140812202983
6664762563895359153095424421023811400942201054358733
6021765607515378419898361830584080950957923288723131
4482568811590029308704194884195222515577142536019696
0007116347447439350056845634437690033574176028255491
0888521532392506087844859788167278966902631720010672
0647207404594248900508076282339720351383460410116855
8951501894545922832586999099358318483888333349590571
2543219713999285566983682627054269275987881498938895
5532775219571619884675840926692371227242945252439825
2383641763866889502393648767029318151507672585976784
8492337458449430892319353106512504708445945982293233
9044945520691951947530503597462788335698461320222129
9389749069342840233552962635602769933229272508943020
2063972785105589058443007897205048804412923489474672
8867838915026228885291287429527504355650209629422473
6793464035255126954479331493143201237484386524665786
6338104602907702516740675225470354471781686148529224

5879861896172324465895281971514061410121927327569484
7104783728576946092451816916879953558040498412411049
1975743929251100959314280976592191588701463865514029
7759601295358470076861168292264437488830903589618611
2660684982818072956009457321865073950634395701253548
2591503536895771466916509576445043362651251199168204
0885348821839921490374688320638921958396771807212607
3278856195131355765752434134536657084865934159844354
6453769453590903140796550810660494682247214518383466
5013178906741103928036189001441926004517726990294651
9122625302754503806032488518836047936463224990548258
0495776019740854482383090288704159731204703163934836
5489265472302515290804077270773036185182519612016624
2493551339087586691042329502521962227426242566715929
8289304043119006795704120100840947826597883650327603
1030284843132564211053477207583503146038617532938273
7617103295213812756699397374441853371632506355839050
0476440504318465550507304089903892699362886204694026
9584431506310889786433825298791700659552819878337554
1777917628876197775025249067296380621847945014014371
1027157594365404088338258179644194308333085998477598
8917685225290274578421689690036843713892403690325632
1745803965288188275818677509035040778567728215280 37
4417828553924639303017935575994696195281418099816035
8562272455412759745369637536195570980535234513266574
6446826235548397593427568458345430580915983825089473
8928769210297688742998711895415173204358065191585613
2717351681114909835419221182834703657288674488155998
0262163347287343430001461760387865900136880255812776
6026346001449208170959073372666103607362423085988438
5252469347359660993796975691783492560369338961564602
4653120908008670272784774156103600087897533676639449
2456229509234066383358361325146393239122021141047583
7363815388273912015906594748848180494898363656362234
8542146894400775250800230777863764248086281910281820
3471183174373034933442617797543695377194812821564298
5465056109643559380311477664303551338888294851698137

8638751307499107404631767287595323655332912659476404
8735628441175079713904740055133239389619555128569785
9802047769760052520606805483520547158124294736280121
9088311512081693681709227961780504089455783599130934
7632280801514996988913569688136111051402341987841705
9507765640482463118885901480833776111552376819383063
5714319409642574226714354981847395407794813508778895
4885441300133174612510004307332540107531424253624326
7779801060578340073046225673594886885619126692356012
3298435577265000064743197514702226705282803889500570
0215205390805702685168817909665359775812636888591476
9419797482970011325857658785726696799680032089677189
9522050292034907180214016910200562865477296008692638
0272911146940147119550677284396877248732715812748042
8597810302450513289974410757541535707520610369102909
4313706923000275848953212511978468846423010680449037
3289192260583773886130912107117705877528703187465304
5525558154038878565241732495736977488760767011950492
8941832459182180707700622495028789984059966097284353
1403320074981937074803208342455636186316599741500158
1892543723584147565843571193494334859756354649191752
5117456083081918043863357646830719241410750040948868
5010605996185014204586165368969551147587396066060761
9517162625250236781435059299927432365421787159533803
1759236278898782949132690189601796158873699583536315
3030503048325861920935222145942082074130397798738379
0906406212477970343951141630488008217868194136392839
2496378920442456710399995289756704861790288062734238
4547397844728193551281160733812632791742199283209721
8939696111281497697228426437027540750347640790134533
3133132456496302881157135538112815499523882630906800
9825704032522248214189545731194710504933927455593513
2042065621536892949366468640905631095853659861276550
2965285049085025477459821566929512737113321745590753
1571802909943258834053869948164099672325312246083210
8993175248992344381194820664316769432057280142533520
5262187828721489083506561087419272164437457408832701

6759943094565014599538832556482404066112833332293469
4473822981475820213179753550555457762429632551222563
6139644685499789403443974936816101659194123565017327
7012122574966526646311730711573483666997513141825911
5846143281085787581342173761329838207675195559755419
5717239664222676454746520021198627822878202403645259
9026354617705964647041183230009657954902487137117915
6318995621081675428352688809117989183270015100894309
3350226550988041726814890881229128708295210976641544
4761833098113691278977477122980202133212746589852634
6719266245553883192771449991408341010595248725690306
5476732185557814100283844205889041468210966905433755
5206532913423204111494441007724598556415289012162258
4263525511470785955532136098809328950696448376187763
2525942710870131467907675027632598732576679119043428
2627051475245364942311690421570359226975853210231171
9445897478681723215611444092844987448122559723161679
4441503445868448206178121426891723482568747787662783
6168367743249408773244948988691063112600380478865818
1342462107208455447364215184728768745937138298428035
0920827521176899384594910962462437545367749182746541
8467347459270497448425473396257541989940808113988809
6178870114520143244990958695292621402193397300316065
1735918939332358754137954654082138809127473065435451
7356644909327584900617263889764584218459134590451977
0084034522599909618879193398402889543235627906082493
6894154632361167752940382683462355519428633099565126
3696800115648880489294626619364652318434098834586027
3058541333874462705453538350203975715425221285252128
2183013029933266179041222692387061468284264574806844
5855575824866887213502541730937126041796912427648131
9974611402653413800502024718139556939770266513648400
9479262014589759138181669048177231659380550527069076
8397746184553099126023123008399979841076047829385141
5988787247984223909143764543073187654518909532650231
0947742636863639679565102799543304550196299510289841
4290991155800188855761777033734674454678487614755250

9669944506590337393015003131608383205997298445703580
1649539935756873406560221992121815670455920865803054
4608103653688805197213057760336977912821309760536858
0630079515176056476299641062473983496958999491685187
0974989444093459835935232634681288025988542531965131
6359589443145420068157243560539801859680943714942865
5669441490034911155697973461530423686606894504700045
3891921643594562912220420454155483397818361155128983
3126210250769674369298551683203785418677422376337169
5145553084726091429391925640068094877398787536390431
3284521095778702457216803676542536975566933769168937
2159682642757537449497418005416994427085677468627943
3457412781934597508929237014499253445514353396620616
8461628694631883411638168732635660966844266181685087
8358302201726827279515949196298894654714150300669941
7879085694450273240803777842715034138537355605050774
2961538376582005594222095940727751019318804644245589
1294859364538378440501595920916918099209322085122672
9247492051440151722161097921221026915723023656934985
5124309230660305936577985029687173529133449972416296
4734825673764078783302942775073652195965175395117328
4253848099934666697011854131164051894952655495008143
9314826648814446464020940783235393423612349538986248
1501646262872514068441540323241030381292120284342758
3407715871458381042016886254561952959225028937503794
3829260402097400297659130668606595275532751069879445
4143399655516194986920940659152382011727678666162834
4383124816662520388365762541266998454645663056269563
0052414535081644907959147990964778200244813028767072
4944687100703971599113759347014971086244076651521471
9893306301860213050480860703451838006279564187122853
6242328066541232425929709109770029297212199474442622
8543308684127337911626984752085482300415690572101220
2655469803567154815773525190469140430937889334079783
5766389240794583105380933986171461332131500267957972
5818962290576615119916240999193263215059982468159612
0350861961621851409553835077370973816551296513520263

18952939243562395145390317544223633130657390191841231
14589429905606842925375989779182457369869473304239367
92337352111541335847620381814269245862206816051705773
72233329961192988887052130039696680004463367180969237
08853050360875843204027555756772967005007093445629219
87559009851741611710104989587970844695219578481169471
99891507108449626352546703160579177421169382207722093
23526005718234236239343941116792458409152606168641300
25591430568111117797082777361518323078893856675090928
09708369331327427004104019029461443725910813483549170
74104732302975622169328016927704919328931914111490886
97243890420492218758259799706756439265596940213811426
71289816729574280407646858336987035423399064120619089
25873490116640570308240076491226128501018729351002460
46182225683153238261980699821183840107194822899896073
72010907570340147326678500305625838998049807198497734
23190191747246598372272074210855769543250590340703340
50482242261430876280888342016729941844956162886958921
34723418576797210819805461815839733317831479781392885
06482582508818965761697179187141501706091545003996639
89704484130921254691232004758442453751004184464771049
34773844112579849619665499434168902643712951268056230
28208527087804399988117787752643841823140576165246619
31188606904051324232395317331395144937095686946093074
69314164276161660931859705889566352596023775307483043
13165254572812065791236728498218575347455538754768260
27002266247481546710531782539114273717207458271650962
76400991584750433634318265463544026404059732843711650
72254032651366553949818905099347937981985806978218655
02181327039315360049815369763298650553658774363415217
95877561245185803483099483370886025489091525828237798
36683487395570987066276329805725192187914536028557177
37671378299615779605186852177462472797744589456125497
43742831904815350809536033451762018622246012604624880
22681448903020869954053406814154385184939659903845911
17458690409573581466833603032345644628063430524441222
0994971113470241456

0239693118152077552892874878936396189283729487007933
1117133257969760228398317745577545861299311051810723
9694876579428203755884462077707173234133890415195554
0464755763270276533801079265904501991399911148631955
7132221245993272985911849735013224173775276724983321
6157371180056909299700801812447186896657877268005455
5115368197971988268480789452275424334870367231884210
3520350485308724115445938475496586200922220756979932
0338036937197011862068678541949525556332466091158655
0421706176765066648772023067335212871229572864628593
2926867706130769255514137249340019276522485394327005
9561964109070391620844032832798986619398759246467689
3911495031828083479399860466583725069502132569217564
9023887656152401852170122116212746858948178985028647
4477872880522352356822852035937224412799366379674282
448396437806808494923017209468877742518660795932453
7172908026139941096976765466340943127143436880658920
6992388966339923778147478935028094373660275900364624
1613009187349859596342804784736956735049651242784358
8542222945033800113262003971584611455196041542091202
3544705338201478038741000722524862318388819378019216
5821037441670975857353741134083939552285787470682184
2943745044891824751438878123345558634507762627021771
6576308624209174705004017086640127559230703925990257
0961549547425743327576518018968833828549557132288485
9164029072281974743908432916254992336031309377259116
2447164874142268157619944324616057295606007775572712
1477555028219570885987610308616526997434567736849345
3307531078550952036834215382026867636799048087551299
7682554833269000521950313002848256936888101150090070
8334010090541605439705010448801241231425172792669780
9424662461608953113615416805309768800790622553127495
8649556987999188853625530061132458335482857506369488
2255737304037192527978009324019657545394260194974997
7794696391673674187369879878250594371175061368451255
8358007146559799183227867154283537193419549062224893
5956224872350015965515958203602782874174520513574548

4097432753825755521956807347778912724295252854753774
1631054371227392203054066316539460792954280119497228
5986886262095970577136576745769464229858567404085499
3161468923354856786331441819221221368862997065403171
1152797921389343628299637884132778293950989205495568
4786747301183738455056131478157054163011814605181258
3181526609932668565644749486427236111006399020153193
4128825983210736949495291608627029779390422363625159
1408247034324703681924639827518952691022795031493373
0899837799279245439411604715911973315412320997178133
7228886015363032800532128349392282194279859554554166
7925851085320640320984885062781566162148812846676727
0416201689843688069485991007452575301982376453842056
0472267979845632601505549341618925533263566771752943
0341119018787056822355471201598295028936095633683611
8560837692702315069334026594230419562045967244077011
3647316949691061304972832176408469533539206481500583
5018255851030854903488038337481831944218813095847742
2037695226471444172459939593411907054416363079721417
6855938286411209471656201941013076569395469755996400
8297600535418866178823111020172715573022550595290335
0137402586928542771974669844636030298141183451832193
2934662119104972061197368362956703341942779167623834
1546392166558972067100211039085968668390528173719652
7813921345015475687862913183328433454758072201247546
7487682898189234688032497121578622185846523818179160
3387622054450261053527730169177366478088241739088445
2589138524041891705393013605620502532289090913146599
7522304654649626487270505042798563552499956894354846
5629053352838905240747682823469442512422052432146049
6911515995027451192115682913619877757442997508199457
1497877404754366466712183671863998593864775848049737
9574310313556106922648893323794188621034303009967129
5756250603303590322176674941553563372391168890323669
2739237960555947822231835025757695690484995783553419
9270800453710290967308048719319218734905095783279092
9334097901123050953551778787974257302591266151473309

7195021283664439457963959427214348479182656385965253
2019504389724159269468392524242186038072871266166423
7383521768393446568942250005575813151840308505972561
9469623569650725848509348137829283722170682289794264
1654257844542009284748645428513828372778381335135819
4082795357922839889739911377359314873156434126855773
8938282797440373940385808905475853764692026568181661
0003762245923078253690283197605974266134558262895200
0747175198661393484522858606709892291398600737672164
2745021841899068751282105164071420660638958780283243
1977581088443726138355462961246061281033813110128147
4830649250015655123906492637605631557241505944464475
7897193493099102668095708148423810488418579255005136
7684291351141125175793902226258740088453641633871277
5753025819623722715907425554757340119363083529254662
7694571481133866548463902123036916905804954067841731
0559465918598303503065831399068316976483876101199519
3662536138889762266165084667337594784663042039384465
8903846488347282284523795317069941161364108194469699
6724440746928392364130567933923015291910836075357808
9544072265784231508405883419547823568799736621842186
8049687643075313951636703662637544391351854293973863
9303504732241669528654384395492271047001209217380103
8357609531156944974648759568577239395946125495083017
9421864512497606909585532848141303982833219574415699
4123083932663795175897709125491779731184593992615345
2213996141098349106184057736741482444413357246529150
3100508044625750253745275459855153929486042330214582
8071311737531913940893517121551804301704622055447973
7669667478931730962714817804320524153739850169540905
1168830681253148680904395232324376023162252504388366
8986857066165306618190456852747466558716177265851650
7341550659451756136678425907159518526492326611989419
9276978330288377794111175014474104229080890236608339
1940652111393237267613597885389234282889008977305473
3631268033251710912983926148497000542686160299429609
6384886444060049102945503966529170234236346606504222

29602109878588006944215698577053669844841086867310 50
69630296090619342316066387489900188289495125615591 10
24044628205315937709684265648034603378173148140538 45
04499075920355090644590990909144525972567149530282 72
64521527396612218260413461475010286492244572657657 54
52903240547143706012344312177520657776455271900115 39
53342505418075391219150598379852615733705162295441 19
70307888743173998914579482534998715896943787734373 76
75161566395464762435208186531597510366689707775832 83
86505752358050171402323620418208538777467983029508 44
23383311037458065364190794474970628477065476794829 62
18866925906821760067313916066581607858328425138624 68
33062603366795467912492230323889295687728947612151 53
63960309406276531197669010911380487405493778751586 08
27573236354566858067490627165972159902992186708613 89
68951373706831731568009195548277920465055777750668 88
52118669297652110390778631844336959067235202839999 64
32396387686732036137916466867950311217912679120494 22
68631969522649415260318384601653704509259257880042 44
05944349467622855508545255660206942268773218946333 90
27671535820222547033122347985668270282524598801234 68
39537763551746891225367881238189199546378441270332 41
87373763325613042233271184445074974242237826619704 61
04011122653275889557238498505763648049409248040111 16
27708715552798840790361257278835912414460810333685 67
69783495825697333839956169239890008502935353429996 57
40696377950214085190300644954777473548411684055102 87
76110864198567266226787228785939335839702878835954 51
26358807626541463405148082978002699115811418605476 29
51288199430838665320484267405555510253322746134221 99
31036194777204952433882117850322504622922142456955 55
00331377584285710265052016884487127931856374126072 77
41141786989397154226727632286584861696775818739529 46
70929434525373350650056960134607318025776790991551 34
50348415839846752650573742397471474812540093578195 35
45973458470765007447097325493931035952440878024276 66
98463084264446386736380162865119392598513341956303 92

6559571167112304497992321173606848975917558263162239
0226224898462286579357279624910376403823104080481974
9428592702463212739006905074988773173218854689364243
8942096492856644246175264232707272793308155715588664
8416319531614117369090813201902738952842515499672243
0348902376328054372916139822725299083690889249738852
2995923775450126780887326119546163620900490104565727
2381218607340291854477995352899592119455188052138295
6261774080412733037942090367807531125676008042604950
7406190564878080502343735589161553899274656330762008
0099691290761006320537323126937279620073905433437462
2102589957213978891909703175501973559378692404636111
6027206338338409032239894189167224940658460386883599
1272521420089279274309866163507957283790514385513872
9527112983667896161918139486407368579422616671178599
0514038234729140090097922491302499787040038054145640
6429066379039225230449217160958905202817816159901170
4358289134091358894393049509659960511332429448250983
6510749851489015128230584721525802518501199967290923
0195341591046634975148097657585419544419386603010239
3289408190947927848078304524045296572175086034722585
9417577247107124427077986907209955806822251978998647
9329847328307793432374088524818024125993607628074995
5223104632199106829980811820415770202403786723523630
4158690964281330364662533949494024268686257659183352
2241354102004887850569974716852406728657519434257062
5734736675473650312557208374721707773173242611280307
7590654692395022906758282061276734984710635197644664
4271672093846156072556820379137520949039538816862720
3084219274933011968906502185542429917425792226204508
8519366428928772287755130094773646300064241099299000
4593246677353137444054866848758777157886475428825132
5710249958355230509904068164634131457314484939386983
3134777427352150116669767950323492087026670633342677
1215264880879536592993735189924206722511948073132659
2334357720796452977775486743842476892616068173391692
8124635913549460964501115658857292538244691171975572

8510786552984542787165830234121111553000746635255504
8245234541048573187791003476233058840725783349844254
9880584587627134845326920402256838316423222370976454
7140509123641344063619592205000038714900195783359807
8100859888605597284847911653207087459343563080218716
2111489589626800897330821420606706376919309905812237
1610501633813599391835931871303044926180189710892820
4261003161499659858596679590094337472123334438951788
2139213168961373456008847211834055370417390886550221
5538977424805978954172343536586149015842836947490060
4308895256061113875543858496272691509108623975548059
0136238040358262600530735745124871077414263477512509
7739773093907252823363210026480288229727064109520105
2883807867694620371563957604766114500580804771764272
8780013131277413244302723408851298968560383152465747
4367982370460127907578060892422712832226440307990061
5323141020797123800533820530512191173870930198385367
9081734700155973475132833216025461225027248671258349
5315739005620693532954023597171539469246442743880275
9614864941292475277337413495071487803353778663631185
6230615348675776706325490186829580038258788485176455
6308708777219208318388986225367791551404036491281933
2723464641041042088110931260983961840185584637023543
5946724857491530908048594183126096729940667484419440
0842598321754803345986174002004313785082861832997568
1776571031196315818659950490672379791787367219075156
4065343776447630621705766985670874922707502808767246
2042253538944741424134363110166436639699880878573350
4534672454066921810426881156695438451339196056976913
0745213394121929508769111217525555025022707891471738
0068542365030323737486778702558389059893630502100871
6646978571180516617315670294795858844262719841035509
4621510005467265326668617442935116210475673902419773
0966147242208704524111707433388517866626228389759505
2031288720189523544821739478219424439975899983500607
8041238792193065679325724872870197685892465232506696
1147397300880047710443386543722688439568250088571648

1309468446127095654313955754758050793514267563138193
6217024229731878481227768318957407048462335383621659
5868433085796285376160313676474784725658660926676629
2576087452731246222355044158659706369989276817043499
4116812797928212969226927665070866724480088023309789
2820355930828885978535341197850211821400468425054674
0254977141733336689523049168314443762977348272050599
6184279601382438264090054050050543899562203257694031
1410729669385541238618453416513571633768620415488226
3119603753953515796511561579168752008167453622412708
4386082271549504444160687673507152786736164990684079
1020504384460123082621949618692440030831429304478409
2563118632077751264305941995949177170641308906867774
6385439177938197576164468111066389323583618409159532
1163597880906332951732610281710148666948643961131017
1755365403322789044470100197631212341076463636245830
9773268932593277419195714992449855964697617304672915
2699247405557698159349663077375470407104034797497551
7710849312135392429329973675722029345975758802576403
9177569051215528023145210293130499653339373023200920
5054096856615040586803416379342521973861293658971856
9331253741457848278270086059892165403956081731874297
5871919106932757121799520873208269818635984600672764
3387648206815611367912284097379440355613871405973903
6867993047423692350310653421466641916361401970239342
3257526942990274693396440815930157918795154429053346
2614104915531568845012558555000913322061112701018443
5006952174617306595986416881093816304249760877061545
7839544937918719179778214050225790496722122077816023
8014923812940085463237140405297277693352142122684618
5978211744358524205910970289419846246899952324223981
6555904667713228763619601838487431946537298122963352
7406286439610417591624718206354197435685713007167543
6810628538358587361141488327923310460061344514130864
7000247195647216798552823676900196362140435649853331
9768880648886200204617790255802270949683423461083492
5724753544521339268873693931289850485958822328147510

3811888094604514397803692352923365109003413934584678
5041229174135049437061960586661948266564318216656629
6456063733475093289553432517497273791362242708239356
4558244327306600773155471301780840169319607522344183
9922700118976889848446350552660726916423783790080479
4161768433438322845853768413811825116775638388925839
4020120613567446919670757527804706934458182769157795
6860232495884116028959837840511525597978388184102901
4784762345089670289667288889647657977823069979881927
0883513493304458981708646817971112838182002044197499
1177215410342051654270270319998466349146952076159865
0684885220316328732259038150648355010008451778437545
0743614505922194240241872628249035634126526431192655
0304063202575366053150918626942304009333102482986091
9247004411577693505738700428590486697364230705586127
3564020540482677643880903258027885087690994809257329
0173311716530546783923878398599208449494428235455111
1075567141794156706197119864199564089819746965038359
8560001537763197865942684706532072913018346587824492
5078198783011235042612991308041293778671171441743649
4170413095998427749615238523119521635415716204185542
6172577328975427489642524558369008052069719897834053
2795249628326415693415613998572867309874226214028851
8128600190846032317501333078074831236164178261380511
7793371447041481211899591909802237467877728128904530
3579959340056720430564404493547802696346463971915660
0092636864217347411761939064145402317939063173828518
9036523986290070998309862522489234371226060985810304
0957638851092438950976336494774858416747664414121044
2699677685198498165067537953954532870805499052982680
9162734102724026108419569462737591630570306936049304
9551760291677145256732350336713095707817696108902556
9013133837451158173426087208091976896237615824723279
9696762659491769689035000181087114738574349836409909
2793369454718156290345287008933177864626001848458350
2750939465657435197966246939627896713327530010336216
7058028060203271745531773549327571281289722301694317

5659529582131377364076503258241209583942276617489875
7281848476302875186074619815953091451776753761422878
1538324691582039685564713187068248641469435543418642
4089865340226529793504253105307455647445859877651882
7085892094219965186703306605540243235853057412498897
1883964658269498141403272968040098567409447220294441
4410268192913411094218741513282873310798232925145066
1078279210121801031119704277690769430354497528305709
508933635333318510969023890698062935716343038815074
8514989827901710659364412707476112985186817773137047
2344540374560028621916791322136578270702407950996791
4180334821962558626695197436112488465946582728765644
9897715944442118234927551749896827448384642155445225
4335777101571115312646908838680103921379271888579610
9842454401760358927360092579861974341024609362898969
5597260191902173807009755802375380835506211199233147
4090884681435199033503297762255410262247665024469857
1713535726528821856861123930221223530219051627735697
4128634983540839032773548105080711638965572944766845
7921741120944840199655614355306898308612084948698645
1710667370769181572544604668202820526485233586992499
9080595519278288145869413682298976929469666575230382
6034310488580760551170830091971011984925364807283615
6976781877422150711037399536927661524524496535091003
4528895977702974277885415339785881066071568949262130
8277949265494362107345696787855068738556113335206511
0009516722384444353315866717983535953576451196680957
4527400042434015440049062666137962957470652740757837
1460940023326556772078370794501025269804793497599536
3121472488812065999504473245405295694916113543765127
5923571260142803889143997132062841954681968858841752
9292266886442621434063384323544903587183699050534709
4758449581539405452988391465186318150939736233214054
5096908945353116235023162645424287888032188619172536
2579809217969251069766315534866721289025535342766321
1799541889440802311171241078910639117913801116908415
6147104326412032435201946687819777327163070488510951

9004363450847138672677153789281655621250931993308461
7355227009185624196261048152846404552681817698323603
3294487179905087448562440525846670570782029673843869
9369354654364564447230044536227375266531452992196223
0988117645680676713978543791658473209504694430438488
3707718650184325692701151713494961736980782426940804
0645009225068239337956837176121539057082385854829100
4571562854052309447709233679921158488718861524422945
9047803961052354775294518432918217764704012674190255
6257847710396875503365299622144881793526405867813332
6215920913490441894123125521539959779526570682574624
0764954866113924814670065722771960249573450805840870
2213078799454350725560360218895109924837033130076618
5308525391274076920389698099676917560701484757577914
2527138022094159387748849404964683120257215335558064
8888199949024491248744387026964819045617856715182153
2142145868896572634255779575626527006292978459645061
7489642530002109224798180888946251675732735002967714
7050327998757989962912792874660757819647948475239203
5222645594063126283248417539754817457010657092602259
3172320874093274914009797818697026014374214786317938
5313081557116718016184198850077060828811446496753514
6217195521385239251104198410816414048220442494178279
5380173574309636944364950635187482039605091070892098
4431416784064461910936771361505772143004751642927846
2760612587932649555864471865902984818220160210049977
1983618037782118959565168922539341826350103713632115
7781214602370936691063219118839169527053631999425358
4490860239808314298724363262584106915366179150117100
3458273988450230019002557747743121433421560910214256
3004729295216805128221564798791314836123033461102775
2033829796984764875107530862096284647678635312023068
6050987180875128226550528198916999445054582528742745
9706340229603356853886949799382828014710998600870595
6848412151964733434766337627435135600524136353719011
4791202992225315331886602515377068331015488909995369
9017173694298447066677048279322448234182065674258791

4678111518974774907255848915546647004332922203989506
4945277075532932299426267433298828142825568965733266
7865756193371959390806642919147503111328446751487396
3577287934298975641336042658143050921348136798528882
9257251595246368667052647189839619635984921134156953
1986380455814847907178090786952927279184045229430102
9428367110839506348425063820855586539232765452862272
6648834185074874139385493074176302795550310290813677
0694302897012025317701849418209149792382963165602229
8361980348204057952319381326654609721358839304485216
6287214352575113723684505929298249561147422682921830
5794265334692259325226372031701919998447675715410489
6916082005560929388043856153939388610800038893141551
4251988072813467570939331543005256903783714451599730
2344291450569986414675950303181271527278942502383466
7263317333745588099435443950036775670310919094071611
4115405573408848342833535993192144371060068593660264
6993465720093603162344426382151424428391715961465813
1742473170515632534823793147494492006979548016468218
9884332759152807095398821268398936423181065489044804
8927684892185471067697886135033352449811386163739573
1026802516423133941154981062670497502013612853271679
4506413100084685100521121363995839031402261910807289
9098163070206354131115422666078695487177382021410547
4014024032631964563930096218551712914326524427459441
4418275602144551534404540028743249577705844940809840
3302725818096317965249173507014948466841572546592899
4947623637005980489039657631382402769423666262269015
1478947116904980254949025258782220685372604573303597
3090775835393170308887085423495312230318022928795990
6405017737334421306274314312861522090518089923909090
2879120512231186846024133278609750024675079171260577
4045752925643648232591599931664450989019146148028600
2548234387196903373987053600376311002385030885515627
3942623157390714593247386985029721347782260711680996
0060440749909341853400572171993409130820882341230507
6123526962124545508422240149861662157316029979040149

6699491727462378267301789409424959606884377389474315

1877042497676234221451309179368774606291685183902468
0313643989741696135882918597365538906457751308119345
5198291814912845765152299640726298338964187393802463
2247073639276859512913289625226756903128989820889247
1349570142944686822655595752774582282842997280977738
3212732530514169924655225066614655913575986554748942
6015712380877942847279114372236711153038274775334142
4240717948860202834057759385860474399509433265113648
7754324221472663136485905777194353727858104371833790
0951798179631135089599605570568421753214407989559231
8480852487808367550068345948393424783285656352476418
9493364922492977088581624231502049139804359044981004
3149323301059449024747184763989316316844434740475056
8117237464392161343183812685470940210927957187993693
1386000038468514368458211668002182820129991748092639
8909895665394058295888600282744802699916980106424049
2426581075921272469063099245607421661169418389720402
8455317073315231511142056785094965756120390076831567
4762721070457356617187840296262487211591769263270635
9138767846553619998598318230451455400969742685037750
1200832133206958860197305121870854184388061088381433
4713432566689792627702443419921807686310016485733840
4082407585399136557908117839085420629330681229472933
5597163936794620335391527090637206946734233815138342
6947141024398224388308667689896244157150644747905669
3233180852464295629231016005111864095688006805225939
6549137978135636662786584825404606873151675168565871
3053052115870559273837534786890808148447224684033527
7000011912546712330388861562966960158721813302472824
7879162197849116433286067222446352883105320858360458
0514784985209442235498471314113188173524407691698381
7007803170683363615105313355153309584662365539260080
2125833620966795554446296223486922443190847463919351
7496015959286204453799127924439547619127992304158443
2891273052169124081319234088661680918543382424029454
4610170882751608810427472192492094539469702686028526
3079403107069099130922043115964504298191085799442609

0015245142428334158154882276758200364152510502790358
0048053428092302369982089414594633169359389907001555
9835064469082244498619717044307628693201445951564848
7770125948501151405397979633930100438405319067907436
4712995847183321388711090998235667221263805486838741
8940101075400137881018393159148808747865878795001515
5396440761625775081863582179310117261440466688848182
1945604405137924883197154571243841934389672475514442
3665893647185182260718836387507045929451981123641178
0972307434095521941405244642200902221666865483560810
1851443171555087452030706681358203884352154987944960
3949472960654708063982456273947063974406405137892953
3572858668999008599086906703943434870800817607140504
3256903200881536947167076099967808378752608107329924
7949817333213953183877832823591673701258312428906473
5474215088782551498414142350130167267508217911100255
7972517427600739373192142103926032970810769819058988
0389161813825799144434322105291569057878847458211103
8266054989466937643057282009457407529253361163657509
3600636734985028438872658171072599410784930308130117
5864861158329066057327427060306999694824005860864033
2505107457077136203520688649462344119056316806989240
6954125655639819424718889220352243226097231754888531
7282667739103757237101067393376638089815158426760468
5203620367240789149018792735561807661755878181538307
1457220388177297187593895662961497505678534508134597
6592828371823671924314009741054372978445101625754395
8762457903820890293493150148104836451353888928030438
6036901642658923162207218966651935591627497955048009
2555159519046186067918808187226970297962202088800248
7787177912581021178942507276131979543109624664019772
0972322610026863745530018619088598256565626527936608
5407822811525155118261704821756956154162785139637782
4799523943593124697524369059218677942638333073373920
9231890347396557573966490982052618777270846169558736
1455929521981268933164433823973142384726159453088525
4088718865776402238510510879731017225228452997694761

7375824995263247141181730160201833111918112228135088
5698210048181611461676048518736999561147171048969514
5284969675812583453612978034773231329653596522390633
5742142020499047977949714036872153280706435798771212
8372358671861232848101428938668857037620723484479675
9291442195386422012108207798876551102503115293716932
9199651538878042516256025746324085042288449977623894
6926963152865729374134301693600760353293697064178277
8773779306501056973945611621924033667640309078354145
5138316434889449979237557815875014455206486934713086
4983300738980880568943460652073487595248873814528563
3118869663717726065392901299624512136216182124975855
1092460860670329278198602652523263228264835722732391
8383492582021275938940571530411273628188361027150253
6481549937009962498261244882629179867282497649487437
5237721882327023286078786741446467607560018209489067
3465740757676455255636422430212091679637261321178552
4325445167266178454961087379037460663294176926463458
5023927554463003063866779511564310976617009500805367
4662921490319228931546134479522051756730507822406654
1216430977472513044952288499254238165437964069910930
3726768644369877159674558387881782369993907291362951
1356585921262470605123799207098445306806751021923597
6258038209179599403622932241600021449483069498884798
4025711575886642601488261724776092275527824931800612
4546804432703694942805199693247405676562139868967945
4104577806793818986787438960315493080875448508708413
9693776973877043177428658169640187113187555513983743
4149480436738506910996892834084554865883150782980444
4292178135196816520654596288592489947217333292060737
0072548385216598658988907709883546595044585146811251
9578727298313561434878739357789016506310758644069444
8397384650888544124079935717189272387604339267541421
1824046241390905429387061282390467449182904617638581
7767365973869331068507357006883922614400496752244109
4393867947431047301757352214306870791138992751983374
3170803283017996889374427312521582135631706351219458

0243208618521125215636094841086274928276882562501686
2899405981589304199588164286788710762011095607597412
8723052213171260507160053682802813777234299991773005
0094327364174874089319317228940480334599855226410892
9822154491181471363488518083295243711560960510757651
0978703876312812439595878309461734903983475833694660
5791545220589591440569186560964276508524589570587748
2881204132612392858948735235603381043140003572891237
8562743250250463650074389009557274063694944618836528
6660543726152454172205829830850965936482841958696169
1157259324198273583245853831440167068161487347508612
6487519897807761879174968270011712481891846613898797
8076114957261378239841534622457347391337669061778806
0546347492480215010001624112121713945435621200002729
2737880906863891198335160811628709104432373070159783
7680528428190952153900899546814560480934772723899921
0771270883119994083118190220414788829239645428650979
8811164285982169556628812903107065000093202373656256
7548443741799311021817623100802014046166684899134498
9424122449701917555764928542399239724820650415253703
9228810588634119197203443941487016575998455338478687
6567773477091766714680382367313401639828407880205349
5468329590014417804615539401292235930446998544589414
9744390975240122334235496542515475897214184893791435
0277894140168852816498143955298295540112830729059554
7761225000882002512272561369373743769265949008737460
4434475687479171702479007356271678183376663649010405
8900490891903746456201786773831536991028400082287598
1974853524725020143724210479284879675997055778633111
1404474698078827521309539908080311709187392213784736
3226332007664951635051082589788722053445147150516363
9234558623587477446622427362128110925795828761191662
5415368465731854067726266281807464561236527057026139
9929161653055198046526503055454849917940375541108398
9139053949669149956241308657657440203839651663557873
0617890614207972461756714788535806936170019844579843
5091165510200227365198453484798249644618744625830476

7464826173612333863217678883125049127792009266884842
3818185480897273944618069227465490467321094749474591
1730963142655993105118144649849164657348892990956236
0569180745378234301662743162359500345076371867998693
0718629506983311087556768767520957403143485900228811
2378236242261019467416930842492958188229097115975301
2490415058393440528022350651030862184459727403529392
7698964695385469623728428517188700622734337169764549
9882868235530232764350376352044233031219185947358945
6042668354496126899092401308998337870135124523948795
3845969039342470609246933983359445142833816687683906
0145419707922473108628216859276547233642121807873842
9972932575872522741969211008889948349564227696402311
4275391244744324195895104686825138415711722663081780
3408346944469160547679329388942043396620864727227134
0585034694769837848111659256157715352284042764824972
9162557314786408820040147197345096547770301284755681
4101632861813732440985906856266371605889070719967656
9693695577718697000834782679846572398312806248566411
9583559466864633682410147066349690545555892651629622
1007684166189734544232485587531576541157523508244197
7440508940990094515586829666106709670902421885672635
1302207694616474196101248350048832439909968585541868
7667730961363628557499053167283427905552087494170066
7820958305569962494777334336350713831665009160994458
6030701991206362094726162816281543050557973001978755
1337883522833593019170150243774923829395637935896295
4953801061735070050621153794116136924540177933110673
1368079062915857711534352906341740273374652512188369
3915935520571525480391410141160760734932551286532242
0066828708349329470271123823044819108063670170555516
8166051617085257753897094270729668693022600648965741
4662072615047086137372372464532543782557590092567538
4226193084201930141705544579737519753004704168611858
97332422245538937939958889725739655857866212042113770
58501004153632473380634849308781705556606153341128860
7376743432071404068751768453010249189447882289246709

2476454535781147272683411981251579374793859013653631
9222853772590745949489165008906353103732934604820812
7547665544803346897509941733772273008008835138661216
3960876950803327744632752021123833663890757017365868
8340714324957376376131890231795720119331834684075713
0035560035891019144267179484864097507034891571557880
1216721948131742438746984840193683550642575388859536
3540645784595258026348940536848524566120713520282183
6091210728244860357098485986214991107205603824591948
5982982354366474772443187443625938524431934988597695
0847699830301144277916640235362440694889364336665191
2063300227069643142891417815651576241366067940864427
7393150876453278200005143114416938450213602453859527
3582412808228405148007249950595042158198022032119117
0380602272622853152778225079828160184848044638845423
8187782171045347616593762607297088842176998044060036
0950716305937923533727603256790842966377149031844698
0759242293061592887578809375455664416098881032273324
3261160955215205586557288084122820000875492308382839
8024930370185292062238770987704278101256226828645825
8207489374506573468058957357127024469933040752354544
8638482611794977490652298806639691549161945675734171
7414665666776155891302285172493749891423154404203600
8376023193268910997554626585144244130691699988162808
9074859690483068772415009109248226072211953662783562
6888806669171455147204065469867244613418392754666639
4924222757872605386194266203793909264926951308081431
6765928250487647210486358393805609309137306920848388
1377562730641491551091108441252688428224477997658280
4516708878961001559623254578416946031392298780545287
2155596749804085046227082625986099215606656776059196
4786619653501186103387301206818626988983175553856349
3334962036254588905835397049964181414344002208060299
8861413468584239139340153564232472142530142859120565
7638569356277624708771726378207859531814331466842886
1440577879484394185441455390118963375663753705902427
8787829869814914979048998218102052906879742490994284

3227323929273792293589493756346044711741346181187637
7556755317135332869446788663368525299779133888406252
0535576727452268759094250269880262302902959198007187
2715439050757092048947427880811216175370570817778606
5723273944368967597196132409349773380717748506768066
1100024970948317758504896131681710471576013281159979
7966971179648937493658018282931468853266706863564738
4591938163006141247385388455900458292731183045319492
3750739495359713935327341573558516210856392947218980
9593934148208358495693346918311894798014433218146335
2839107766288844960868696378265107375503636164492166
9180879049103339948482442207474508737919354745696156
4410207232332009042650394967045947946721723223141046
1977817923227968911516033387085969994751284280501000
7507936666864488336267250377590252110262588133484892
8731366640495934702802062967005570773406220297557733
4989184277808485032949241500065621022687299216940843
0237869071121487459408905047676091555973770140654311
5136419939101532216931977506053626463245629378601040
8510259722532517309265096592347969499224833117681165
9480608800082618923704059498077040017396745821892429
5489507168771331001039785153353453147106859570944506
0537335919536586171963149700601821601511678164458777
9937232745484370216719698324035215895168535513924782
0559128279615349166301514580517888411480971867069842
7226473253962785688562186013575514802584119125786169
8154027609127050741951140396563792898288285227408601
4503953521126538953883867793995172958629270947841273
9115243311292594640007261346492499726818109451514403
3606300528712881176345574744004968524233433988590299
1306423665017309231172549322356936462708264736539614
7019361465648808236638568712743118146280242967084161
0218758998609649692350312982446820228624101581183380
8695873215776018223789618015767799892633527484089573
1040846106853771869639844131858461150414828730502311
8066959863417933895317469817706686635251177878445758
5311149522528920915042815749224555913847113304271589

13555341114234956673240438530735724622868469222840837
9373448470602916936013993040950654596190030818380252
7282563082209372697527529895207663388958110039346705
0195502721525407078422203158690015995130826384841974
8777684991692392219653723166544757440764101009965061
7126279590178098819258884779270810165459373325352841
2802007575104825911652513757834010089518476852491018
6221173555348636969841574019955602865124223618740583
4090844768789715421259530355913270720760615089738197
9948118563835674506257294695134369167891826306650795
2431155826364371997603774446201879101613542447700627
3647656549196267616444680772190922929080866228112959
8849247811486715342571553676031195245475615652369677
0385370476886604698021385248030561176382068757506758
1364128628780137765140619072670023309784587766918228
3339165129078707608669375171168917416276908095517572
7969805201682102716262136069706041609156178870530540
7209797684930266376389686627360501139889883673359135
1396031547428897471385416969392931830868002123289632
9599381271653722118529514501671084320746942049020976
6700742282746672221568338634130232860911046024015612
3049845591409523921046097565436483444615085503749970
5955336660221346260483996012273273641910694943992246
5245700052391362273333832849727557393356495943155230
9012332305634341523510569628228677882109308461215686
6114443870760958284067151875349808971433102743967023
2072868301053055854740611207109035935222453189737748
9287212239388744316548020903900810810885266922772461
9314382529097258631124615779550275982207061668708048
7695005424198755620466992746639886393484868401422335
2872821444039547842289174547658271945439737217167460
0060853686206659734305152428332728939125107799228171
7140591722360207385278926420803737929278643557802115
8357573583590602073781354710814495049045918321175530
1545019836835859869923267993632243604356246020284705
0065834538669263343147898299146785381491643329373377
0647285611977754717802105720656241678062049249605068

10145055544400083691585386399196405583027286332121027
1592547804919890670890668715121943494416990158181216
2969164529656672888226173464699874380606663650125491
9352995156342670614836862978791948285828518127841377
0269402216755806828356762152712510027225321755011753
9167618570824410005697863552015457177509609839035001
2825128730291388517597460971546374785158162912190028
4214937284094101651557061686206462220736689554939314
5435182266250177371479799689206120653252754184139144
7611392215213461406928641935582165447967025369448854
4417123943334998602980270985027141815954167802543245
0545151692226883222768070584168571088248940081080598
2436388069304940140036054058825147274558825823285297
6555436600609761418817454395068351899430644076770973
4158274975433460741103268881973632510295986891271391
8327323940654677652057012685602158373854658756251286
7371973052008354520571151156379671524228543756123193
6717685670090063735425030945510191851196229951403259
7437920935437420911843155116834985132549158654381928
3995573831259453672482506711473896296729908788858488
8259970107422725250403548180496285225705208236348512
2693831562806728367147435055983927973260668429507263
5138275104197468897676725913690173065123657317195324
3101286956107258052833631589009582669841735396384872
0165200490649064035498544923722942741837841103308077
7386940393450466333220017422240536339407527505982935
7491086988185410522731299102223339766125332469073008
0270049025891197866095461412213341213891962638826661
7865561463442108132633077176078663806363424319091832
5419432274732411523770487756138092796048357649969536
8748511308807062476127325750376659107831658785768550
6619883979458302690860270196411220225736605864073929
1438772259890740735987644733155270684661916708199094
2108608514033280601787851430149967994736342366222291
8206903244530152501557857601522884736715182349449265
4828857179769926316352753845576557644067754489066824
8996613455952180097814220867268863056429154013829653

4685734899671972420171760414562999130251414447524884
8788595219222684372364532964338807885110699339769612
7782691625463097323535451462264137652139297506809924
7765973472612690939787221392807455856470999815894296
6851556270403366058762305340310983459322900126983226
9205022015814343028080002236580154632882552904706825
0736300630781189568719809407671585126103344019382962
7188086008611697072396703943181261309742471117179019
8574301239907540354716506288051264920095605513453336
2029263066498091540875817634021299600161949111034937
5157655887524827214237679726425642697736819889088217
2269926555304987247904798100065046843426551930595486
4635159658341932560199539427254246373171515176352439
5140380742442385772476565961524567466993162860282648
5504068591224634076966397694270136693882918560767582
0086694383375631459329417145519966697728221814706209
0620349763171208646603060567561493310916735968692629
9803365159852130088802404618168331860877929665002779
2213476011617481352000602598951502189632641538513117
7521071464107072577934971822527267900599337018659615
7932780066176911183167614715796464874106952245376114
5688792311508863036238407176919599890035658075472701
0719916644775922773065889611193942758602466572916464
7782959517792521728407860744329511315658116636186290
4473631555156558233734102054109705539810551185877748
4725531770169456197732740007627466291601557814989384
9915480067001588028727139647788555334050152578884671
9648020524672424493955186437255844099960063063002898
1857811273531940034448275340042960655950892253518601
5542228917642946353097283905715525192200671822520448
3456957283024858158777835437195126859654267081696079
8234661679259260735618939895590289542298200690113501
1999278337057617540712583480582780681238942856193111
1141325049331642605280233039814169006798898037376032
0995164384890586993696068307101231796152669245304649
2374632284627990824233479833037403445017864480398812
7559225902046536608469102137948423688814526296968662

1356647670201426725320564435031311714441377135146736
7169726545226546510614403820274351610225460111162195
7755376909012981481368832862481603668992773908339405
1471433848747692011642619481923964494632004463847468
8479111517044843919386765303110082423870498152452040
5573962058183162486945032953509198580564589890874425
6831229254345074737341519527423647641802819954873173
7721290125242144675404294585540242390647170225700446
4248759363876636706774796202442768094377854825766512
8437769009314209866607101870293385933104377328530343
7168895184802547799127333139633667250697624004443999
8428715482735400213622800363878357808619303281309994
9059658918893750753834426925570332089037583246284679
4994772016161849244806707260664239432321923207176003
7252313600260079381178885240250457731492535024973994
2544051399099724277891851189238955567249088332253210
7988627081564000322527803151097668662283346777383494
1746122589454209800002909329690743772601386909102791
4062598515250597766818440107816706693400075149348285
4055561430475539110533143757464229486209744679084475
8764683631789277308598851113550957947449542186298667
7496696159474444092731132006697861138085905453176264
9459016781969714986139949226972233845270514380938951
3504644375512254861117089139668089366777973099438488
0861933190910435786072084967307382593792396359932847
5100407272793862627167044754075719202503934191972545
1894831172790246140496689551790799869102537648909648
4598167090171393512067828395799612263119733149133778
1834338419318523675386629019004633433285328343418249
2107191609276730801323536947264848927644297987068112
5654628061172604607332191763047412150640302017893205
7956876051027752505304361222709604675845312738661652
1424194086834083758914009511413392954577009473174811
6885364094183528610997103416723558178602823498202031
4998679401740417191402839362105194806048190182451800
8170137104219745921279840402426896300533553685494164
0086392177456769534458650883286298216584601645078000

0359898521290123959007042026607449887850627176077622
2550606390745319477188925099058100936729891954099576
6335386583780980447935377991877121429774619764094727
2121402353265517772361634196607437639153866019450177
6914006240684127316295860806350676293775125293436759
6073427199675135201580087373954838973439901238256568
2907859114425882852136616920140299004131462136806323
5265086541121808474998758441118362909790891983411875
3766496048093626489329104390597725632955813887767446
9546181579787041915975583350003772867246073518535088
5495464202930971585636949785764048374150555809918840
2093433335128195514555185733348881649167694427782405
5543369773119201437225415419274505976013798414437359
9420287605255571893780411489261257983036183801077110
0500423923926441569227790057027947401364942291847933
6925354324840487802317774451841779583558257547618425
4957539587854945834308480441737087984926741952893230
3148958991600685422879578929481590006873537333333386
3813908420994872551172617108729440888684478508116345
5852692141540299672098869371588623222033148558174268
2635950689362344813422473786939460923256600754721256
7414103421849013118963967654732916742471899342433028
0290926985075229490797094430092386728774393625511131
3628761591973832413491672268261228263127793811750807
4089589439719002542641266494798017644201645415881317
6018972659624614413913192067850352463830657764016522
9732047089201146917133722878630370384531341942644142
4966498447440486177385135293731080998594728151117365
0719139881269307482690614039286422766278494329558828
5938372148475070778355119896224911955882370450645820
0561017020532484449021502294561765561859214574399009
5548229413751544136132915043046914179941224609338165
1362627902827884213812207758942403884062294331727989
2594183668293679259604224184594177065301954864394817
0335524102874704473117150703477508343237333116325876
7276368346858444865625391872394608273247135004489570
8680315032503867640173531507940034557285503700184560

2769961506156829141611706104616174082484626703351891
5255188248212726007259576567889912567870201497866790
8471066310748764673098909147197987925985906257336497
3498235031260983098734686162739358057907354908246830
4972840977323811670824915163734680870510521919176205
4169882625476054458177111799377677869654216992578557
7204263442443042074495489703390434507206119007697363
5174019526563221392825831205282400374669545909280452
5968798460814087071953425476138836335351191121441431
4850552012358138062649231353833876758091889237975855
1573203658831762424116916752381458859207516403523668
4372679175906370539219798112659770811394735167196299
9970520901709075898569163890664270423570074450277512
0104903948148329457443809746260351057898195407639870
6078758177407249118450697884299413834206081242839014
8814872599854181480292949227878324356105549156410917
4887706706782011985910958909838688395117183801349148
2549927491426985525951776536126242157266244889609614
8297970308420210516027967147856159406461363824775890
2011025199234215321006017523025742122375421074959187
2867518955215532994532268942518840942282675774422227
5528207615607277103947182566802471066067738312063031
4662847443386204743259568505689287162653290832787843
9965071672422061382945329166004638087256305952415328
8120937009922989600698139627968627956763519874182401
8562931984922623336431899309017048819952588273388079
5326588514493938124115432732058916457864229452684117
5188081840509569046913813443077848902114879712116283
9773443054846149807935624354967141294733326664224830
5004364454547027491723207839956508680187617327033190
6565795475395206929252015307064350471306881289032053
8545563987992101095667298287304794661005431635846234
4874415554071338143234413117884241960190370286869624
2276246502400420813713504640159934720532755679503533
9067127216177906030216239978041857633481005448388703
1730716255256479959999869353196813010156297097871167
3069649402749166572674943432081323146138869255334949

2941183163877889903259401093411157274298013318145782
8168791351424395578734205915614587736898808079170711
4551154762661682081777574724842787971298154823852036
8959165240543730524667287313030192582952498763932098
5731209161056135722017639271699598686108867606133843
6496169518654848604616848482407438381180741734222448
3947893998278014734222305786541021772926482091409485
0350132241515599990163071478148527051614432258254780
1434401095336602393626424931385229405475364168364670
4157430055837298470401814557100296962346985059977885
5736998021822303549238318797629894156977212363844088
9178927527752847144776218920574254531510374799777068
6236242189906303096282211773219708230347745455060238
5859114038989691666220778768326012396174199976236550
6363266016990661168908868774133378091951614977054921
4171181911014954347869382062098082787895731327899060
0208633911278471536843043814981509703047708868816359
7419253413412219139628745781099342483271410917827631
7654209631250713669263658821357511998754011579028028
5078066983338418338266739455368319656571703194629336
4109272667274244992747322913800983574450913407993085
6836048467786653542105866800521154264867497239537936
4215665358720818554125981945474275161184169366255632
4193591585695088905353141218336623998780221150424566
0015023646910815997419202544378736102267129412282988
8805336794395032940480496167711072878810316146773037
1502183528902414648175430954905944887541042501680406
9999819671937982082773346485581585910776178828497505
4685679139904011384562436040357440246191237694402200
7604214335733872603925372466408966491471465473007219
5292737417679613565233067804268200024230741218667486
6946280587887873829952327805010706610138540288011913
8557309113805727558936819403093807188637937582154443
5162655991203087848205239374935189559715815518280481
4807831310535096823567161235509631615286658578590195
2871775464250641298494510573035332433489097997209457
2917009587840832214044050147411438170956577125267157

7981709767638478642064876136793945784423858603564966
4463140131048667056134627690174928322900432711208192
2956415894815282655842219261890253051381591180342310
8851491226653917944817433904322700506874288819967061
9606885006733283536146804985960814028791196311274825
4664043703847828840079815791344523210166867374112129
4496813451012237984294021947946916648150312431873969
1196021567121142279430682010408767568093738982054684
2147856014277688201880717174141274936746781751322771
0904673001530946777306922329095349569449160116904681
8182723477816197512572785153556986162922029188474530
2698301566387491770219476050941748668411972766044408
5483297504989078258061539020030103521802717920995081
7839121233970070027832906193713368218336961540826520
3927161518022791528640768315032997406092804831050625
5157226287131327681824780116915493786248322401723837
0727160886419294211817185646748118569400870529961131
4105945829387033052862979188264932474201795512442329
4762903488349527229283080401222637227611027520380012
0937520885624064531125037264015399643712337079037439
5120320285350254748277980930302021966325093290944881
9057451029852172830699622421954710165193932976937065
4576333432433256433136391221083529135557700812650539
7931646814945134120751218835054739685955538780919473
7465662790318681617136416152155407745354380414798454
5840634474745085680280862324126969399311229640560327
3109279384916918793335961485351700181496982616480034
4027822813985879014862852128674617538485034806838652
0168074767440211476655696438739675796644295886409577
5591541926150719665373410817064974822204191350352239
3203692779239073695588057799755260190308143550849247
5981857797980298126412519269808310756574659469351128
5627975905780034193234760013696114472901311372932651
8877314672141074127522101051515586571349391276885573
0064565566635935253545094573789688358002770721080754
6851979021567766355960858995244927220497752998154586
2535929556885865521908620348857429443548834017661683

0553266024583599445433095297362056428294745396641017
5938780787579040143195672644502565389640520071487068
5370656271174667615719181064228046438268678064178170
1710145579987859492981028674877471679017435503993169
6835285921164814878612862891163759927684710887436616
1776239309829498234064137828071121741237834222414406
9984863288239240656377438980846317043398141901704177
0458564678599349411376710185104828834584759112985991
1326180631166520977109326025457776640478113979812027
3962790521780796565933483031963863726360548172617544
0262684357264514741114361974748853281352543231321040
6553747535609140513357644845126503159949939847591233
8676258998862826742421410656317028268420274277535478
5278498062554779673095621350107513570108143767120973
6106569338981128770064395086822635241957989987524453
1513268435771252695511656842606249795300564134236794
0326869432022127715528452294559060131663002332978618
4272936970542564089572247314796084227996064312115572
2898884134069015706079169855397062706949869226131550
5954581624066542762653609898824692284240221705109208
8452264193800405064723514106544629397362950062687069
1857435034689078702526688990310590307332020997707019
5562670800784214175004024877643693827049667540691448
9563400092985235499129699604654028321299512881718293
1733802589343360888639561881453054936332102246702371
2349079740681256872488335280182087968683767489328659
5156350015843734542714861534249712522615178086471835
0199049932178199155353873785623455087143139050384328
6965138769032905340152384282439639132568672927224870
8477969680927365604215169040230075017366194993585447
1440415221410647497242161523037246612553017965334543
5408898008690163071891588084555221984311574223129451
9670375862605877626333668485075275284790344250165376
4916331792832073572043631496293326217198412041496939
0021578987584055068836313189782803270681820620114709
7078246477602762963513245135225192242786044766017845
5463586895641834408259553254334088710161510703674682

4529220467808485318529560302603933699879125296023073
2785846937162703749144838891633750052621697332332152
4007087580984856462897791809848253427212024812725655
5150503057004181198951860921678780927093724148047112
0320927521638219693005366775084683589386544528152235
3475575590763802378521519460182724568236395490568125
9327472895456989344711418098131536495974851541068546
0978633032486429877701863553136745081107676386047334
6402957688211528898135664493094931043589036451413175
5241478099253505914813095314712262341354064609059803
9300729540969213694564498578781314849834710618588580
2583353800866348220745875475776920674100675717787021
4802490680993540340497808697475231821368788671920694
2034418621529900468668535749778710147528458215215350
9827098635277035756293849313931604238871689907398239
5618988722188420611902770942990322757366871860797707
0772155923264181360169717277740702884134571773206253
0298816172640649801780192696415503872932862487451156
6944009014691037146658952807690750630991658019040282
4912396345679839269565571594761366943873933300683738
3081812309420441592959517187192492679792971295029491
5811852694465448881606448909108574738497225266256135
9833739178319495991506543998379398350900517433878817
1829551618447325706942293753652551534711785761849774
7797331677160714994214526381704752272729557050017337
4401356029781216871886130729572008682704728150994493
6557613530068382030072487238918604555823405676318762
6236258904340796023955291440818688735039163729934716
2815765862853465444767189779664009434202490164341084
8630536408494091288771199650698045788071379779515912
0425380257091986045596738860190163441482900997297139
2282778638712630801766921370493497690438081956561880
1297909146507544068453180516940496248454952518633632
4222226145869950092979045626886597039309839695602356
4388968632992332376590811424860358035451697820676948
8924866383306879176594263820092637789031658270502911
3785729750974004938675358051632321698473633858419735

2048763394553946427392885249745167489904651459473884
9778743657589470771779435627074325722855361014929296
7565026585100600321814615361292859106934321487868637
4342977020598254510958054174455625081164881956344323
3555509154070228734506176634898004928111484860336673
5880522429802762450658064166164609330965832277412016
8700461864708735460759460023553734363668584630008298
9115932174654321735117553157555101863096588560274473
2881743617645365592624617643877740379976886080184145
7644764303571298399555691565594231174530659483625020
6036438221735440965535395090457436938651043129342716
6581052146477362856041042715237293978124871060598593
2017807577384367065631135068593058886083620234778064
9879758531784016729059787657028138758026032333379883
2030389685399914520291291252264234896619763397182681
0602370959759662980918534819012504031589962520471707
6766399868353153196608687529125423115375034755005153
6740686846995676773305775107923504903457787782633964
7387642905640164433316822456509039570066856800982518
2733058190424322611289485184918822142539686131003371
3722430186839559644173467084178320400757151241680837
5541314824792400452430812536772968970444067964317974
2681593592031358215146396789238533146916848019237856
1366570636094221329625671822214085376580706834616858
3496375655648152132788002082516961807984840474867542
8648124718923245727848809951784920023813214950559768
6097804392272136566791432558387293962885557531898513
3904712378084559407184483283611533615780257058364385
9337129234780510580916071579237173588327126164560656
7945034809907997205084227305474190778110649402160166
6662950945932469680827885873826896084877445611837984
4865730454320536892684034914234508815351016785757066
6874956376066630729363649798409105284008454623326178
1814979761412181872861268653460711665095513547783230
6274118114820717164665265427099148478807776892823457
9548611067150494736473997364806652995373677999340235
3274964801034460008563976973633732229471469730277640

9407753962883901258246629396938987701586208937991855
1693308463815996539591339262924500564406760974231690
3078027155970074441832783863920059150419266647287527
8979549277034150826391547896926883308166047956420911
2379312512452382628913239114259259248872353135341403
7167337993061711996480374076644403610827781509371989
2667899588375052554760929420018430744438964934968142
5354244228754826346725993997879597972547224411728760
0771260884093950481192159674887395558626710654903021
3240224770792652497769449552051080564104822068021729
2763561522377385106360209720313789483890876087730444
4339107037138713316374909282154921296609435599576542
5050659702392398001339719254398835335819326042355334
6407090728580486521415339035222970604712201304941234
5535778954712486300047055628894550446086222664852384
7310733172230385641865311802179351207883678931632431
9258064485005114528227697470288129755672304193896966
9115168403425219126668605613003845205389336911210081
2050252410879522034623660139062940958680638104861567
0349189528899594434218075363312957022412607656628378
5776310584133249700802109639513491608642051940551379
8037382651472598999021247513597282764870480302409576
9246243692551113922353178694829284218555724374736182
6577956003518901587687598144266576265825837773537988
0116858111611885458719903282377257659747925555015170
3522520432959456689504597798593133546030580462871210
5099200265391999022032802985513499055895296203508118
5191162393317684154510610655162478821980694146953140
2163854294846509930575901842547725876857641474091708
0050765864163376935814660781872244505837942437576287
0854540335011053709267102160168897425190172664771091
1877340196587613192382440167786694473160364332359105
3046392692548647340576221166484269938330305048174397
3119350053676164505084476304732563764216661398181721
6715311128910245585162073197007003112540505208238290
8931287575642187541678615691797009700938050142984666
8521200159151132720592321337362451009271857273494135

2894812323115992346158969097896744541306276268725633
0468233531371596513000406153320056264213843223148270
3547358635091988446533835684004841684653835147529872
6685210475900910516731592462648535707081943035235987
5504929773330753172203557347023651707931593278744454
9816445932473938377943652780031092775354373492970112
0308885081138511036298562059488346961563623208283108
4250287376136412846726450367957498216344694599244023
2038663609579652731250489120123599962848087517002713
5590882128244592123659726362362619317174993060278006
7270385204438694420239626594307173387544843964935739
4016544694254389183159395967500993298475531832019709
5868111474034114643077490787323473238318431697758006
9510475962534780392905312278972028317103077044152040
9620275360661687750953874913564865489935285108137546
6230230104416440607874337993886594343197397276280988
7255359642183688625252868429209407669353101746760614
3615040224106100657518385781845943097381666527422822
3595232656315600436346060661799295434268127824698023
1107831922175488282484434894215590509624968095516816
1417840810580681730090026315272179607472345797447534
3900826623840807048876766201256697542263247684144360
2996127063203642476725566239584136334669634201833963
7825422544743063280535146449208028401169350717192513
4344917591634464815473816005896456760840579801659025
5748976711877225395270001783834618076833574250619637
7939729071866416577565549223989511605044361810205941
6850541876584715902962594030767718693860546243793895
2917738297414846392692211185354790599095097299760747
3872962647276052104595112827435631270960595023755731
8061954525987653198881915837078060946073010461531407
9672252575778609273091902469192988451739871631507685
7663518649379762973806603165031986181720651187867335
9197652395674431888795336522908865200801404998114956
5766450135714869973922194603166589011920226137063954
9215000080968080493651561790444139936547321194227076
2716806623267993172218125745760915245516705097654927

9875792849661673761936853435468604260079278210505029
3510602699110772925833223313894239808716540504631465
8617867219949974969729059791579184648037035187388643
8202706098049340045567860930765382619265263181120218
6385925165692586594832744831643461459403696172358059
7179588829058381523697885331682301331452292999154059
2302274058550374062853590459765785964573806521600263
1591774288812480183076664637707719985469471432460415
4308710677882187690707969413136784605770183828392830
9479160915776228351293308159945269760684026594233928
8587110544086830636638581020473632094496534988776275
7862699053539155327911881192633510691931273397666740
9148613974561452545275688380518228526608713496328388
4183684848013626458289540073065079156074102889083454
6596253808696255262637115923594826384545608725836860
3761981706726337448128763726829988777229445404077067
8994016438125002755420169830230695451162998313431071
8899098625410160029355244154862591376159747870918515
3402568888130198921884456611164910632688832423448309
6288804965797427103408223736673828255772507677153051
8606164830633559801934071098971630988408746725910445
9888759596053032924889936385211939490968283710864571
3523456015804386429710353846683059689879680835873497
8576267855658883836495162723545207087795334676586720
4916172087981314914169891375664645456974216663388028
8709311596552908909685832466209423687259552095876586
7639645258106276095395529149176518173767291898766779
8425119216807552581600469267289283769650286492087987
8222442107140464673074851974371397605307432234515525
3212123587085161628521453800836861246295154803945332
9683773642486296457024358189231686226499499218634983
2468121494673751199481950450744965970432717604013383
5522236081534000479598193685223092588145754597956696
9475623894957247325248486607065428825628767359288397
6141961905913290839946916384100532188918694325657206
6776523238247316438079769604445729694465028343061665
2216987080261757570512416411109207357726203970411688

8594037610584331302931444829065955684296487296377161
3219507321997164144170256546894610127030824597238418
2836040311660348915371607623286396661656185600094671
5549151974673084232558676741097334029846169675642107
6322315367897998858068319892283085976337509294649038
7910743707740084634936236240083008495007355816496691
6759777865808111823047707424696436047327935182038471
9889611703590083004556851252805095510667699602373878
2353766372782267740520510615320189983806114919859528
7500266294554227352605814894880997940795238178862244
3301332364554325741038837042323980921416149632755395
6636176867524381765566251013374304648082075528515649
1167182834541304723879598488899915627591908215219594
5358943987391987885447878558839301953170347124025007
2072868196663188915409237562982477373635896293037302
7486492513691978890676358246153690757238118890003868
3408930397937519930653817228736677538511417161821464
0630053993407936721095094123328350571426528319496746
9485021225058627404548110939587405372888096652279949
1313504114857375818994915227341352040220177174269756
1032405390025938558957849154940687591085109004456698
7790296358999578580439959472228433554403913845515754
5905101223677094900429072516327250643891141494031439
2933816049214114829869615126075681193004885160428925
6533437306862399621812008375911431730894419899802933
7723050652250270984972059596353606069301554054235809
3562221999017615513369475682897390100935189886891023
0562033456067478159556373737243150513104638843684616
6022210580766055016338443951339275390504719811126778
9378425274293071714287376055437904153578146194855984
7060404010163365311893068369659397595008458513327284
7308800484877403931718396492321212575915598639683100
5033440560527898876611472430892073906771713344899959
0499556528668704313897455638641950652562633820072560
5325025449791258937286606936652747569545855493793654
9466905956264822166011090722271664336271478599464599
9926749049549615126423085297640404369503078388756766

5223227853593755677217005441761522757890068849222464
7258100685483167616870537047140293293794295759712945
6134524552754565682494397143313858049509737206075394
1257415629101102326051852161087629611346623796743611
2342517799340059601494432947964164600310883776688705
9200488620180196337697614505829217689137668547554193
8116074706436461835509504278367638447274620716138197
1191770444487975137788922899588473820082965070462231
2728062105126991039445893607890442906612891216488673
2932771245059557649995316456888885393740496576571688
0913039242348569376445019991202878400273546363108028
0488390398644628663159407840102917796877741898286222
2702853291914550355495243566744611953366897039280307
6336134227848130080590245232298645365347593792470097
7123725148559789622335055037140823738855989996357257
6815282498573230291020966336552980975186916429280761
9274983229652194481696043715488475908552723586440480
2017581425754005293438794619643196734902934702696186
9732830560212662858359414291417643270438839971403798
4862965074407226526416346342898291915406945822384005
3228719781301203428465115000214159693875605989584674
3378832151044581733491029314084926196254430106947558
6326736133960131549360027028822653755016913194817216
7727277488797907130864591251768962270234348267173744
4762540360443612383426636852909793473026221774344093
6100473438325947055617962424701479577599383961282821
8670404659473671346740038028868906734350560654196291
5439823233191663415772757772625566712728756776474035
8782518632401429351230992652418610536526239068980576
7959009378267212512205511131583478239251417189326668
4869674104933862880589331412583648730855968523648709
5735227355156417316000570435683375128586486798777626
8789119701741989262975037390169668114553105639638913
3491947849112600288727202243629554113302103288831252
1790386091534116785776233880138563643471230253133127
5834921496268168434618056550643768648443216916009932
9793588697132634594804765801673028762362637540464177

3711712333755529076168457609841203148149067126544788
1308749669265249507283763395824831279554241699844491
4156090821234201446635611514398786928367644038199996
0736130356587640683341102908782368523045177162181480
4943262467842034037569121810020485713346838603164091
8904933187028256511334225965139518362851792326534003
1625303117768583043905855303143470009499540428993106
2006909384298598494625764243642747550200929598219970
5371385675402423995822493614688183578390529256276625
7814762825490521531188451672679258510629964112189047
4290326898290300195391955649073481812466843389938765
2291244231198145746620093780807965110820809202500372
2176564662265462851678069756165122984690402876381965
3602313563613649766389022197923617233885491819809573
5229701423204549139476002030983118265134708319579082
7421779732291100149811042092919972707939131054168843
0564766806828814326393122924847528035155632827952732
6560834108816169796102396306652310043702682309153399
9011018517840853479666457696399564386232590725663075
6039470551358975851270259376353252345454607115787656
7775171323140719849308707274172599038347990837752226
6238501618900013696653310225295738651936909936258963
3379104117758605187819207087776560999410980515517605
2433838149857884415802148300377380729422205761142218
7341912017380974191603908969457081286919618687718234
4133397780597593991704085257402459527953589551683671
0461224141484882350704209894946405334610298148181498
3849628748546950040743248430100423702377026935949778
0909390095551642812541683795122263408103402400601731
8562283671178791286350576512210523464875470892286547
5797991986634913433674573178080001015922803245804156
0620405881497336905468838305683111885642692744666825
6282605693182711497110790802221674273725205608523980
9705192876172818294917079681084956342092215680052226
5765905027081010596468960041915941397134432079748752
0061970362143653107681949231466574683330194256177258
0296562871057463566793619642933509041759154760564372

2848231520544883264267630871114746063603473283432300
9419042086748123219633559808929813901374321723190294
4033802138350378466582478492823716028457090185093998
6308768974636121392228744643728222306898144271080876
7398474519868597356568324758502379022904387433865650
1635430920494975139465171204239963804851524043827140
0499873660922159516794366552097162467190287849723430
6879414622093306540351514653335343817621214097499529
0813343668874219608546300562941871857198979654903892
2609940064581345769898902937525260315947935205172873
8371667007215479619563757407012855926075633358043611
7856957032715108609733266002110088271231991637593409
9712303202613810988996422209731755274142553634130615
9729484372048567305363345356618321960256972367743766
4041985539084586654773133442430617212600862976052967
2078593225622923558462628463331892928319836100025921
8711370399715055774932048759783197671362863201844282
3365393471636379457127715884098417687777055114469465
2404220885382028249892963950846704657019347597460108
2109002237981389393889173661995870235322219629414991
3960484299279341165867048778571113372653492856653689
2887508962608405860497308118077965006010068109796230
4559548011897685661967624238931275652416559831945661
4716758220572051507331874386499592077292171541152707
0251573742932740603038899701817639249284449614498921
1996008741455920119711061917813560083117939437002185
6788080126118115799464147173035479067670391626448036
9152300163809750954986478577153019332075870767280738
2416272997985657219816684445193115121472153946260624
1547478844925314745222342898144439356689652081728352
0184910446850123635346659443910177128605907892320369
8177814414065765795314169342756627442756479249572256
1771122441508205905726084814714045923953079597060427
1724592752165506005137153967772426318259343164959994
0848813489629443980192805044699030743329586964202347
7994406085552980260168762745948165247001320475095699
3158251056595655451124681216741044189201497129170591

1320464869299521893607872571989543326870276421120797
1112580634371790702853656868188130649315290358649123
1366257190281160804499253317309331886197389137211839
6348879203809214817739751512155506071421237847824951
4649493285850008995173231334179449195719549373810225
1620433558509640442999506549482181749182982098028501
6135764697993067913222478498512958741989762340204826
4287062020511304845432074564363054682985899790358840
5564610067458997230028648958952414533037922186076757
2196030001998405346104236967394846598198903322644713
6417782885831861564108998106333256695880149087548911
8552944475487770703325428165404870114529025916253351
9801238741401043603630920528139960818578217505919920
4440716799015338445864015420542276603949509852531714
6476566588668145884224627681547177489770602016193969
4773855297795179206393425938196355238038842483958103
9275670564005177458415453647792017073426820789537842
4193187181125134001913549075223018853664245615188640
3026504152062258752697464483059606000866480846739758
7469403100443050711008032857561852600337068332178341
0069730737036403212982503954591566451625180163585488
9250188649041767130862347416451200097805267165914094
5866695297194326626136275546662763974798831324207660
0247082565551686346541573227303798463349743541760533
9118055495848517100306504714657091740008220220333420
8427162102550699829516696232279205206790124999105979
0172386421758536906421361877237109133881499560018522
7362965360920637316839784737149841162314047954571631
0285083096492946561428907773603095689141375737242200
8431953676737207516723402117432726467077757072056644
6538170861043049130199330474798596351648879646194492
9528905825730979040950627720666835694427988054961983
7512732746507601821215205334922080851917626617083613
5723334251006968310417820797186065027973257761157165
0724027365119908613767038988665032755774389422758156
6173076972183643680194219584768114175757657802259412
2936937628548376150113712094447772678839072637650618

9327449474109264058693198959058559969600972046356710
9558642594222507134228108316107878545620838655268622
4968775589567428400965170789743998364521940287896992
9676885522509924548113167981196095844999451283017439
5658645542534924673319276992922114986442884274282189
6234377707149143857760805808485642730371713537642453
0679378694579057523350764388152181065660792707515725
2883991851571052600561883391085362212756884261536867
9981807360170875673642170133324263530619346354543601
7260388552556774580621314638205488997794379948659252
8073247712770153356672430512874551113052270769226215
1065261479813961301205482595539849829038221658540002
7871828179452759345759816062682566935359109197025317
5878958507864251238169022868440885550462338617680937
4387125500306276779114460427175023690482605349682200
3468062797939752823723519731070856662324732554815669
1201846615653934486047540357405735326497356594491899
0736160811763133524918574724029491422191055346694230
2866373906308379579526522916582302609976507409187334
0013305234310103037778507875206351658346277844835368
3665590270558349479612983505612394339024757396930505
2295130811046709132498820483153789422613223556634309
8114183098668476810282571550286827483833975719258734
7401153664671682629370155268224421277110523909179693
1923912773797717688015422468834962014009973227575932
9391772408646348928364644435387618113834143988823534
3530623717836037727092220679014913157494612352486498
0368060944848857198938411919684715248368996081472224
1875431733533374236949008176264368931636788534545428
5713963603275099228957125803241463056342872931869971
1759456583631036168351573519241152917458568654839747
0920699815889999713360367051251783221658704694156520
6271294188303672861697327759547489832700933761709103
8055159386077975183164791131510918328012752515936515
0488607939884858609930811934938012778967091084784084
7094315489455770550455036182181154602650233359866262
1190754572843333406546559236526779679009724022559037

6662770157178745301067836845974624241194146092702551
8165357449640803707635921820480703170780049503051527
9198046841480292623759144109388211758454223002587105
0989743405380196080967060017294461315943539319021552
4986391770306032510939596603820623564441379417884276
5240643899787215289854873908254194811268497400341380
5877029331600234975233143783250683468323902948629612
8917710704498770623085845558300725385819369995627494
7316552089561794462764776882775111576147049901948228
4323868027347074673862814752675333720121010716466861
9616006337713521378211388572121554773972169330440194
7166534830268285098055310036760417985335910839912135
6009460421182548283826080594511633337459993309854009
0956755312822504067646007341266354406248261660563576
9125279981301615925942686486995004724775332770649993
8464441139295589691465220429908122439062860004650357
3812695217022563921682177336732020915867504500478898
1687036460961518460048557414398817260973844721274644
1954501983355932317688295578306458020898295276315490
7355005465411111371947630436261732978590284651319161
4666551676635064119334575866713781810984156465929623
6576908362716072233814788508550638863177490787234919
1987679896746362545471260440954168197799220710427640
2515484331964773905581049610001647059895157852943991
9179851978060893956786471973638900985242234000544223
7631164031518642793802179526568429978142883832138959
8243939588770155799078394489206709170655203496769986
0541559021057651477157873786751371254730446603349942
3963787796271176836236896244596776440199697869302789
5704351674431509802289865012877454039340739724267160
5002558785498888940993842091001681738748983584562893
5275170791170549000543621962170642940432780627987378
6676730222852018370415921352422427596283750372997349
5914312511292396853190129753561162967230848617432317
6389298817540247039152280625468479103097962829343042
1339402288412931047594043460835668343232495613647542
5798625445544988963516713644459379357350070277317673

4884517488709966825081043899155677099848874017882990
7494844233167883018060597371518899642950324199139880
6477821402001041117774789688395551080742111648042275
0798779310315115110843383177288725255853668013231927
5233969078693015925808524191590883768045362608975083
4211044419117285563214646123720291167406146603573964
3190133267351138318086854427530611568220064205077627
6243163090763395746802731529177990810125354943612591
4719887523866639937485253979788879307769281320874570
2729612199390812546255462541090083430520403871083838
5722642176756302904930176929511528857908547895413427
2967908455152662497710743422153765699549216931190035
6755887971969593608944405602798532282974592300572929
0225019186904996051519634665873845766286165583709262
2954905255871509887801915359854417167213573070063569
7462366450699943859586530458169557668141518863751672
9582855194702568148675216121345853932624252174469525
5070076927008328053505369516383702482620563452119086
4360945508786374555688416514747774466068680273209052
7456817539515713203754621706982626148475907106925055
6887184036300053051386810957538874806334632999487695
5374348108245413109810837369070751045263814615713736
2035056374120611543698395643113673185097174963434058
3229072358340161829060871658710160130421591917139023
6458970590273481568107228986719530257968376456398971
8769225175933155867973340773733043216755653910755754
2389161607021227173759207005630713684374782121352889
8979866821776283257253334518323039274761865403919208
7965652589286081930704035739642080989321089432200851
9941676117175273539788820266187445622820220531528314
2784830338686027996517517975581988333102562543823041
6018942339170386337835792103203778668218061890863264
8674196620083453772672138030048353496696010662241386
4272529253000537291773706116234865438736033357315122
7139300475257703092179902691056833698681470070046680
7131900900638769068942035418676510749733938247585986
3862891901184985160567627206356818364125792244982802

8996536186177766867781410290398847347627179023486383
53671988418153776428435790683635879056700585812134 27
8486017342992731149852582942638130658497932171960271
8883988881367210302529373730687284284252676818495032
7932957470704537523032086408891882046920257811437470
6774210439314645278658963807040853074482407805394314
7968235436205409865242544613096095689729925320004693
7227651926452523471043868061628441502632729061257192
7257767631402242018562351489059030613122440819438963
4712598106722230807362487403882344642804839175997118
9069304948686931188157335894633373137830646307582206
0360727221651203940831736077127229108945530997277130
4131061415677302487550303033119745936782356969206929
0086840655178435506990979613012085913420825236538319
7216400257843865315526851804559506527621829780729270
0162623807539845317874932145717674428422404980630487
3363455955714192165559906931601168545680475785694458
2581465254105690942130415186078742436405050757001169
7845159928514363633141919856220392836363769807369642
4802317505295501373358979603598744425903636907172462
7520840312123803089261318802320710301404611955903967
3478322147276210394285191545150346992392868606087894
8272040131551785488924211858750326012076622758794866
1078061994316694602312466700366406945698803378941916
9279093056380077125569611138551110682307186263508781
3016591596657199561663926952401328193712268673839971
0231302371860614424028167500027852370132974280057314
1233771076730526300291885432184952037726251725196654
8985863027520198580555286556701741242478139844114794
5614036133723592910024028800169542825276370017260558
1740844534308051145690107092006853751074980056229956
7939370360942165585240126826086201398966399798182668
3814642688762945498686986956018358721332468705175195
6171604085036070219265934907270251099475725221910842
4455952078308514342748979583140906113813687321865624
7805199733098940110047570718189852293784438141434541
2750827098497919645792920802355036436353509601251717

7168056554998278836687203057953323354892273581439095
6051222929425451105961566598998015880640054229431876
9492707622211102847618082615964466027043097290549291
8095775775902696247824342719684252106673708953912879
2695710391317015524198956659379628884094286905195234
9190754939683374338510867886831129748407742561428880
2420254564707508574033953987674644706472412244050841
5745987731692742806579384510808293347136974573171707
8012150465560773087987578705024420182513066251328457
9379669342674659176754473212872979925532293956541582
8683586256392962270161695814361047964633701681690383
7002557364944013958190229025904302917933014131985196
0060539395930151180348550630214863817390059278659377
9628397460050166002456192425055935165239389938578583
2922591711647601833158673958922691798677199264267707
2280445165745545012821713074850073677934454702487148
8371884766882437818598556832300391829250722212472339
5438145081249592057273858216386717412145544407200077
4625667799993033885814383952246841860609946505511749
4931262474754541149008992988802447511714394147171562
0484316616148349019046001190929625568516287760436851
2092217653725203066326102792604571211234638430897691
0057607603820506269489451831336812975700849465036278
3044503424298855803363561984544505418523948398908251
4808667971595308723718732952461752619649890592054696
9084045225197466547763406553670508396129526943779829
7198225507400872467579607375592988511922700401408992
2309976925072908243725293025365545849629334370195169
4483160099816953821753975089393183088183490256194526
8977264011060413490804531314501071393769710546347665
4293873327884965077889115704433899875968620677669411
7023532572196970063355980127679632173540570173738734
5600278846155575539188888190157937805544173152124711
0485252795976660872618979299145615755209797040480867
5605446942831227454025633219115710445542963152252304
3608263630844221033568337100347482864619734312032202
4724439322933802978839273166096657341966481397117329

0576318607759419141898287478329113668433025352925249
6479211019646490652178242802165844480119414837560870
6264682217052788555866609397308492117248898807055295
0041388669076368399430081887793480377554118045469519
3374369307401500438691562902746936145881457045663729
7627949440616093119931117418045204492835456146997128
7682350175144537389283837680072041680696395364925057
8693080932586432379583979185083667852687093943396098
7834815131423664526254152492755877990653256163320812
7186353644050491018131486487972806483608496665024892
0735697536217190475720554079269663844354262130994352
4325188309713530823407873141549480966574879196207042
9813217624378108179660003627805961686458583311024830
4331251029352663857765397147351005221181616567263513
5384136775558921388335613754616690150611753776496372
9764127572979618546006530588154337466463750246766187
3682860135938997966104887037321299898807718930345582
8424798632586508329716478813687836934043158699441584
0560731051700807409062423229834214836459375406668988
7990245296775038063088642462354000679205016940257668
4226733377623470073850861041947110699435804534455704
8916857268425464900093712562476105936696388997312784
6586244379914413995156946681083926325223181911728640
0745587634104562885751255056808152522939279259781486
1747545269477638044769959550506070304057230748023470
5742344672410312296534950506517011654313242852197592
2325039634918024654316346612918022256977140890212339
3287886041739410135263115203691114539200387541090012
1740080037076406806582462780508755720410542581570457
3154793095847220829190461794537539554705589553490462
3810081663731945502358481019922271929120161775202444
6686059009664014265572476843653318670352206558014589
1365214214884279565586627059338874918786393912310949
8561212629941292195705509821590414591386112526656218
7945691785864140841846629181231277170185607642984140
3475924859795364139295700295399600044765241741180636
0908910703299357601235674502894896672436831132348273

5638137007848180922045444878697136394407140683810853
7502379067925491990743545391118754416978797744588070
6127946791026105972678506875496861910266428987266041
5540003559370620961468777480211955901297443475619036
9015032298765074421165940749484642116358360774214267
9643983102756815554612105716791531072217779302545732
2874537292989194954644443021474349443399152619976461
7560500446876141445197137848176211789772414355467354
6702425073478364221897884302406489528463143881634350
7529653333478207439874478442439483437862158000529594
1201695844997595662195314594638517065744940644541377
0883853227475395462267472178164107859715062639482911
7437331691230877947558401624524346731607652792303833
6108403393227859230437069614233618515120201630037106
4733923601594093487614139315780147375520999305714396
9093614178086869200812729995012689406338154594261804
2524870705529369512012093385006182670089611255740262
9284289397798199536585334676890135251331654478486211
3897579395337638477043789600054374631526792261265540
3500132944795933203899504403691234100895651264183332
7189635116513065117671202257937290610148423524674378
5474046961258221106399832604515812547546779293221601
0039689183516686733683039313629853299887287731336514
0099200771811594958529964781866067889568730412979681
9333868640365429982502489760838987278799683224799486
1290621538119527811750653500768127034638069985321345
2396070408503822031138373124673544085400354980559889
7628482182550298417519242138199529518355344031257843
1076669819823628904985956993976147201950407415341828
8497163679555729981157692899029453651754415265328609
7253112470609774310064281021572319914392872471828409
3996741383269759137019206039345242446281820941620536
6883589174586151994610289297586116332625393634857916
5089549747556108870430826578335339548720604361931381
6874211841752981800450246240132138892154135406316330
2382161565464533991219188665880282830367309947128978
0816548788972062587681901474865764326474554358647966

05054419314007833058157665753933917660149690172511013
5193032764125437908172466899998107870743298828606321
0542872923068461896542162578575560161255234030058019
7329431255534421488652069980970043014298345079053809
2344582436794387491628148340152942878299190846211322
3908762350246378918770122675476308791945324149114109
1253487734704529022285676973757021670272351520368322
8144986530301933624735824640022653017817862306131826
7347574556271918313859370386942322406078341585617375
0148103203795100191326222685916207583999928414258360
7550237036751437347250061084565212317820690269323271
7071701807175007667107024383008821349562114209666627
9257629475353656122311963034872979923961295610014411
7684309914435980655668473318377111148982516686052882
4315829668157736742930175053096196635158767553864128
8368646401680329010209836414539013618201231060000077
5796607677144532374490366645381903303565964053774848
6377056891952143606796432870265114198205871409962188
4954127590552896627174371637870154631182349958080516
7263878986901169705743325252330668811410230906371609
6539338291754247405722813182485286562201408429483589
0305433176866938862725299013743012988285332041492596
4724870842887086812156215565905552011876699209953492
8858695129349256205885299759459814735055255033151457
3498130249635666005715621294128820162299984814529158
0272745294336713461339895499251972614295467134110655
0874868735678990505261841864946700362615455651785900
7474346830220335306212633198282054810388329437727848
0154352985423406909915122071140819165328421628682826
3663561083432356662184583496584243511408252713418446
7043885460564089153311183112391186053987811676276761
3703658428481807313920056249047162328232991080660606
6470626403542859190079697347301887058565429921291410
6508310502492118115256100637422924329390473242858590
4369109978087368701612715657608625272563041379461868
5327159089484682802764387819688303469013986309935348
9043993961413138596462884385448545178614698948363590

30250233388861386605788253493820429754205348196605 39
43726434779791355298709044097516714256549173418657 40
28331056378689930376561815662351820475549941943497 15
21548912142520647079290437621284540363565400426443 88
58799154465265045884415022083481899808252376263783 49
39192990877418633119502333553507950955019850442946 46
00379207703264723165383716779763225510461508344705 72
78649843104069195521190290898190955066437540036659 89
15712966903738260586886220943158567621416946277003 44
61465588370961454183836781653264671798870475162420 02
76984105681049934797417474275556623834482187239492 55
98724808179028869791780821130782667388227175699536 81
58100683494417701411035314110709827967960457421557 34
67737293131835742900988172510707001027056816105909 25
61977387252629549670643826974876315270996332315389 78
02141958462837023585522802977580214061912503939000 95
49720689692293515487099579621681865172941430023651 00
71410275558943126234999452621742518340731495182266 54
13670711205435950477052984055164144082316040094148 59
32276718335610410319523348484607952364661069986963 17
65885028193317090927754380975029691292829028718271 86
86765605656010388362597476899331918150868463510202 04
44319141590772364683025539168088913774267233213994 71
28849775979539773479229678936219309925112006630115 66
17390657573683767636593718921142236849912021475571 97
22305574100572545257245385556505686460537112268226 79
50192963103945844174558065212238893467377781174877 11
08535856365104804007023513841408394344914958733631 70
33745247442076903104658940280706520404901026101806 30
71440950890118365282910010426311261217230160730391 42
78299826054825937487197078945590863342170565376911 55
99133716964152529125865507192748345937113075626822 33
14419307505994006735363650372935500767980421512143 51
97253256056226439454141131674729877836871409967589 65
63070199695608246280145049119912407121857480537069 39
64038921234646978712255066960696516150681300606294 10
74008047075701309161735077334755648772547369122521 50

9813503319299338334382107885543833236187356907160805
4558098436700574350850753199409765965336872104349332
2280618834992004827873811252750304920888817484642831
9016539600630466502915620890532294731999667891042999
1774534128768910309099767188148030932716269812365720
6043315964964934205359309749955463534159984382386204
9425069393201426663783689481202199761417986083058983
8293900673951455175735499971707538924803152941090118
5149261408759447372115939328926370506958329918100862
0840410286489923463260345642370424824257063909129380
8017046596536307241449281837553304794862335063366363
7588656105890660335316291117659790023653012172857536
6201890101649815977724729054022670786268338777301117
0094318514035776219454816762064375319687841485184258
9344814052958399638497025395976212635978594566911063
7060133227953341910530379540425978304137667516749768
7874652996491783423364270007454754819494713598668069
1566664585329149030823205989028120587812681576743093
0721017613582263189992887973230932010149232126373267
1517964955489689247766118431545671617574939384016165
2290784408942315087668705466527579323880554912666169
3777598953925561080406938118224800051350809372046686
0200390550091411653944819959417321871903409718825590
7262958428371977008581794185070102392475998828961357
3778756435885496342561146780783340518710951312575999
9318051909219022665615695039282619479016017023122433
0575443126065446462500868662274804386674419442015394
2603811557827540000404020712181901315746401093427335
7483361014694036854512564432034707528448389317545617
6463157220928781027991202902189223828247035054633830
9419447797469130882924572050292206249855523551056541
6430457629886417680620174113480892342272883474543201
1980726606895692588229327024544775347217655283908061
7430025428281598287674713598313924867754531846844608
3804815081566354194625658132963595505948290763301068
1564496651788032837772344264957347620939757595579306
3867100477743934006498340550723621811989484482127172

7778513899568490447627008613126978157757229546476033
5927355634301851525659582920138209150226200880031749
7285389982150390653559331282828353022910424849910104
0960080812922737134515814582959625143817175466341676
3065800124676193027393974968282716096494372311355756
2907810196124240298111767982618671404839546685238558
0727594056093297312405555373044799293256493279811483
1833202137311156232404092544419362171293573215519555
8269307036208693910309562625792884944657181752916336
1050844013855636792071157188857988513496298042756138
5500828166953039086989620802903035538006255632140366
8177908180211648438779482785129378171151953129206025
3499397876175464080018854911265094773779403380813851
6582177365474212231932820826980804014905348395503964
3892782029247237348271696437467813541562307929349307
2403367516992521889224056696146996814738788518603142
8976353797624324269819101596656181862939220488095891
5352336772256873643846681697674177357449240277324427
1852172458249037944402883067384456400185365607465417
5605371082546817925693646964418020722076795015297467
5083841352311356162549952917635141688384581887938427
6920770036330816674413405717150430764455055708620196
2187509021388493215588493146361185166819219333310687
1041572192225845136232219900267083802223487321575797
1191139682680384884040281459269592328395964188662679
6149305159820371186283109450197691335579388158951141
5327250422495358886450529525188569766476775406419896
4612563278225970170233375585208862221826002853592814
4630861090706741716126123002253062294326943978032683
5730088162486844963806183381290631720666982539273397
4004947585695873513934547695113481873226417526311905
7768859219802373209352298817249598232180205141646542
3317460267710479573856951746726028070968152504373358
9820550474008030133017558152271695309675197201612009
2056623087754287106964586347137428066751678319373513
2565221483833173672308119816534223980262474376558947
6756921634368776691564947998944908953335482608637982

3253949166726899416749983477046990214640899683758249
5058290814533142200622637026588908567589263050621772
5045902749909993279197623786666529191863955876879356
6387776474276695160789608393162352529278325036415584
6780246181598805142926914406989652471948193396323136
8546341865090928413827172521695383620063230092109962
0624942506081118148675129816086548637849168389142024
4074612537349911807444468004565780723476210113068446
0779794221320441751848161601019084311857783736923028
5339399275611116062550093838015931151113590785216256
0485386914323812245904299772946964322273715189525802
9733660453560070753438041266967058679143693280921830
4113925179378607725904330105369386056453122825753943
1737223358522116815430443635849742772083634227879617
8301536250280185857848442597198671328342485127694810
8214828989987454317220977924036083261987325361585956
0141938413651762588232316649713661871498820913042815
5101622439116045249633842565407862054039684759841372
9509315814877737124018717977838047898994943654297767
3257015705381263785222746972467874241300793642328497
8184783841870950009202723276546498176971856311594683
0120997154727353173557025264097429486253814814007859
4209375626383946867371632446500669475670531599472687
8112561704600640744455807429019997012621053694428071
4092216615182130793486989728370011952930810736667285
4879857871482235316879674787357526619385423624000713
9113056755503388529014237718541648922941556716384588
6141110633338312041108527645828402610255558454722593
75961872346364309939638012344489255565298982720299003
6784391510418687495824429546262126155525198674509444
6529022196496329554000070385210563219676582484226507
3312536205462602686845226609688804383804743726232331
6166601591279368819959405025719999329176733931027100
1259536094694379663859680832664319316496584963397729
1944145183731667573688536451820230238148263770653011
3911289136913462491327912603253534591991632345277581
4247666027954795407043305095730585710512198291209443

1653359432408468138074887538729708537544168280660988
4935066699637289794431656792687636706605922664955035
2104308343571403351838775617420963086662250481525113
7848588257002504051583861836164507635919756645647121
5061989852062746107207788534090089741333788845290560
6432467837237844238116245396078973114333370609605259
7301996509374395456246686628124327527788178576450978
6865491389233961453938720160995472773168875997332167
7118441995894848614261038915187575363531390084472167
1593136105659055230688227016390060464654237193404309
8508250773501548519931747183573730449150694975724800
7908426935982914383093138985548754942322744949162792
1911744168176225851532923090627028862627327172713736
7472885363862123221521565981401434617442086412238248
7018421762113798480012891846115029134072351040099682
8163551558849625934702284245296564463145812208779641
4813974052131898287854284269657822434892186212453402
1422918738248798312836552083881102235045871328965119
7529723939560011425296402196067696795758147935979993
1418498120411115428221651323925590709874816323371019
1611397919813113343628756085191368235626377581119520
1592830109802456170110222573672111807537839399705619
7834785899801039586427329468431331010946879646342978
7772268219444805699924553685460167533539940861811762
2592942776471534124000027690102741176192399224871721
2932198828213214581558330810327676656821576223286102
3578888664955757065519767142309504205762010616009270
3642002340450972083564857718166824151192694485068151
8629351905611712803659267821369850750877498220731933
4861979007258977582664142806319759559863145370970654
7766476800725878500630994010878945547097206380389483
6930369700269745829254188935682769767592328677671016
9803509732769360272883121039870404729507335731757223
2713912886823089697758812579145493793429853594473006
1655931505590245069290180294939653193727641307597943
5030186195182849400682794556592773381062149016449883
4284750153040835721432245892652000275214846886135202

6832020123565045190191187232355916703723297903459787
5506946927721571147182379966807463907229451229908534
6532617467973382168394091646111762121075682647338261
4614878022120885461094160724950146164718017557452315
7572564916552016964627970618347370461961194707193166
1127537770751989446486891611246406779934062221965068
6591370531036233918759281965741610783982987951386477
0740645142244930292524076236866433594986611393309682
0043150156117174402845705892934250488997230280359024
3885579715131545883284619046631488813005446165010619
1633925396324995668918855812076832961375023195402633
0963046940512479868495658727645287697099156573617747
3391905312486844946258722691209550207072170716218796
7918776070558048101519229507432633782002791831155368
2643767887579911981167968873963177814529954425665773
0927567143291248470230227864752245335402514079094918
2546253529101439352289666482429845599342557559177758
5758242352338973747484463340047259313572926244839848
7770176548131880669706150228893096106568820300579282
4776845650575916401812945586968803264581170511288126
0506082220110294813788676234788832482528250394668865
6047914724349593624085753393149412855950658025762800
0890635688149232095192164003188097299919802546703286
3218567448770476679740080524449412301925770616860793
3337726166795237231022560304560734945720635603131707
0375962721464901942508795425966836567349678477964017
8627750094170839259651792113718885026960580859287154
6564185389115112448280957513693327438297878402591292
4252701523801839428277642307786599800714990011271862
6725697797493558585882762078441921011501227555741779
7638627058527065459889378332964951877011526441421644
8752086329424108541931861114096827628512901437848023
4391248047246668281527727431154896462556708029122941
0470366544127961945872451600591100008959934764826857
3468246049695113680191131432130230724663416373425090
1921361990587934861030116849136703125159321122314328
9163231551486390389204907304675379060333848146223790

4763363720223176833541143297333311418939424737931987
5513336595192369206055641815362927457172495382044574
7470974338111998252069390559257390707774360691544834
9546454394555065175130465938835163084824746341099051
9949623679710358992713562010978505616249952318905021
5581468446772463615546978316832482756963135717558314
7078970917287783371804851989545984382859669977686925
0594382809953116380964438179977603501130870739448519
2851625490589031108723329631488255920742740143440238
8147125455959070011936654709673891258727027356852730
7775193968862038706430537341996367859408521578977244
3560955383209737122234993636234092989901253196033994
7214295787475147685543217321671274622768033233000563
8232707227545294942172575194125105391672186433585944
5349268769220823873314300392613546630057296367331556
4909795980489924726693864564374620224968338000050557
8982685667696711428018767262177865576649157700352461
1009327946825636771615396243792771294838137972344729
0968522053123318201578958447957484049665426250203956
7468235473353258966746482118862207730244164139640057
4395899344512407081002310766672834298600575549363747
4986490855563048804573498089745784926319775144735958
5777356307725467852982333254687020095719748961900232
4974468772535045331727309242308890578830620727285549
8567467904848611804926653657381113003182547299877874
2263224505235417218301325634179543964983867938294564
1196752277218079707950556444508380435789200441015991
0087105620865457535861988133754425521273028942416530
7580303308079636039796066642428157932448605287249560
7487409036181364062430201765226239558683682892096274
9246091188942919022126987681974610643447889946335490
4936758430485143499806460954496732887730015221544029
5681234534489469325923497985578607921934790424873864
4206792284192513273004963905388381579374791162995933
7347104425865737219183591342311892468121510056417553
7835673127527720339420845773309322939337741474629120
4143364237584532278001041819917548416469079889666380

3490314042085797512767023436973090178204120231201633
2370681609809619376238353166428146780856607220894937
8140585082658256120641566908039133014503787374700861
9163420786896584131327336331436338237259869247856701
0819887206149431011600932643020534519436686730988893
6587361852746283046848650865893166284417428158134399
1205834318479331438251361257686530937757477371966080
8241387978909568631924278782136534145351715120124156
3214557726125717115423757523043561118558919663144423
0869366705119913815340322621062159439742712076546523
1765198966424475262047151909698917445511843743312560
4112060004810628341774495189990610221341264453400653
4857615800633640466882739220226192144757111594145475
7056524353886708217499556288908901677082397310205194
8388718354983432880888607110361769033231078137963055
7273898112912770386827993216590405314689632598639194
2915207644121837405335668958199426520915206072234787
0111507849459926379425827381004560937340373847005260
4324004765151033974426291589785942161902765461243710
0731415331339560670120992357055893555586437332466239
3682733813766609885613860817560855257518818229823656
2059303984802684689256483157272038196342750244905381
3871227283653813817411890381862937066879655274018351
6011067772154427487631866938516652689190966932412414
5765251754771386167907074687690502863083641493189655
9562354278624551587599337404868888059363509476404640
4226690237394346938131980945539830596379527850040281
8801715189687315831068254147354750483209376687988786
8201624977207965462296599869709286344427118784043263
4846582416724416037900486290633522739626326436896952
1863745855477379277081083209980065603785497861792816
8238018443737773965967582913220601255392869706131183
0757497364982101473139995137801195836304657845862188
1944149784937009840275600685498026335735027690350133
4915999410634061547078733290603707231161240341870550
7883790008981769470259740812636634023320523475949462
8647720279625211368644837784212775470890607510255301

45464807254596276113743167930832271444495182515532O2
5306889930585319483180619097628180166772303612166067
8136951717648759790641767207827490044745667114390302
6184811709882218889166496393868513193469112598949862
4773419222103929318565373462824471898957875867961128
1511099659370100378346036699083660549953931420999423
4926019517200560034985940409635399105472773946979904
1357870226320692025454984165726774279463961297439136
8521794850340618077285098742812276901141368164028467
5288754276648347172854512389859157160621106210386115
0414532497879130121380688937808878574113569402360108
6117274749076863458679625973610019911702986963967536
0563303585985059024044857052314430822798558580421066
7606978466858561399205325248703555458537010107735008
8649519635411914539954755706552627758435765745846125
8415631074749536770190450884604517896103563009644983
2567980305263711009034833683273800925855783963976978
8757455040977616660990279542918858930891795098681231
1563053892116814233795813108294191301853876649838432
6660489096880186078698996946395195235541223544449510
9149042793265931689165384524647612102977323804949102
6972063197314457687223018886454723103520336088032734
5189251460428378272357373813799044513964614587724157
8009918294527691148925644955445782081125854974629912
2723264649039750796423060026194526098123860389417285
2569764474274292893781644102259782780808093811884898
2866564507504699035503682364045702878619264007846083
1062566896721051510787599806260533102100122358089908
7864525527739274544838949424609773097681032881810483
7755849053575436985478248720979623960285646337036722
4447256026531392813243226027749446017801180064353804
6561772418288441537932797330316564262549644575247852
2515140333292180299436366892004502808793014583626559
2745303693208581992368582405824492867970470048929341
0367432496807871678052871557152697953873176161393561
3259300309851874585260746042807145302953344424836352
7863091514581116770338434981090347138086879895128909

27047103603336771077814083484462468606147254988365 47
671435078971650144527699386002279228873124616118886 8
651454875712458758319486681960713512978097342890093 4
829960916137217184080568891284832030737804399029801 1
977674459526087245017878848275822031416079664684533 8
401373003633935223913261746422839714692012729341080 3
224014862978907413563620435519583015124060878235799 0
545993705598339963964272542888444321935083739604441 2
680349885498678014241201398479946947426725134575043 2
841611528318934336257825555762486044698920810829515 8
121308074724848831737971440657552370929620468722922 9
750573904325529358481198026632909340398949735809290 2
735650265209686282787669265416618779365146499963351 8
918591982811237177058085129088914798950229698022775 3
697818326436580306594608092400801768907232584144129 7
192824247203800486590762483298432666125302685103349 9
026576578579963305728578158044815065341472910340118 6
510752757591462138595755579846533174652421247970356 2
965764849716996877845984143281364150626588032373537 2
672051002039187208654889404916333923844805717563887 2
509301231379069730344640283694891453218747968903689 0
800919107696487522679319688397308313632855164428485 4
543895511923516267055241661011619193956232121540447 4
588449317760332645864833269999532761311560819867303 1
798777096885473206973611106968735230086661325732823 5
545132889270735840381580895934095400477668633738822 0
989994775888725251392894710257611581130581373079256 9
508911106036337471430048180745447071235856967018761 6
304402485928102941244816533043001976781251848873242 9
146546537476530840785629909045373784763916341069713 4
948316414943177259257359898774141042585256506469922 5
753338667022937999299024833844979636165826017703760 2
404285435251468929382667737884010050407781498010651 5
655297476502302530481848290223516634907089449876811 1
961215106508070884528940298303191343694788607197230 9
108790654545592904426962102412650896208002377548637 9
219005822137541799451367737550293442139478343845408 8

2496830055969807227427147523461863844963300372011058
4884442628228471835669510578364180709087118119763413
4679230482799919294233444343374526571185195827581724
2955697536554765355857153715878867315582373179120358
1943368590246116895493584548438102182098056456671152
2781111528663148797912241046715034541226359174023105
0726757815651690794975354569546610234350635178926282
0073876715784184485323312647481696410450093295263939
1072551402638097234507421452663467912487992195042495
0639835057563167002422209788884501423549332644770854
2073295642006479990456728288273089736342401635598151
2716558570876026319347603574797113673228442544954614
1241031441121365329590731844662678111062740199008005
7408513360217113291910191481723858110359763233621594
4590686068858174582721066360783247211055398822853111
6230830772249320718760314283554972399990954386747911
9041206409581635030692153653952559398291083644405778
9476865590642695907855588927801481153129605282973963
7830522396568797932913415829562315501056197500574788
2584350683894802720131680054492417655413415536722967
8186722629752193572967621597457292998278457699958185
7012470410652755234709316760210887462018949830999056
2680354732391917438803528523945196855902262991234055
6236681684202614065946016614268981636489632267056714
5452761552084031997755218112222830694640282754583090
7880095255382670011748089440085824422344840901443930
9950760458749929609291944867872842446552942603550401
5366304857278457506789033420637548565180260586465453
5049231546366067214955979231996128389256606862453821
9662137909561418094819626802348373704864453909857960
4117131070124331472277069438162426160779174735893060
4032216117319023629010812414634682390910147478453481
2647369393780345086690201178540922691890721139084273
6264008402256095279536622268780363107499295189633493
7247807842462077385456462974698567818779464133277559
5949591985874191668447830186688758259850633580953047
3059262795878826628135669574185663518366296779633536

6162569000658834528478932612307942533321209343130986
1700139402245159398630153300275885744826155520116321
6558305401109024799131744318609478889442471917656585
7393833615293891646315787775206926098992237784272107
6226754713629677265283382604580315488184291834579056
2055852313846748688098747574182995143608952757114896
9698465913305515718249017264063469473831622484570295
6882134783218600502197579493279575371401946919871033
9192150346011748954935005764831184800383210704551000
3197299460626690817032846523263802422485817403542298
5108136931116248203781568240267054026506092762957413
0039459067445279197996647299332750032387853445521820
3096952772541183313194929837454299674345083494569207
9455833089572194166974255446825338646561066514161947
4027323306891805548871442397014824881413024333221160
2822892880315913275460406393144756504510247078478270
0310083062398337903505920853878236743848746907159107
1661383527097558515843491321035012225586592857429605
1076914689252902235938279127110071779878737897902060
1040537927126554260278572353850665444666703866631450
8709916557153712069207844262872580323455856531437854
3259039018168386005327549657925726069037995759888134
7306888807474475471119322401485057132580064528148510
6494506217879717366536758124185561781818650925659150
8967858838537346006935149766940200854142411958615773
8199026180199575558106225483616404106207630901758560
7457388473264507133385756546041732533711302801378940
7938579364339953362780923121083970023302836469423386
2029910331376974507251928193044810605361488981272372
7553453645151798438699737572972344796175412263854325
0152317808397255726878502256871216813534253902834781
0191915420343242593460913311364251073193994337038432
3886137588056827156164854936538023785090393483362248
2273182074099684532966420066268037268782026486705782
9851192845030221581505303564175689377542106313879430
0816908619258742869875467550923494473961567075925019
1683691129303774804955199097039823363826364891350584

3039964819572362115740772576823369907023744639354292
7020534604480383102314448114591379538474923915729431
2780741204406080309758729669731071819844106693340826
0496984151577165929551948655816486757794233936898827
7590035789421276160470473421167218792048876942331963
8405449947511159325479617623938498538240196477081852
0948648559242287732496687300301024626104466769085254
5609760525276754260972217580423153215767353290753481
5364111375640334493764477315701074417762358231865170
0354720537594043178245991335983391817819096329898151
3912575431913898337254405281508154570310228968029957
7227508331211659770422199826896583246941224511283294
9898729700252886927820088561271599497751356215241446
2120118572601525246972290915140464075319219928173428
0327618599645702580211560174620906358907518617575300
0042491967200921485676401596976159704057369425903568
2705762172698082612584580596167061886633284026338564
8286426103859307490073261464768345904157879793049982
0505904321300238863933029787387361962755332185958012
6622163258466407754647657936338085449649570965549194
6816120511556943910807968135393846059750067179506310
2561224795271996046570862538431272191945200310509319
1236947954585612237929584674483220803838519266315323
8284507062223554163557520160484585957877279643566407
9289773151778925432460163741951653324854866212599811
7972463576280561784916758392325772236684392313904636
2506364023028317230013785538397844059603568332684279
9266546160672110959606388342429504300233636510873394
5914246608246878679142241097259955121353439193233952
0885700159038044698266311069545853638464299547406804
0679609633400896596476123350118154718221565202593800
5625906215027290122242604041472615840342781626054598
7338572456371477792160474827443344111296731237676119
8629786698630802417950047309905486807868392900997525
7836056842325273451249388464642611743784590176544695
4096515947087299765071127626661175786960190328974286
2638348064602583518365195914728710254160903417107626

6775815046516946254495799382899617866660985727357260
6855066958690375723054995572090626217139470396850395
2363129448708741524764225060652166970795528304128108
6860497392295365924315417259939327074860795667414593
8706841232750041669086025761487588798762359367347546
2930459683470578182433923747399924993182137763038375
2990385151049213862685594862282898029098604098401166
5073222899953224056223484778426908827773333002520957
0716387891375728361113161508282405867748061283780997
6588122450097507098479143992524014162240532859851784
0602677533819228267849527886672456893337594516073195
6862168560635120350246309928374185078076042047738430
5325483876359241255753163645229316351178652228232896
4448319102692474163633999280535309366592617986442495
4948846174813289956813731394830959499581124735554883
7387112521903370051546804023677832615442406566906224
0436357919958919161873198530482800619742223209883669
3647084018123214174762767322394733060040517262859354
8541188246139912254599360462896971413416520309350257
3559361261145139647646649021062754457374977144715508
0588826860313596615759788880825134084175124084231421
8875957026396166667684217885015116650295955978618415
9605479201785548116465418583113141209127828459690444
8081819980631438903805220749709996445926804268473445
4155781033449320595015632019630539762141807399086270
8480643032178002476090143977156423120072263543493273
7991573155185911006524727748519971019847797665568944
9196716486168670790375717066483562808032596486767340
4584205686501819370242695253648169558145909093659078
0768681243863993425655353304650567850655486555712821
1838173965998363357678030887104057297206384819470482
3330793522090608439862801838670895297949455903398039
7505682344935367837444146988538880452810081360625583
2951931121193475177628452066519222252738669692630025
6656057137987467747226321911039544639384612184558577
5692692112744696565407157141418197949229614464039021
5239365217131970168237910162539056307962867690370366

9998720101055197275883904902666193971648348114974239
1173870422714295049803918440920353535056464826470908
8316271727043914121423842292160092712912360067010295
2490482688952858136084943203538041356964515930924338
2773010698290740036378191072142201091906924187216027
7555804593264901521505941423353135286277882691268505
7008770944176708211117803391613231077884547695892356
2860626769063681154808619448865059856349824207852248
2102735571711682860437427930019053242147326980760359
3366348559246819120239471480921233284186050062585379
1085553950069351432572141821734241696582891687104773
8311059705099813494096939955335172453464547253019452
5002197042367256339419925959139030043726038976533972
3331018927319980366786966546139807538001500749940589
6396569604444634104888910187725401758136482992701898
9318743514757195119016432624565142006231074765452195
5306220940907622656396770318223285959360902816252306
2797685291067138807712746419025705602446123783078902
6485486006490087858777224582811024344938706614774453
5679373207145334080832496103686003164871160455763852
9640009288990565903880787201538168686057845360262395
3761584365876413156895420184516680425311250906128057
5860518512612612637270196042362190166991290755289148
4255000203166394336803204057114116042375798035856595
6312474994695698495135867617171614861342118764921998
5740236045258155314008760929919647446288133829073216
8822691315735551490184217278167785213038366037635612
9007685445315638810905871806940817781907895745143359
5938390351399592322158154787478672620334227303779525
4810134334887011269882527069457297044292547385863958
9403162524181768010338837389259782856379513349072256
6439821360408060769512299054794220774558697799330344
4390441999735760797502802035252928636562271663753010
4348154145433270653391650995946735871134933382166989
4856705573807822002173120939153007826526128684514242
9072659990570958254860430354523299996515557814617689
5285778053665489485926317291170626203829798016084121

5931881886962902828715797871793688490402576691969478
91119701664230794471163004796916602963366541165671412
9266583864167655842583466265066904169951079819904156
3897096165783606780345803858875490439836681107946259
2280948439890034735745765722127802050766361242474530
2728327791975368709460269211026412850520464771511319
5909759784752009540677372174118439959013507272930599
6362535527518365125842604722808130501701636458298879
5296387473523944228984041274230766926575389191293792
7022735871589067800740485921648396309083923403988400
9512873222983061853071494441245528974454318958561187
4000180952752096285137117256845726219543878785925627
3722400118928592109735917744530909913760185710513365
5268996097982796691301664713645669707323708148465349
3878889810099483298222045461002017204361275120315381
1658299101511869430491144759374415119921482776988466
7239091980921508515824456105200887185460698730137255
3634632995644554638726445232695578243816895531108965
3134069842041218638500690537990290112965560297304964
6054821960184977149958716016018632187928577655514826
0986889146641786743554863988576721640798099307846447
0041589252981212008077546940444869022853477093950868
2132340733873815211764044476048345552930529995893020
7165021318455760851637824162907159794086617952632265
0908706250000978529459880341211082557760720141887727
4907101283893632437344755327661099462982029247845727
9595995866906046000101825538486745656582757507158587
2968677167511148726521220658930514702981699114593575
9706240523820427201676090897193644031047142701890112
291972783281602113559756325548699922433885045198582
8262910381822612265599259159708291978623319316688106
2975254106841171086259870307144086083889081617340183
9452178676169102178400070572151153318183463981090485
5059841908065379840393196765261825490144962826381313
6868301905372776290221408228253205031575814491105214
9693400800591718428867474232818892082210987075100960
9422271796061186875231086513054879407303247558551585

7271295685166711502658575372152497020735665542874804
8188381689438092947193924970783426911981621047130830
9570279144044887529446522288091642693221208914373532
6343470189414186549875808544389783754276111604821944
9857339919464163451058827596415074970868687065333087
6746855170863672200413355069089822703363808724832477
3623842767127299827900047237503365691359648505894871
1797177358953752351727045329880897687060371716893813
1149261179907638681757227782503037994810530997141332
1067911176939082497857595017347343274863356684315499
7425554342879813872888588408286655506303116923034262
4006519246182085129405138410943833830994337323561184
0834156109525474454266787441539452743808547815748171
8055429230172577082542155145523116898159841576303310
4102486785934451395633720568858890690571190839996979
9521334483952287592283977865693169705450275524986037
5938138006589525705654871539925942499971731716797625
1830780718006517305152956338263279321271806269594806
3416496910211160236464323589047724347766517250437018
8262115843764422666620475127825235188702178980272672
8027711822773152103035872642082452898843316831579166
6499256821611449903424669515324639135704780768526898
9984893205253372101246744855945116869825165086315065
5902335060894613325964758574047430716451021988964666
8576509308330045681872757241680767059216043554681009
2925593956489533295027971567218920273257081506277090
7173871031323842496009919676665726212488713499205697
2358693324073811177055732346090215798472234521382771
5795803472205402589664415390534121374470050890338715
3834939078365114580792027210147906297989923573034032
4996976240607889732184310882449308266190802622049990
1128597492955627721746461146899392940105949420973381
7594382878025044409968753401903886017717012459122767
6884675715236965521220057724245036539737696902851785
8272010843359833984544568178406257490431917087743824
1040318671414204172036811429895036772565460710501286
0183188433059026704135914468999688471783347040829146

81241547430930099815384258505632760473908768904927933
22492409534991903416211544608213898364543627345255421
37157891521320603765643461287733464232652794907883819
23864447557705950091194441422576047487273346424607959
24629817046415346065133084095714764871552128608657847
07352955176812758232095338163731442904741798402689359
51138362336190365221936869489539292986105606543505797
02715511242608643085927282057320382452863375360040537
15977663220991349122262298215524644134152945655157701
51918986924072986258052706984831289548079625435319741
71285842576611402820095349691802167787509064096388938
91048450466408657503729405221041128000954731966004650
04394421684859422090144524292722255849059366382440277
33283626450303053353993096987770343090712332576249414
96016107924287316685263884527512670603005707641095261
42989664881812039606387340849855500941732198779667532
74996201186977256904144690206247765655350656642375914
69173330727489154340583008070722148098316774819302173
68453700212286491519944808643918607136506738526628029
01568680073865848269567625421052986411659338692482538
33787875213492229698895497703342047861740980716210184
41516177221481911004373337116936743087732669730859901
03719739779496102842916186841275756991398649211424787
81435310883070128710377476645242406304328370837731525
61194473055755237910984617735312906442410447089451669
04880695091460731826894492787888493093950009950702290
35343539213325589476525032807105285424124294687830663
10827995140399264853819433461802956014311243285648469
65317170173912538789746660971453942277274101144239387
95895987891054566410426814789116268572803599678303987
09786663444404744950237805374940891697917385372097470
73452739794605724927590824927825833506825690837808354
56936366817395591500548911712294589342501939702089639
87204233608131092995278188504027712873857443547058197
14490571304159192507557151185687124513861708461376219
94813881555082938848375691452590235639700631112601221
1800437038477707

6721578829144725047038571195085880934755063748382935
31777538088558941174192632937495430008507222483922228
7543944226262695981911896956305252790993462946153609
3616354944787917503777137415907062317596724558368532
5998421558810316256244427655499025327501763136294312
7796011916430509716959532112992045271774496546336026
5153656582884193638793377535522100075233062499389170
8860305513498245420332860149882251892444978971365438
8787607325839686941239838808204334626611722043799813
3074235136784990845864816666286201110167877285493227
2589731796290083893809378368862243092816036269180732
1356465371535050488691647787791858427737908188613147
5435737043839641610866845447786051856988364354553941
4674488362256907582297656680517777748487429731521740
7172224089962520271344638028938433145251698363807311
7992146783653334217478880597728426160203584308289244
5143907954874192059946703109376776927346001157263547
8653303787050882558626169169547527360426276524262803
8247711129035653109206137550104153516350029477360280
3042651560687044503456586532737488831388713636012808
3391312685040569730582623675060661853976157041742174
2759409404777662958560993046444536969357131235331390
1844731016702853668941803502361632619326787257731214
9724982450873030342047625852256587138441719279864813
4969900336823663592588206010751421305993540511823982
2139359584242128868015569496013163426588392290351702
9638549314312026857517209243416771675953673854595891
5630415154340888500416067053346654082813070083289776
1488249696289240160421141615103617977793210297134762
6579979402154699780387784794015988160852142135353041
4979434853189195283011575006555852642361877164423608
3078973458250523293243640264375924591800563290872305
0762970364843550869676213310828770177163493776400157
4077061929099773118327157053720920536540297077678672
6653277933085996580010922173623890413584279519335088
8221454365191134340143428094289321478884830738725770
4974253324646609483535166578176707446537836480284709

6926101318645029312296165994373555820292882023445072
8277138137353108738681566684680584165539524063151157
3679215491126324775012652980708686532078718046128679
2428807755570652927825941943007588427801209591016637
7504140143144565160939204213995507274987872445710941
3555504501572289709470592871547857842375495555306882
7716278606818246476471999729800367332877207475226535
9103975496408864254765241431378861906622713920337483
0355538284344476445982436340653278136319578083438991
8402072956331227424040632911369762268565400010623898
6047816686297777840312794899917073967230675221629509
6528789631503782846767916109674558488735934754910866
9196172246037943326536717532667964275759439960499766
8066120400165330285049499728607042754250937207738586
3700415441026059524493707585003637841381860779567606
6806461723429500752397765145943294896719001839450853
5071152508262845453452737577969122483337192342761582
0308014628017675628250240714602509716056309317672459
3076048688248790053847158052075074482030526496689819
9940251579494232115902784104325425835337177788899239
5174636657432613728518110052927573466418481851815699
4434040881803768753519740663630478337284405581819299
4312492820699574561423826653050981344664300969883579
8899933364824964722667678660948382182006917644779951
5410064831495092340231329855502076012976597654825143
2214277265806665755782452114351183573372377304924371
4259570295855158711563888077995670992404167839450785
4184745979089815804130829675468152034660396987843938
3081839237477162789771382844434051340951168427734092
1769072524763108922444664789116879432513222007365153
4562311855013874215233544503089388539861014611069074
8910956635966294815046854675622928941510892372943193
1956149182318256112906025836428781721949291753453882
3527993628588963636795180669943553947379606788386394
3299218278556841255867799549795461330287862146465157
1399165914028458734870270526106981706256806358660128
1644979724666816416196987988677368474709635196349675

7677241171938898911618410167572490652812301639681693
1500308252014897457973889001526824023777983097240046
3108506499741059120950157077584141489180528324673061
8351409517491370788512597534824769265004236854623584
3601597072961828044309168026370057859215113722012165
8572233654137444476594139549643955313251616096580180
9920877908352579037577267506223225185883060756932318
9563374733180715366869988498638691437039379137969021
7509625869284257169129054200208155469777726908846915
5248117449993657072604850439508091894328530447493989
6300603185187727458210524655774108122548587461398410
2455231546972860561599004919730133874530431602324644
9926512501704259895981825565238875877954291090967219
0883553577724986861366152804958762154417326193041179
2496997142364803945636309930720835169224953962020083
8815119183724445227526366889209929576677095667383518
8966195961605375350086023233648287212672794571825927
1787177324843095828273573981941079466564284153735209
7892288900210373172027586642901183835024010382622694
1855451746236005539053368444799037802314935009104335
1555983611865012604956960259134576936587622265577063
2075180259102970974492951788854202119297169126631041
4209140107864252386132081347634547297753174408567322
1032254607393016774189560202729920316247975376862780
7845971052132685726453736242255319250951861718760134
5112433762990535554619591752548043038904417376179131
6962743448785311550195253699780779344712638424889934
3581917195672961221652053462230855341045461753516028
5303531706014088121363733345863747449007128683413621
3074664314991962645876623832479456855495264936646501
7373025269911908734889383808220490313999050562754439
8890446347223256585859879118866887408069701032026820
3991645965393982763139767575867563066119124852892485
6994955452747582595838059266980566909315279118773020
5978970631378708933883220709508976733530085556152945
7849364066391245229601266959600962624923623543500671
7356575093246211009787638785940003781112255481862865

91913026546612160749803532560234435636763124118158 96
46973856150829519318006571541263062289942986318094 47
08829996977823600837319327593124537302930803197083 91
67921123822037759386912820570401257724448615797828 97
03237209824314199521913569339907013764865723794090 18
97310267731464770491543124633531492311649828461191 22
32044329798798865455048855028172241271964129621425 52
44081433833184835548531648915655594728454941595183 29
93178851797857233334982197167452209398227947038948 30
93870068880950095001760186510756826190182122598791 10
67357657410237188662724699351075758553148758506068 89
26530110183871898352111721543535127593826432969022 23
39493804597637855809073171634956875032708986570450 91
63882052480794235222753568708256026031828129257727 00
37991853012635237990106194259570352197978694603793 40
90907681690393742403294222448564906242924069041355 55
79781762809939784678087508737889272173505415721686 10
62524353129717528701700217138926526862619668687123 82
22001808199810556711287217886692866514086857307314 20
30260294175800614754339699614454195780173471307453 25
18466502487391866312265912601566023637862665201617 41
60913156721283704302383185801183603595637611440219 63
41988517773587153802798083923962306959289445028812 70
77113265979220538364752490063865053846735265471279 60
13927057425335723992843925668211900759207612282727 23
86085621382765349933921722883132892944445541855911 85
23325131830035127913792520760388229329076831775728 43
47671056843279974476200796878545960902303706015474 43
72426114673976376141025089181934278782304188311556 79
58045412431192103441351490269095501799996042072322 60
99427493619621259445847015799426738408269168070342 00
37334281735882422048034577619180553844186706109278 64
75446871109042765649096133678966932061865099048116 79
65986055034049205961741584031837366244498788910350 47
06165000925499423394566216562460486362752367571958 46
21270971010358678373762859455190356948078418442046 11
64274326860787480844054251871227322220026214347989 54

29528267493240381500981848046570078416191838634925 74
77692646056917675531836754822789203318027266531093 12
71164367933819622800959541330478706842758615783981 59
66786353122126491074162834036375848986732387826613 76
90359301362324296156031166345795033480696041468095 99
98514167331564166366106381359333702910558095186100 25
96596933585381337026672093038574257421678642906364 25
46277737124516142500896386538595275801322289186537 90
60793284411531239571944333059711167462664902389432 66
71796113079458231044119190079783881715223000670094 40
02400638759226943622851083989082522875583396750453 54
32742412656453948010664801879337026477967864356301 54
19592850256243936980244407004148981012850384761255 76
65321967998949975116516655256803543642604731499806 94
15767114957179087061914479972522714795293279630735 35
87611207582906805645927075287717611731266294567603 66
33571335348362257492316090884997339500082390802270 83
40184421860239715622689469759011211164176352819617 88
00201355089860699803555039550769601752355717349136 76
58050366348098917662374727779443889209867951693893 61
59532508157610426498077187885831916660258754558558 02
07124960203569014360241160676509441751163995525117 60
46355458354558683733327378353855866576517556532384 66
58898639838629579349798658078391202883732216541805 51
01504518391046892095056442926187130727188975971985 22
46213142951936737845472512841913917280843195020814 87
21448681809122621419507962485836787251200849590524 92
53663525575397130128252266631931267472711708740168 74
01981820530095690121079949369324046386341005226062 83
35978477750993619741102160991533412417456322272914 28
98453154604467932463165850820599008444304452923568 35
61305958572769849765100357637380948890437898149713 38
94411215534047828538341025158627345220929813789580 15
12526795905016883811079779373785231407545364238326 77
53725911397231685879268790411956752555832558943262 47
46316959593237932807739721523413165102332695551004 25
67134781656878944325901020307612993187061171311147 24

4684738098761661329821589523733673235671671545614175
5516238807971200015234531119945417833872460867591965
3156964945975434352624147173770273516235218690845881
6943326495436096306903278636782721734337872342122012
7791453073615284429176459375754527665511736166249377
6258341232669470386618010995264161414469436866655412
1465355725108823132604443020594782957064696420840506
6397206393173650735131730038804452622596036548486365
9694269514837219194053748866482246955138001416617534
4131094397669630489259497232083175309962614307845299
3318016508452080323635826442101627431013661298558705
4229184691858535800025965709361068886707408368708490
7612579947164130996388859772806057254383677775945949
4879039592655876319211032276505838730439987043541540
7721708153371228169725347084804606784815303398274253
5460673613310625536480782351719373197638292319760687
9318639973939709329235846525925254878475606695873672
6025074964686593205347603330267022587893997155938467
0734638222108716543115987110832363590150691704625433
5142205615002339931636212269316141645163467224870089
0857559784286720908705035059539110285561256499135175
9644938784685362718801645021502346566085754400621293
2808424563547806671370456568519259335922474643444968
1389396409572880300774156674090238261424501639714899
2671505880621404736388771776520916125774301134751954
5220374908961631221049043534651801900090993521519640
4714588647735602146512247941780590042351820766796507
8580026385108261666788485589507491586451266193098383
0583482629690713170673107171972808077684848057965875
4441379929449107047471829728017872590170340333000923
0392538061041675484669889573135068038686174263992443
6900050778323469824069242703312234425338168695545685
4241317436885767212185233924551451954366292887749040
0439455946655074084270251388154225801277993624948276
1255301095759410701557524570174019788509769995259561
7208689434091597258963492835929491764187707351188755
7335239097533861390461174419754910858237894535948817

3409788639450617009227355835690040585622723780649116
2179925752635876771733782545081928110810215902388311
8495295970302890454635523927760376697180464550093663
8939527129375661836498772135574322758352117961769740
2893745677711710209601970860998889197482278183406451
4963240191341985954666751863043821782298053647680859
6413487848827651335906070316166008073856450236740863
2651047956555552390569324136590617673191577439406138
3800816228694214783660581227966013786881036422393449
2899015582775604767214528875269651849840975725311412
6969799968358313111636230166796763740682084735220764
8198751988229863052801822887374008463358398398351191
7977005782448199586712374407288542554258820673303865
9351234884997970262313428093258461206187314490573933
4295261855568315864723465082335631116899039298074623
7301858961751274620110241350253016504735987391299668
7731657887933146811462093006882223088720615833096858
3647938117175872322118804407563604562665101608344367
7231341484618712669394355809447299222058095400341747
7116924948572249761131666871509490601304242787861170
0218095472391317944418416770245763014022975018319380
4488448160477701404081374189395454759655273539614603
5191133930122313329913329256507989652308090598585740
6951069192032793347339726614955417056166160815429521
7879242085018926158908427708999081541020516351796459
8157545596880694260180013035578554702715987600680733
1451816194078367221223851733177317836585699996221593
8487588392248992911068712777416931374725695651334300
6306522737924551056266921880807811832586731369601120
9077146657944366288330014408158630986317264190633304
4044508058228112815488916198322932731818799958851273
7838729385096990500690958150559968623777129423176197
2940408139418136027240514820820737951363148433542279
3933427765368610352540772583999654861874606811085441
5993782634421838393844858485290353045697151099125044
2313825249252954008960027765387386635741563882743141
1023599533922274661428089503023576301847389969569687

8579647614813385274626000172840076610353359970011074
2958861790934485688238614022120262810098784194778556
6957670602082972972849321728359478850747856268835413
9604520503273390011597689975923488237896003704743509
4401262181010948157151330500526967347505159579302414
1924677188377990789709674198562062060336577626066200
3934751458951212313889676805215858321485875621004939
8274337792910008563459608478727443346055618482163033
8722910556433093467219264786656314689814420038187211
0934389738986303817212957909299112743076719777291578
3767485462892852180820396714378507933637369500772766
4266755854199958856972588894171885473245704126545387
2792938355542304128413701321311631686167649503182078
1233528078670360575316926111114131927428779044645590
7453108303045956119947788929764950706463971681451462
9453124294363555488686361248826561240064409327046272
0991938024993538960597443351284906593630428301040581
7460612320439445967861994375881821078870552909724031
5088518044697672070493219326427327474947465735270631
4008460022706069559676968769935816122080876142890824
3828225082716265451015076507112254877678312710489858
1258903148080115323482235485756250163613432445755489
7084018525624878558411615590629157064001861938267871
9167145422978427013185563530075923391743715042252368
7143080286973897230659408747261558320028856056430543
6754730126975992391214093078205163473056839443576608
0505495391291458645641895198421266755703261404016526
4647181916825561780500043662639868126326051567703737
5763228255723722403373265634399261010727774913476271
7616978002420241745421963354236834210899820509839969
3090096683351729155605180032912375027974138336346557
0545278348537476157707297860282642316989566162153641
5061191691186958507630088773984133144332352216108648
0753762756279535732619276807505917919753822080505536
0859214370642558820407949945680532099845216195290348
7471729158788043378980271991757552841249516455389736
6903667977458510604923731647493348513891310767047816

8931509280056525080740033602625587974658733895886504
5549105944187115107898947642772391413200859027313482
3051471147108625684424430556126457313179759466908304
1340349698418952641280813039237953542339551451643685
8329717672703387996607050275380884466661324952084705
9562360351328876356928310404235142307368983022576189
3982210451471164975086495793133102305164007027404187
2292552531255505075196263090141904288158378608018722
6885330334260324188741994743094870980900779689622152
1278204036387184986911556892852678851436920811198966
9919917593944222765985852960731611027277193039228750
3735276746031364987286014859087801089027696481017841
9250513768363944327798589783499670751064591608079978
1498674621295942992949510345561239528518996648389315
1341663425625387229042078570031005057667238782499422
9304331876293310759921633410348818530789457862130438
4469902259715109537842483964045195613952649940910541
9348027833381410505497503863252134416652532578346241
0543137242051343656100709750264749750601879329897004
9526351080384880012217081134423115923301825065523290
0080532960950529674761782776499033804649275642460784
4006224775102298409044645760882840084383965375725163
3090869365950113166805903176065399413046783186045006
2980954641179322276064394545097464031190004046101363
6375665251476339220564085394881156566515641449954699
9798961551708519238964173504861644019602822837023105
3195494455715777919567588597530366609882202496886496
2402373769741490896227826267171647090545458964342513
5810721550993127870018109479337800194288116971384559
7813034375911744983050397272005088385100204864527669
1759729131715166904145869909482996397885788471241582
1518191513119364659265502246995225206586123710861707
7752972480096285573811172643504397723991591353337279
4371214939748639792624792457463092688060161268332810
8013730760678963885303524762479577296233036320607975
3131170032156315773961141417267096845791988967823317
4980250090409276983988948694658406197088371187171626

7895059803799281868790835967674438673941108953719698
6182976738458954392441121744326072814344156833377839
6580093324162653119758863863145759699162822606077903
3373953780686154995433439666253404137239336058376135
7136631484960228465675347318833395976940602425850333
8228269671602514882648376703139366150304655694403590
9979588086530647752720761812141539410681855315173588
0056317978950810699820712326357146714531162758252500
8062009835302998598765345615877265590577677705235068
0073472116044980648839123443149699456629688381099231
7937596151519832650120598781441617932890058633209563
9044967327409677561565792169512569575734306501117727
9446769975484639636764740951978746985740500256663043
0496930378286467308890740676217208162910009270841088
0219803664690320066048982952654419736165080870899913
9972680652562279641804946454156448767012800583944430
0387771355707804656432991321870606322081514275062240
4635739633230334171292053079551861594221891341743934
6426986003963040050800204287041572959373530995771667
7615251624524092900687648409445244870829637234794663
4083467678892429045159840508607169534168927276857026
1007189617528775251605114575792348758289688272500582
9643483257250194887049167329327414414693861611008212
0730148945094220915611353196601506009515300421518796
8005009256702774616975269173663358847895310487516279
2541522915736474184880129370760082642006574185024170
4070543171728758494534967559692560581068000917355323
3924655397658056115062668571458886231368778250270750
7793892985221289837675733944406123896019462548014968
6695371667569202794500414657284567068530994311959103
9872370249270901043532552598050115712122632847074507
4590340596438552820872919118498395516680440487215475
3233912561994593028105975250334168546567651849453998
8852432775783503720834777412746381874054586392925447
1150555432279981820868187969864701046561341186800539
6513134672323812494414047414596471311007233355202812
3385221405072386153693166089460593075675862667907505

3701611040703409092628267488775028597523751023854526
7796205530602752777436781940577953384076593719123390
4161229938965877540504209331657806959653412589292769
0021058353472340561065145554293200917074504394701995
6698366277022470228830426650690455121943837056846708
3823573011816322854340018430494979858262176841561991
7919935124153094915801007937477404039276368882693841
4993758057175171344755815931032745774187576371187757
1322468922722424933777711395755848434693143205619352
2817599764563384463523667932694166365293829299473155
4160262278104340459712825822426701976096925423356800
2538348946563112495170146899400464009118807408707732
9964303094404358183408414175801847899502624738956907
5478192665221993496740541728819110503325279002158545
9431757345887327695863977413383026542251574881272858
1889273284977696911085259385652172391840425825823144
7031064760464814502022210286533308656374402259588047
1321869450144366817538530731589200519065754801582344
9154844317770817671252046119649394501886256763741453
7686456006014215126129242464596730642846488510497611
1918846720498664547599992680177566433790034005684703
7544756287824934898408714007702362965041707501045403
0069683112353172934030263172213170861762029604344548
2095434867636952493941538476675380957735512146374990
8050382956467090205230467398556435582380712001340517
9535817759266117929746014869213202630600393179956484
9444447316119372536942062347563745047826759397321270
3372805781343126621117387222492806333639834519303680
3892593096889952092520352785719809923805762361228496
0652327697097076479039139423140213336869039480868768
4743308012268869946203470823716309724704725028106033
1416470144741205421382403081283471910093086202697490
9153607225275128659774951950701446835957034266146025
3028153384827277644979007768898074458301080580762580
2826525398467121849904439425891880218806297599563142
8163848511209471349952422729096625621310572002786025
2130271652435730201321995053171120419333856243218115

9685353143642809886601095854368526010852934463784118
8537182627216507454141542909241257634281683464248035
8183339986607357730940949860657066158440701673506468
4551083448304103407143306886135064816123133500844233
6241417442522038471620685815778003440744224089977379
5250772272242522163252707982864834239360319936007701
5781454854997327914957152420477718325986255747117217
6000475918686134657668007479138823888252605950369614
5637555094836455249433184937579583447082565112029771
3054890598588936049041984366155680280639101693150794
1239844261414631452720749063421674666562208207338740
5441027555277326425726662603254341416451806464620135
9107463690481409139488173491501909058878658865756080
5463319115395795199226915101752611355353654804491538
0523524309209568391339236641860218486574056531386585
8918038882901004954410539504281908521709568970985222
6604914977034425634152011335435956016208048334570567
8582847518288694326457284709415507066390042021847428
4148150729321142555253848768239840596403694675585587
0650778052888118437054370766853940204366790879472757
6997674271743908241142251517023195532601492920532485
2331644303361410581348081211878445401680049841863719
5377507912250981688395943984374349050092769074304972
1973216682690786818942988655432603934799602791528666
3167424520499357683219759829614090609600277500754116
1182341365951954364768265974353872503433557560760309
4264593717684435256584561894133043626585493339644881
3667814189940353781251863618098614310938046860842389
4176571709198375302735294605318217748498067641007278
0689353948778562080766076462886434975781384916754969
4801333616458553592342187444390487282202327482790004
3809743722240905566579827925407920182719176407315686
8788990869823443332429846171348077941579260789437869
3787932181927623832088562198256262053706157053365099
8666043733527800430297853858257772582081434930689509
9746902842123214805436252144109926584513208698356950
3793412192787701548376684625884851480353282100576225

9531238439549984559771836617746969487713373602949024
32014008936954352464284670390628296129253331756090 54
5801524153362697046334156094771470387833114080455327
2026698516916550545808852623472270091856838115291325
1607557514732983104817451169011854473890500270796628
1357373328915219473513171507142118196223950640360662
5079376610114109097571987519506279076719920841561838
3126232787053677949587209348085310337003707967698144
3339865319473005539550361373716903540441227441848250
8972554341132914140992101050279406078134345274545748
9103978703959255767719730326629205850021841243056110
1471172665885887585811098013622427191783601505630084
5061260958035114622186897039265600676775227127811508
6918240931263983238895143307309208061818367053327222
1996712213824392494133845030855374291395100606121406
4015277299204721624746906019968753616129309594435319
6702138702622427471342529519834641150215258521719073
3435287605089694898596618737246471484457540042632111
6913066108000759019927685277231229433845375373445561
9270731842077003525188819851000091058869702463614133
9463736403629167648502431395922261089143124584310802
4918533766365473795428101980063946754916567388220937
2067955022453974952793604321876041673712529418689041
6787560507158191846283974995177618984701201417284922
5316599764249139615534630455967370036698271392447461
7224053775638845250605738311722063309952246138275154
3842624841830545461614239767958075757928929553582315
8379105998810001363558358025381930496087484840623244
2130196731285727975696380891589114151062682936713253
3904432204439351225627134258357193758075555470995280
5865092748914856061501490586722186301814539552715433
7744115748430146045421047237106590937458945560185074
7065551250496207976349526296285857419106119868433743
5959902767513512428421627873827040736179018226421161
9056545235598213465009844346847498934255194190612568
5394712155093835778399733985359799208044091401401301
9692058848162416577920743850616592988429542875926765

3257646127865813653869302306432495148722915683419489
3397732577238311018607285992138274399553167071797786
9463102412096562992515636907070979765746223022481118
3699337954003728904003552813838791669645146305017446
6783026773391565848513133733221345559120169428199446
3558691199010467025724886303914313189269027234278807
6559567143550810085233287367618883314330625840285446
1381790916491511389868861020424410969493039946188156
4346922592382271542373255618657637319111383935082984
4737578763090888190669748750346261607736201056147649
1958588895261757302986600028449208833098693563896669
3573658315432198051146302918030353282391251226514579
6045112913620481416074263483689048725347741460497928
9686654718043110963702069366180038156492646017632282
3546752577251006348108332460496397458189485625071592
7985140068823456587664328616964994470287617278607401
6278793760031340305365372001633610673119735402165742
7455266137724641467970163423221094839273520539921618
4668443165780514857874873899561173561574229271079978
9378304117463423317231236870629887993956379693234381
4093077303166738558758336589487921922929917029253219
5313106313751699564895556279207732633443995606991340
1234556618660027238644091295776817456745562042696966
3979149248546317615555834788491120313298081093820200
1667082142619538393913519494332455874157725836948116
3492140484733546704852527626699095914448570908316042
3866343355234799524193321743127082642757150153738114
4704648961477923432391883659760417327606918396476067
8663226749291933231611318776739191327131564491305693
7058551533950582292262979366928009889012737440110729
9130758483919483725163875251526812093561550689661281
5652785043774385673706596865712904074504021396786409
8050162871632426642267337613821529562365220402118843
0916924496201670398377240229007751911990173388725654
9451670248842314466733016979564931389538586123981116
6830825720223337246985737787251767301674688527011564
2775820059393570981225869012588927727534775124596954

5250389826116680212877573805636856315644421994581874
0281065680175318556456529558228861695528627420028196
4030614391059003215379795396930218519423268807946852
7914072587719484690411764169152274721106824390965834
9368174072439125672601413920553750443877850971869061
2830895421445090454534852381522612136091363279625618
7136414316494221393554422060053382734515307986723066
8801356293013165017655537704716300915062479212373919
3705754138726037784409421730259112504923861549381900
7039732269827084459330938371631480611283411794863130
8461995943078968496411168708433793344057007952647802
5532429988270212239589071572621544918831198911796492
2556872579518737643726521041961886235970807637082519
1601974822344820994433323650040151033080334509873420
8212792141200420480180501079798561723216473504401369
8114855410588119726087395394254908628737839037468098
8320817221479683074313026938154363684537845761557520
1994791177412336708593571769209592840288800006291720
8716273797741535129580502970994424191380762872308506
3578557022002901343270927772987375157616624748404739
2851550863164215302820835265015575631195908368230734
0304392715181050027526500337086894984288132356849650
0724939884740105694734338056373384023824360725388733
0911137388407645000377344784709100186480455411711002
5614054087836992886952741392619379085138522929789581
0628398059044992413872737639857194829128448347659740
1404143259818850245391067059176832462269483973611571
8148985287706502379213217892695361163793044646771025
3991656844314443202029811685936616659255206559795684
8269167629977734031737378030874820811784876725458683
7334246334524199414107801536921241598778339974669973
9542951868182009064969761036782152809895298760699571
0356909602837100859978989894633487209570802179233665
7487071467773673609724639922157532194023698804256015
0000758404117573195749841674033303526039480973788565
3902886659099013584031964803732938986045942039868966
1622275489436550760002345814107089615093946926452773

9423511469401112771914913692543785853948352726485755
2160208720481249169101852717995183756073284426677184
7679540711162180153539879182477498161858356464522803
0863239977138499283592092705553078665155645276895916
2186920347015645557340287669263841325770360160596459
6904032255123102029559970989164180907129145748986244
3029770711389414628364848487157678567794878143125124
3293550247687268468946933878930968619372172452421687
3927185950326411718319813570892707560107929918514562
7502866219243697984178114140454031960916424784392143
9027833868372641711554943963041940588806752381877429
1085517880020788117679147787731136388027728566969401
1827275547502000935927404948376919641741747437643099
3588281149024158567172118654518195403274463085084900
1867213652717887423627513368543825843572609765763398
5935364879062020368888102701585005808302800529052500
5920409954009559933828156055157808056927750376405932
4228538216945892473083726933480345155440890818030000
9713590313876036950646295255811932331910420173976968
5110785451020588411747429566742640862762260667721607
6333832609244311716326610442814595860625069986860747
7542904124446463286078462792080416316171887736540720
5146931212143251692342945354332472824172960433247902
9063540574426435510165761886887575528924905830732266
8857940636107263471314512803909679409368497933681487
5713298697813211924730599496014027781864831950609839
9949029248482264519690766949366809185292465402548007
7219908917729196093156605486780271484478376604390600
3147146452296723373821018725039173155245869955423886
1364979298934057309118689822968756922198783526788848
1656201217993235571811933947899729411316250382377160
7476012250789091391360073698161644961550771156247518
4867186427418752220983699262511079458767442710260549
1018377141493953846017308993344936047697330192877913
8027031350707321098188209904217747958244675990200835
7116817844883047873914753449351938140117520881370598
4398465495570550101894746331785135447806050024532228

9869595610981274453720034050684331619519836303181703
7478986266327072440653794392777506117384377391003701
5604508542441718294622323098741599261379231030639797
5196290621495493675149348295534255732673405736625456
3878208772478017012519108061380763294191048376862061
5503990171057754937943719814986022745743276873586654
7211937247364836888473863695504723045861687577899262
4179919242888086846632765621548945377584642693660170
5490410696791396585650472314269792668892085692962878
4233165154008797079484044606266020591420715726414911
8427257725445185179225229922899892552417931329556162
9333876104160849199666317442308779405887350839478730
7283099169773983497943684774634480157906910420838749
5354926175917185412293597518990089222621721914459306
7886197603562218812780638813224563555606806941525529
7897496905210255587116668803946253165815590264701717
9900039902483340191847418611177914250879578322003743
1338999438878790174932178328570913622593452398799211
9413582964860423871824067417098622812190185733368156
8570323594309199084131003274241630856070401839676454
2197565982152583422886796490617533288038337592518802
1355788935119163336359125653076196344688595555679031
9313426753383306631643406076972503374612045657857542
7494456057731809473490446364211529985330435369838466
0984613766934308461700054908361012391654874021552018
0743832046374678390499993518567810792289725874690367
7495791338535035133607550732132560029191031551310728
3177116250772551531257304962327220860225214648402202
8427985932928366887990771408192398837850356220529453
3615680282037531395403317643158940563253112358682394
7289210428147770904529570491287593524120513586876032
8451425827346944885456431933445606103927719582129219
4131087466667665924574913853915794463552398869067679
9695355659403400839266309489156370515518332939400050
3226417088488994601744966590766868472286681018342347
3358074816786082569926361457941434157737969727619536
2514816047504644571393528525921757839527856441202769

6318595151992537064738543750734841980458523109952456
6402639452567114830535968678311138004081619234069234
0326098970410456803793483601054492920692731323910928
9489416122528772541725880715769800290232769299265396
7222231259542378978187961729736923126986290413294069
3047792690340796085936969553082870633498853589806317
0379065510234555035981104722630784324378875026808665
2866619701235843107548719344699613511024653823076326
3859894685743530656058275300173591298306951563954194
3378121211180440701536153143845798731666720361840466
5922914000728615704709231838888371259011377511015467
2815683126173135845444955569340401606251591221445202
6304007316786242123473984156636061570022957579512596
0678909491807148721992011513386033420124903343269019
0315247111117705376749037925070477108000780724379721
9997486780512705438658097738366084557141592231311250
7699438505749572084470616176464289701673424853130726
5311314113296922449973258986937933620538231043046881
1672702967817935315147925262277150873178413472011750
8291991814002438516518729332192236713933021839371801
8284319234620686122487712481614464372623846672273871
6927043432266762821335957208657592529933570469301671
3770629372316184028128146654033939299764799450835563
5293134044814997460821573143678277060209036200023656
1479481796166953389820330364315197605893882403131045
8646245390394575637688705927941914463513165656385303
4138055116073830115234965016026289833117526654089514
2764548739436429000310464975634186467063383937026404
3271999344450176536211941839809140543543120466110710
2294914795728223004888150989288005202982221956031371
5553191807236758087351573949415863863465924264929827
1246874287378871072117796093428528221850751761203163
2878650509567859784696943970668574363944173341905638
3504407294386217526100606739944657144525650821851108
1123473497709235330607315603438336668128028972835529
4978346861207930653666229848868445897302368695057318
0717092710488020936988600525942346855824214663237148

3603347088790508346751416318599787318213771136093379
6792267130480840063571274044761901699021280503624955
1464937562857255534722015529162736004842071745078807
9507618459136069577882486187479496027948214683255363
6964805579475195804226596181757545537257578006385007
8828942880154045147286645076936364292616561464092376
0991944230253071776066476460974890130071283206700958
3444148679596428844265653348062698500890014358597934
3204954700739790197624021831864502251873557681953846
6957130700628882926375616032787642159655931252924924
2216132957450406538220167262398618661648581434298883
1519203159057960073366305926447801768242805793775192
5499116138740816555196180080619144877294851166256997
7245150821371463281903651808773347375322139017956945
5469855894609950994481976623160582738973924308531049
4440787084726990640317835935290461384822406081401306
7392119403592658922911969016830434362279715304844598
0731590371353990852305554135850303305994263075300844
7497312132922813900414937292573158103903671573439298
1372366096770703723997564711313836503606104054887160
9861903961849936105390323056035908952716521688244120
7600029377720542005493450599593284765791445138948041
7920358508525299387445006490770503204262460392913479
1860529641815140735430834587703620162613635647536716
6760597478008590314293613337929331917500848550960438
9718110409887752004921222370695585812496257133144760
6650693367882768887310324454243982890773510726773326
6784393233220192621936564466322054855219636285988698
2124590594841453898355243171934249129900618146395923
5638285940616902975647631126229146674653662658235758
3297221661709968921215263224954097253009276090968909
2981539854778104754603970480564097019438611243783216
1330172173520169538667789380518472999961133017530953
6392020079729438433442713241962980107195951189408020
7607854224753470765972679570860437380133715614849302
3257101981328749438961494854878373297094278163362544
3633308269399055796881049489203758750785453764050536

5757571955571940241392108268760908876905226368654030
1870822701883547584723999228940340790795136796379635
0726344224554191280957705903158772555892207789691111
8686536928072383024038923627104980728883128755350717
5113342480969287697201586675189142171434255763904895
9469858075006092798100439092494524081766250952741727
4156016145431594518522171855749552684752772716738913
9029014720235059895496157417316989042855399440283027
8383776236501085909369192015402769486814408826839215
9342531859686296886770735568178360341970518419116791
9396618337297000193822980150272483083508010971773091
1105870894894672305428927151182464154409106645329532
3935450519305278990931728114996492725306034702159871
9613456500206310713361365178616544164970357368209762
5837494386303922487734357175596671101166931279192046
2088982875388710715363136881554667959015531454623653
3727710941929274840416913145419700177540191661941927
9523260188825373033775233601219617116125553889783561
7670837738785260756313420586568197019298042050503450
9501358373027020583244706964639692236906385933110811
2201396771000674772455361451739824657873343641455185
2124685804072880187810597648717763079789332546458009
3215613548419484217791755933093578596719469919505619
1293167993441194597294243460001028579758994049694365
6661909757932956680542467013837449990948865697217529
7862499944403863256488733167600407677690717691102171
3324616671369335776775179668748237981197416413893296
6510983219131889123061288321306175347306459432631236
9021942487650442656800640372373565200124319482373191
1164158611994016234561070555880366626031622166878471
3489649977257144363570318750075329860063330828997051
2791458197779206078069420505494926820444046342562971
5753409492743557972516015720797972660701996918700986
1222418896922478331223069883519302680013335158234809
7469113783697448676320781546648812639240828418651602
4914954979864427209866051820357176456894993560204307
1048929581586506395191730563855938221751430131434807

5986693326504874799038056468350956561639841602315510
4796598859316847452297845327115630225679638245805708
7383359848615942699923530317472056202203726152708976
0668460260887447119518675285250056178297288871967524
6604895413107910513002573103926269088847423715651830
9916354673745723994383500092516539191810184237064182
7844631996490552888569939325282656262824286907604695
9122933888826367890525889629799260043658351292859166
0168162711585038595099200450238288052578716079991485
7795117410714587892789285944554229757489266391490606
1225590184674242048983069603260992416053731980399309
5803184587418756119655169410302588427461326771528604
1675625997166890091974462057078876061938247144306820
0786999515821869152348094620599473367284161874380183
7624468431146227199718354041990857212670067522055708
1779080207642238770083023145963243769724843022813747
4801499459678297652451111647562844948823911265802232
0723393967248735348379683470903721727807182199304579
6170874846683226075483119464636316295504614289181870
3440255160661099604396822797600851051090393629109199
4211938826655136431104593773982337547022348970989382
3833496062224144588181571744869858007680176809831354
4889080730980104598298840671012861381855597791311265
8579462797634402093254046425652321448785499837045212
6786486295967723599386702875890628269927949214888088
9025975297177478906720299367123678763451986595708011
8747717986484510898251955339140452640402757528622015
6098390974367883923433686902937949023880629769925569
2023125042770510894350978320237026090787722102888386
9173065202970742687059235430376889847491331150845727
2408927276852932025683038229026985498309266427981696
2154826437896461283683804207320924463481062482376286
7848195810854733788917216503370931716230082783544609
5355000157087325853718296069755081704358349922043478
2397372708582326963623701760976748450030410906040116
8774813123641572249254136735065969999743514683113000
4790437633894683811507504479836224977568918706331215

9825736569343060981010581075128320284644444466922583
5877645410765424544616023777827884230143241483760775
2722866645863176868757284362034638726464703378104558
0384086990018470013294906720162910673208665601527200
3570377008772363933708461915283204882311403505825354
1286718497691898740183701119711247408166154018970145
7760230237450381123110997102646614140418572610895636
9605831244662510334421769869553123069183533005618851
8487114675653747128425378375360270025404781526769638
6800281490670826847143662704493834208868297956055915
3091431959230537938970909123501683175231569389220817
236677947717971362719241455588808601903804074946015
1095181573219926316308153672778639533982650376935093
1962174930710360546827463851923840108589380482153796
0570375541363419453179102144027724400228595051052507
8853005636254879620396305141675053890281548398938260
5618460596925430923050118448202444057353339948646232
4698616427152552041418394545883390645070021120281626
9037643723678632709238480835704928556654225657004171
6643467942705669316946590355900250098110204615997069
2226960430393134150112385202084330783492768521280232
2911525197379137517432840567171957865482009683483949
3549870633410451091155802399789655537286716719980635
6218822908985971359456599565839007199084113529007032
1279848681737876376919750159650347621049267928002979
8828161442624704549931873587379611446122075041773768
0384233088995124892738217191170599517713453429945729
7252152140383430460734029321292971835990233671675519
0204836798893485785428130740917112127491351889665038
8059503648866001930517747973772006038594794431466501
0410771923517224472581698761357315539473606361022608
1719549477487334657530769762382929970665938113035566
6982838308327669547610966886531811220411325508888982
0627979806480301121722792334181960798412999720720088
4183938722113903472473108515327748366983782486796544
8539605467332451782821837371290684884322609750319063
5530179412648238953511473878639950548602246000335769

1360318382569522393043941627677865022636715905426082
1794162606278653413781787284203815659300744636406739
8966754926876493957184903132121136562023902615229840
6287309564812816930350186968503713109547232937674722
2478572964170819858940521696598105253378892335031987
2584948893728640768329664533840034139733684656642998
1296207453255651661075482537381673969227716956365366
8258137853929592804863934624068004561289736777656499
2644452755616222399431748911099786814130034087861609
6406890919344660106857817399649669192940715197977061
9673556327808303748676242818953299379935017437703304
1267846394107423900040751986059181564659477586096999
9412559689622869881578902426577897779204528945726859
7880151203918786449716049993622264612495868772143477
1817782721540331579087438333180945029353819157281454
1832682112481972325972143223494026289546957476210108
0787425847657147801608833404962554650632414744423711
5967899237058982776656858977194579259926929040833893
2723324793025420327412082794436936354791397957972039
6366395782104958416314403965424093860475283325924343
5007813065949179499077038816705671448564759014708193
6288460634387083087301399925372529836689104103313594
394347713254525111255211092798727785951382708916366
5909520741709275251299026204554103080348103227000462
0819931349774103396993570520081349069420803787220379
0464382890249902401221380463397980242182106839346850
8849301679182896621304254933387996548743861093208514
91053327587732269026724129662567364590687559148963
1205017424394526389397902442323703264903596852578213
7809651504544989318660685590975663239442261822295196
5642157254180903209980449661381395312098534796711469
9342694938245149667475159852900475180527661226007197
7572249670815150580391434618240125218092935634426976
9077593665452090820250602780467149336326778720581681
5970250481840076454284365495003371942335655317061100
2844230300420530529529330867637602864565461103827537
4154748491009312872595644205697610999302088104035318

6694892623960995356572233574757431587861845904831966
4295632227261052716481983985463379437083657296409394
8882039748696069433331514579163972074807234334427069
7628656850889473094981597190683110612001028675230952
0111063997859704188194278438731917954837460367190355
6930383994015483738186262489261815817546723094662836
2055651216917470327887284570315345444858522369606979
8688922493453282096934356761679260108472382625897599
0526379232591641503276362559460774627417043325449345
7444889477216876271827727072947967992940703725682106
1295089993702462171998898944678766862734579413522640
3349817753083399390670296652133803469837272890532476
0066193925458206592897014152612950727426292279324657
9170438371626932075365011196015575259469406191818184
8737771342683444424052930660057334458886905888039317
7484511774102397358775282284385958672028239874374352
9592115562432243892896300272910568728872681616117730
3569527231697743695929142484462118989457501169031295
7425148284517441987133717486576746353974745761595416
0878152194938038219063171978546364806877248861810391
8944897507305385580490920796321483089352318480379090
6668134527178233532246612521949926765291427590890922
6211751081746705000595680931351952840080439007572617
8665775125745288433053553174841429175337424877509489
9335437358359554578827060373973912922693703012439656
8971237739451673185967041931739307423102053944927937
2556695143497880554570330590834011304552420883774530
1823471483685405703839803008349014666157562782972454
3844947365389473998534287543282274785381373116324699
2938367029583215296762931690157701637645970731154556
8276634901948250420327162355437616062896101031778920
8130071331349683386548654072526999424381887465482728
6727227810546989992436385383891710091592717082304906
6727659616237816864044108587574547936667543859698670
9554749996591202364718630251342342866031230832887254
2614846504913339140845571348974212133262795637514158
8593834370232883676361427210910916326438119930711805

8137053205218187168903404084228331495613971410009101
7150993735501625049869802128077552418620045870896844
4383063445989495551440098765192200443482688701270198
3039406922428539144344376925256906037856331636359169
5975636285511685563162452502775453761962942890436619
4589680239185158067914386065545763066653855089791671
9672277397520763582919057664796718038856453881886603
5064543168335512453205238283278277210925962979474365
0827334190694841174771358669416235515154189766502736
5243778292737501090510930825867898462418684947221879
4509286730565647293726572105569324973753017203003429
8462939903576040553269480197521260030669803843503995
3622458649675717353644866229608676650221147619719066
8367000787861952572772496077495301582704019156303489
6276315535912935217020929815099957118977712469085446
7614485083505413333978482613439531495371915024132149
2525701145762701032683591978855164103614737647596250
9762230288111889540471348042461791153854163232840054
3710846426909503621868337287756445555844513212607093
6508889689962610406609727149015825926516847763506236
5733527671953832978920009067298565323545222748165410
1948007407498391882302793263914494236957352889952770
4109953394755282701081943357336971843195081657518177
3136178920372220046232022502571019599757952402244477
7321462083760083485538773062739020918868352510196397
6707841850327839303456164014187654569380004166672187
8598301833249780430668413708789977806097087513122453
7331792104776532196322692920624441186024038239395826
9384876943864799215827507678016075065363560192478163
2884895067393170475081966462719511896879259504855981
4225373534918820225223225452706403110045058649340196
2683243996750270941977257999962112631549818066293540
7155836102749719065184274065659372545125747421356527
4061255142087368319535891534018300564376147556000590
1887559432489987342354418536298897724641429112985184
9531060905307036852909517474704662175927821270284427
2027632422188650036282932734481277819082473719717873

3122826245293390331056612313694376721597019056278625
1023149650838505178495474625792863354846764750561938
9560487127247631535427130602573246197073058891449957
6286610805194016087738399359475968793420630616497610
1628938474378762708398093652868909362413539742230974
0440123377345283506225830076819495350573727124729146
3024293420118205594285875409672998247743329952532893
8891028826238500291868606622306007695414534401415437
8027435465277981124805950108815688653909581055179251
7892616859476198901851288548533001971913658050934308
6513733915671442531069334558535936906805731112135220
9014898432261639643263077611402495957275755180179589
4194013197745734228922330997391962454237815316373992
0532476645534806101436730683257957605166743647366202
3462120548832579620677794658961534666628496225125599
8837366356154573809942398223413977857318118526694509
2193334002783956605221904343907952187695286295362583
4511428833741880138976683348345199235437275950997248
8475499853482128754160212142000716742527322818658471
3024374038012472127577155173543806869321781709846930
4772138693436239351851772094380919024767919123501634
1974983001943492514392273283998952752845430980061397
5570079141708167825793398258034505303504355997163018
4552816829264227963795173998262569721393103488869523
6503388767235345917921388311578797662440444585686266
1187618660778544234578255621751391515121750699702826
7121482353761675339029972479438694009843980337239260
8257591497122524969990916251682241883027706483153811
2236871275612260858402325217728238991975461696687100
4680666839513940546830147066324372809717308526175004
0540584635799643871306025046653245098513711350478406
6696740812062280849524708273677848967506686806656952
0461593590640327826022810236552083797774909998813393
0572493066866543878693836289431253517516138530476569
6084834268921637953176445418916273051752216789720804
1102237228388620965663043269375053812605807435715564
4252030153606598273724463194200272636840007290391352

3216097806820898002503971154135638074184333838437755
9456889934327573287635899539343330132152259001208386
0051252010931866882673567260499879953512265864607687
8454488418338341366254221969714632518921172825009741
2198388943799647742461118287564927400801095806810716
3190905554406637684199248303038244538612047639180787
7747840955329367731266650623046349154209455030131869
9283858704049776949876230868160119822506037840778293
3137148693195576904124809600289428590147156300359952
1187511349609284646438827763668164442908723542365626
2418491311097071158811075995684882418627659429311553
2643553365781078624936606809735256728328243880471495
3365163044632204199356237763659235469498248612224043
6933050644547086698381945613271731670628721129223308
8278228768566112936704043109736681582156652530953192
7357606575536633813081141261504182742591979158468609
7566171115535926504724528901397973074836568456676376
6075030008038868274480925601952501822877867751168351
8513900923990735105970327066961915407328728911684660
7505209092006071456463839356591565542668711062586079
9966340457758882769823034744991771278741658923779796
1170443306654990881949703719928121853092042455010101
8728097074432904339482702886320072929682300716061300
9672672956269791838625419239274660390007121099734961
0532335584725675941583353650389695788836122277161020
9908180784994235623092109652020074968190970233682046
4796210938523210558762150886567676848354321163469821
5738765508378320373381431990074202634978428101168489
5754010218975450709832665427621469333908039904675311
1520241502483200566561580635979236181932300762882729
4668878379078853974048969533931470031131324493093227
7053261300288505434290778900634031910022855993741995
9054534198294838865764191083821664299146114191054104
0721718375771550615135153927714008775520012284097818
7258166270893127396454647725980889490463874441203383
4039847054748264843421660150604097870387197604533296
8159456039741792770386611691754030605604275947492737

3175806087531129660798807172302188309181631035546996
7899967681411322384058487215045110977754130927334808
5613994313891956459179137461296122743264902894505828
6960183976632667681848636729784429610824575327353237
8558101279916957607536616328445715479677570022259203
9479012456471885952733235380132049867071615509158782
8895672746134396154959024810526757899163956156292280
0247341472909294565424144238427975134894573060583395
5546620662702100141027670794584352116489088168624369
7965682341977082233313015802821876841167102851911373
4962555044156500322013218780720836321567275832894119
4293009420176277343107493222163016969037110211968178
1459611298508035678247175572259523376464040239924499
4117133227064814092208903934067741659079335822479617
6127195757906232160753334804425925247216637653281249
1737879135545453182838865387075647639734081624444987
9336143123185696534013864422093057439128727633815813
8712550673712242883009985818632101563534940227831056
3117033107671249904005132001293489702721309952154923
9150785904214026893130046098656152361405253039272543
1314097867037236715981350870414441556847409342428580
6826691887058701331464632050819150562484760044352070
8075408782114949462115092792335641676736833501642284
2786529339283279284533215152892040943012000817086185
8410750441576216810260608335682836973843197136510829
3621246800257976799115539990764840380499281718037565
3459518384595099340093392603110508797537641335490529
3957087659913428997297701816142947608013283728437159
0590628796866400470614917846595143380897979017472288
8221305314151452675047969517343623347261533030009304
9742656539457947474078856366781947087581203460486212
2119732683985031983988067512355607212314224839768206
9335797025454114267856882868576218146166824675502952
3775266140894926217941023421516354117757026907294430
7695709089606441496588167174212166831811496370914477
9139340867791720363604718337590738200996909450123084
4029782431989830742991247475096595054024321134629833

4415639386846663845113041716880468008283501799096954
4557428581327744030140033636837871236116275322391852
0093151086954785404060385142967533705144922858168231
7546757859933248970433194748116313656876224420921196
1639847807493990632550658610472649946278570911848293
0764005230239571694045302297748433753449693479104278
8046497550916892848110273355938094404693489578483196
6191619156876786647442091767696146015961430071187637
1159818435709487634197399138085628617818195168335660
5131978094532258542655251653405256418983604191809877
5647547009333545646386374588183707193089927774747776
5194007121001602129242904288437751885569693798413746
1959487866404952885179702994403417092257126983643477
9234200044508972401956427768435740004691806885788296
3825556857695524334810592353696323776654121361365941
6589648936012690830391218907966934638782699462568989
4323842694790019544917649079925967283332015020405505
6395822883229865421520127390385712551158338946014788
2679613070593684462714073176635850748773515360878547
1059945081573750376872175758668974763714204508585934
7552037159289414490384555188824778224888605676817948
4488505424827117656020412756251081698730294789916929
0417780732082029453912387288505780471150279434067819
7206798706677346899175968570170964221498843862131723
3301403648409062296336613973120512678548019751401068
7614978678223829515530144775438800942091958111908455
9317284191284475424593024043441560468496036522322070
3918197954739023747928894306287558798955043463327292
2942658198189938496433903901748591900745498519432437
7468897151635061784044765817263836980897975093316068
6709027362906796736528277031546320116423755537799298
4746403323273985355161097777810752126229689498605135
1716560241028710377241294082780755589199253507584971
5477021490914683365543223108654748777138622887576081
0079271785897902598759188635196306045666633363192174
0794453033459277301240490432328916988631072549085903
9501306666592730117026037662981068329188801540077400

6822293021385957645423568417243649753033910342475954
6679776970800273759435800647152486835066819946207850
0178103542812825835286534039521232796603536324082231
8089825447710520475037042522647972286991591452243000
7083320007429597732272579503765299376768720265918931
4667887983961876650840897212071621470805053296553068
3823375864780997017362177525182662259448897555479107
9002943280737776954120378881938575336245355575553862
1513721579048564519552478272383904392255558608545983
7832420422489960586622158423688887828188750328772057
8409778708991012397962235928130415428146206709469072
9430442763735707951946382406385353975389325145532040
3986581318766650671801285529209290281388464944991314
8962151109657353827367110519461256070483211206288125
9687496905332546516609855153284705020721844897915130
3859961827075525350830941788153307371333348324728777
4790518106099400650621846957914160902586333765370269
5033632515901240061077265518504085743720504028696419
0345060154341482587482135948966805169716220412921890
9013651942661633491015177709354187823411594434257301
8458460479674977341123967367460769375849063542999397
4545300717430914967401452185883758079084100939528251
2399394188780009800085298325011797155246966298052393
5942605334256683484171065964689960240675931818730076
0771656964600274918478453928375977395610105436229722
8330796712427595819133817907834096214082773026094598
3024116813392425402102479082958427192272091231049877
7743600822820404793982383576317324431719483156971330
0108285253401780917584652294174735919734937221873348
6503775663769454517348058412741929648062378847469600
3236329645607187500856199400629019636181432769610791
0102484774499481774730369751899228135576935750144684
5470381796435604192748202966411484264755388461609284
3173667326095117141455146642087759372116066400571318
7183194037824930356602642114555465509638156171759883
2630660635415029101213817574070546034758943777657343
3474433136145706954995855206815968719207955264670235

0322893258869265521158374045717679269786983093658441
6052175398397969141646921305288712473482152684048633
5441603666716454520572878920653903968965700988330392
7822831246398832259368184889730076202950191392217469
6391900812982244757810301784071241371181333474206915
3806373196342037227001353128231561282092730787333606
5731882243303526775361685144012848142160469279280062
6149042372647529552898067238689801124635261708922360
9419514298318505493877642205598397842354960683843080
4446309187389828110323261749424902459296870542909598
9327718867278818149022051859424964978643722019150845
8725241513773308659016343738990369096183941846604804
7641285774857339602432888485615948165391309950784326
1527612427437304198132191310971382332353652196256565
6841321009977934658671253098091631236945456552408670
9902579573737869073570795762333041520457760151388345
5847419623747926673163943170811046161492806358838918
9301292776504366428922229524865961964742501563936513
0455542218413698115505602264569224268844270921908249
1387974604688426215352222321596952972046003562844801
8051430923514649064831558147073373909909403306351628
4763645243070399022890691032269063357760364860551940
9027826803159378088265928386788589283339814431210743
2421057444077972553048758075438271808973816058294605
1048302938386321120440632379853101812009680047840131
2104193172311588019894128999509449051823520285501747
8454727620598636970700962150536736780071040186618081
4138596278076915303308597359792227429779680644323689
3084382221616134450290924444241342868204598923914410
0586494855598206028492271624778702699558974228142701
4367258362020191046924111432481136567823885316616782
3059101302957723739494218220628532292532966281056278
9429374661505175320710232540395606954202499821431539
7713255432975868552527248013252592049623639186428240
2295056529171734982073877274864534744992666383346808
0472843102113780927195036693983708889807928735328153
3984742606450074080844329450210486602352792585313312

9653132245368773308954167066148363110682779419010528
6544385525475882138943087838755469743892676454966213
8072288425723934505208345421566445773935903272319758
1717659160914992230053647772838127334166622533841472
2426299424119246224009785447297982912784403926399816
9698249831998810282024020189496060671263656610074693
9708920646894033570492380927010705105350938561179427
3021697988253541628015272720389796835160423690238188
3598872104029201907105608751001679037111105179391713
7546623683832541447178593865302970564626260948159605
9731112820725571828111332460761042177477596454839111
7971361887347487868253984586689749210617703503217366
7065708216985559866053152727023642929621060332762951
2934921752142974883617498973053872979523177137651695
6560760094102572096526413672280471890194845365773052
3247184576856434531334180912602575401390394116388610
9277630735614771042982037148558809988280770118207688
6043581805556974289513493920850270360998529136566724
2000409340681562664800047750369267010067187565498302
6784039497790284984080411286490427373187873235749233
5157779266546405875527019733174152553436934333587817
7439476676986534109034241828856004682442717355891956
2996250797908273745606776538490242262424341394441051
4766832862609811698696299573291894803130938365787720
3054406356888087392173364316856301974958824077856564
6911005844850632217485602786987082544923435009462428
7811424792557092875881603713337049894447935541357861
7677742593003501994878818935464570699898023884285943
4023529395735903638779955485801844435598231690842488
3535500656784002486228853977905919021010083266439140
4237858347141539211252119664412719679850014037773781
6113954694556263934396372291698397034431618022638018
5350772747903835838776038137578386470121363152065502
8504382092685471604804944672487705111529563991984619
6607048019909225438759193604899417764243237878245127
5097624575285549009194966343005632504215824785039438
6565734703026506980027224965381622755412581238300224

6685055927019280952766320805532391360748859854952576
9899507927614076643457646429097181815270408434116864
8751952915242506986869720912197272764166399894803529
3815572061036528542998042279339909846309262878679188
8447458228183849154137902575761730557372190989173358
0870609521181391392283701733047688180850991710905044
4013024907362722374529981247942216588118589638088129
6789286027173502479061364226666970965563306010179050
5226554257504425919798843096292810306578173471637056
8213331695467538525704127557040755862586832496666639
9960771750714245424347638073499350925572652501928764
0864927618497173045897625148716488591598951255375228
2229553578092755157717734942544940646365328643887343
3421753070279721078884393578408051947775698254173932
1292803528190432830422660811527761503372809322221614
6272802255172759402589140495678040276685368356006483
7512119565560379290890174970568828924953768000551170
4451470904027647709526612617891164352708494766333633
0375047626681813493831684698386036717861807570903840
9994416681888508857156733750859452381605828850594971
9141157735194691638266329100093631969376262656339971
4388590830624050022106856248393754516645389779526455
0145438489917421968312193140137299511841009750979419
9237468840954250132123647670753289568716490593510289
6844431703315138814848447542911565514954932398604734
7014309945309592965732996406961790565718915571395922
5222377996616334929240926900212172351354308093758013
2161231377545123487290335614609148164275810499520023
6087135985496480127596933372611487822048271416542861
0972873853149810697327536968559132688931246588639737
7860527964731457744387037551291438363101480880102554
9750185056343837423264942884038079814325557800163924
9929690852845898363913292517937062540150226681735964
6575985941633224415157573672662192304256955666018622
1382619088019258120372388154600776003859044903886176
6421200157127662440876305292839786002937708353109004
3918658812000874319507410289980656593798887012309731

0069701795579926044738696361160786515983764865016794
3285933431962812865551608452919059142677920493990411
8967887787866743039299757616764655468134247356325818
3134122662690037483369850318720011746032135725115548
7510227692201433444170414759365038920954549976990021
4280493035695458920060089088122898084805497614642398
1241706535745430437630406316824995543589767797872906
7940476948112679095135574378201752416467534613297580
0098129577909972707119039363809449713956296258726823
8358947185416563847329242758695098427673777219233217
5276539686606177282371596570821130355810363815182791
3256944611930915881335319427340665428840741599708194
9240291838849762574831533937525466736508284396554011
3896630167639046301783547536947865239365547983209434
6814763378984065784724561420759944989774128751744945
8145346573837899796025595862692807395026992775699077
6084920221558974346739263779175303666438705307148525
1947022095708928286369504935855848932090955868667023
3457554319254514425181288007849915141850033013801828
3582919167951015135063258915735687948767419183823306
1591376859223498332250831999552784862550590848674550
4695160242052152523566763826664320624401134624618483
5538231919030789790489156492881675694951297606022618
5159575090914427722463313568357223509941125528091622
8242406941132232568995320003903020219696821738613098
7969800721311638620390327297191995557705918946517770
9310867334359701908675463750777494181642136624587122
8362571309690425221424092741444428629467641111573339
0260689928399591420734459182101298789382713047503738
3316679578727390062834881721234327931679007801792778
2057627942474196395117775094573952744389586335334736
7966070550527012142544299479080491346447357381092368
9307860356623664611507441594192307991226358053752635
9624935883885499453578608882349064790166624557209478
8231038700999172108651627001741277647893143063607031
7268396116681796735997405243721801206234819404677351
5110115013575303935536135039838765404636303172924040

0443934542228866504375595216796385599047414810736635
7622269326643306422466225136196475599479394475167428
3039388495387860866631365660876086740168682542438595
0930983900950824741461151827971514792538782363231160
3910159790837053267613356530501089255973694782566935
2228981542080144662048550197653232102181616621953034
6571841288016502644831777530378575721075721672703735
1922240314148705332812556025237909652855171060447447
9118067319707100206860334909523369289943517346016999
1071845070490952869577741317941205305663931582598080
0945415945684740167337619934446441859524845857806791
8467182672145793302716286484233254970208681474069158
5705248301426251349130131793189738382452549317175403
4351063059441518585179933228918463985687661286780829
8214112906685003925620774769600566532434851625148542
8270485614233397106339613269314052482118402280208764
9328246007079295186711877074641645736456342222618471
8124284383548826605654175590498795369362959566497254
1183719335697984699399826697082328312099109341255994
8081987322038686457497615007315013080359405040673405
6012325709787469629188299464967009553229932888316237
6227702344608416178629584181003305951772290600660895
8130305831313955885880482762259625175518394264980631
2004512718100192221949705766974884459659269299769162
0797266423414339698096085014545299116867845278772258
0150857428597643180504071622549465121526895079761409
8356924309417467654518171956674720444984286032696803
7184108259382733558497438685580513595228455287596361
3860275898194531001707608944273324746872429589116478
2188536209845829468303040751100830546676612131694944
6563866233697314905363048789078832874042072673383396
9258348281353332462611966397276729576987444036547136
0016591674771423861819916453062722898155773566229226
6108971779771500834627946440936058431573206378347619
5170000165810602100928784044356820652019452702856826
4322176071358160175222197346722332778027439894359711
5597803812762780652604670385750855602556081051666778

1838263691621120275954763575092735610337565179769946
5779495961144911621311679160046072342568134822109174
7041610254084248399240423509621969126368120916434903
7926693492225463510174034101546543752831620706105390
8212266935397141446701638713391957245620135139204050
9186147323218293619518912364549208823904972248828791
4257299133977822247818652101339141437160107780810007
1612966209368046726337703019405918478585965573088936
4507857936000878628686633807976368902807806198570100
2322447725140393303221195720569671864238802321863347
7611259943548649924474751661178360316695263754160436
0063566323871059279357921756871199982281111089142464
6101546553565962127042099944031975873515343983331956
0989417309386547454651409993893976935539224586430358
8762315761562586158746287281878133711235134587883785
5804216977643985259927890499624290653889621582122218
9788781162583829659073632484969779876120071306818833
7251900370377403487225042973496099347660728486400016
3199296066954370714271183192140992442394695825665420
5419145120455361423648841481602605377486614986411028
3759516290999622091232981921678339239642561390727753
6570773639372511982297369678692468915137165264869759
6513443576712282685837531440126280418404622869358897
3582463737878484474816412107338167587759228223079154
9822108047959043483193387634330733994992519424337213
6321191536581087255275493903497011795310565274368888
9440854746664727270780909808046997309405202816129582
4460656291655076968023561451397859995356144918568579
4960899022815409931896227352107475824485672452619510
0482672561527017230093439430290688053019264553530429
7709997828918871272977567312250801215690898921046206
1030326541953558830843463311842324326679350246989405
7473910493235573872862497868075144049873814343511838
5255825859650848619765348305165577465354849251847062
3584636112896123106347561348782919620808029021541889
5671695470578412636065128050970044865339456921267610
7464890218260151360021764042075934304251549604712165

6063826907266881060328142047412200888667349415179972
4644431487852302819566773009243452527803547237133249
8112111447527196051728526390932400074100850410135349
5340437750708682692909589645056777697550518169976547
7424907498713768970805422231030736998621442134449704
8083338862903619246209941700659527552349945608408883
5277119951256411788748775542690695378901665837170505
9760688177908491161857021874607039200411210127386634
2584584512364593848810490498891712468573926962218064
8113410347799910285334504336352935599469991079748386
0730938322241856554365049764945838570957547589241810
7954937764316896260734364589130152744656521145786141
5577099210493343713276548550207819757756130190263162
7312439722426741466016940435362112664762066895571814
7149979122203905936045317129530955540122855108112121
0265729697565469723178282753589325263383311106965765
8155611735948679547594241902955820124457509075373370
4603536226154971039544311293340778323908570180904613
5796288485527163402699650631291909942696959662255894
9797562872128775719854241967347530158746347073305073
7845796145354542520644230878240841408078462796736873
8878895852044709279081010712119877987278359242339335
5755619371324997959387609841477989821381826503432240
1306385745448854935897070486251427872429652100656649
7870701568388386453995903650801711143641042662786967
2414199832654733411128487933093439580866789075434070
2026793784774345535265665129290134337234628978010782
0115522188723937312016093709704068632553677092619190
0398941884114592615370116668658956367315052216330410
7798488677211494160046621106586052485041079547948381
4251404996368249707396894144246667174873674631685170
4563577542423105150580986476464173910628745990473792
4888107274308142629524890931985495762898324171908961
9983841001815466443780823753328402254416041489599319
0873008472472855748818376975818953479390880198458062
8942112361133354632540040466535717316673238652078450
3080166195770809027132411891654620198070895546091872

3854372611387496629976665825231471481880032379500380
6670118598942195994821840489226045885487688802337594
0142157915429528712356822719394165950006243911835934
2678164048354162384967924208087709963757338733972713
8570051017216390102814006199640295555510920514823727
8889395862250635820057623558604606700333545148883088
0320699079612318449237973339003311505882948380206876
7102409164859588883944324656344667574633584953224141
6518803282548820899182410294068318220651171221056358
1095080526760824300375506483382815036583507743940163
0802437480979848295664872770213417005589113966218009
3523043539548882559026670900065711885335952330130700
8761904282865576790762169073394085607061588936193013
4878025952499670315914362442199778519830043798444145
4865516590828204141139376731810408771959657893608900
6009298571878482454316218024538611683785858623451078
2849182169145864330972476771131495903082734618524789
2215957636176194158034130313191841607759441012157439
4177453130001079505644225251093067375232236885432427
9121353055759262100311221186203829750275381784254173
1173375517119681652409284367663014028433123610329234
5231891837022303444549741418863973986508372860686169
4362513165598064294041596095582815279474489407960067
0401560663542330566088822410924622558187358804282793
4696642062741124597521522027613182620967360927370341
8630680429620948015190112156463233546199496049740834
4435453142115006726825166877950667254182169648683330
8233724458549241292167319098180229151717027639539076
5960845015945874597607378223637182049117249030372172
8475214768189382828585241866401407091409917788371395
6921617076979010060192305266297842195100992184012322
0629140060482178194901561947902646631312875029083244
4837155466248494038615248535327366354838102400072695
0375367248133315649581319529832958248945699470154098
3863588375105274450387542374163505345484390591298010
1603370240650914196544064089309928745635036122491584
8602375132873373130617776438334911416797427956353582

6432656509343897438769712540489099543976103094470122
3017049596779161345460240920721494649710448748038255
0964755792337840850157046965409260993752929750657563
2445478640178182086899760355012046052898752372284096
5641540104304874662760719959290035181390822646562780
0691749938188975755041245526281187695373575052501272
8015922292778267737194613719216516400180711103260665
4657645725683990411126551032698984776200494525732076
3366118794668546807565557788161683365059804799752893
8859344621827274827697573534811670385110008812060026
8451219316134987292279735435145251620662644675504351
3009958875998659114434873302760578327386061081960325
6783353625213985372146327137143733668525942897840927
9847787866474664495788358966808204514776727190703126
2880771077178087594465928794910582812751317845119390
2192043621739992710417481747355300551496807137461376
6826102458229788902686407062087169184080590668402107
1179755073163609712321042997983319216599464118767390
4738358239727102066913868675822234076837140251287860
2433601375456956121012111866856952757609838762420131
8060097300151094877041864750146034719095600164459132
5816011088700342409510860065966824661861370034045403
0565015603210896111956432794011332320624412968527189
7339002930438752682641325237281181873418377266831707
9822366819849255173111404842922636004972983664647174
7032035892511190314552637820369748273764447465796347
0343627745210955607492099970685966310831881197052675
4076084185209907452641268309014347673429855065555504
9583671068719038492443885017162418959241597064389551
7919761248010511832662108395180919193168647150076228
5491546332100294923138434804057097770139827445193483
9221971306082370165463372568013935034810121226605114
9768782946285862320643474121126266335278215774732452
4831241354286604414019190563714456193361673409966378
2714960097805768754169534454405994793961814891494772
8839344493862378544571071502666829049614037984996997
9904177314534985205251546802967974626481966870228616

4231214776292412429658312763661501321599568755630321
3834957850419502366939282044083814911150686568428096
0304485296973825380024172676987094580558823876675088
0469991238113646498178032327138862753999814306474612
4047754171778696333861396561788596483175635190236538
9428609879813243105964107502022049060393598709159547
7942138098641791325093915143022918235919452911502943
4493134123621519818215831203202948600394027669012663
2207066257065165460052747522401019923873469029975061
1631661928185097677922609310051354456493611645965660
2849112845526224785726874706115946260327304926038731
9098427270223022793637175671192685836471857815155133
5203402009425573257024135674979298326066168923747723
7909231564483699398221906223959835129277487404921884
8651486180067646836474766349043546852270441815032947
3466880598025997045147190867269754209092714637247939
8724290800146596030612664416602228604050721331815082
9460988850709805662279815499892792431405323990348617
3808021493341026331550511037223388751481620439881629
6144993711841473065439725635803732060537419376715721
5516252026287912434751769566274504051833871608859843
7046467204976925718626306817461111789650712733894134
3126400421900228426884632219249602699285376809615893
1948902304258339012860852902135855253522872936927725
7311248398058709622089306846646263634164743178986700
0474061566857637857107584274947429648579667625487697
9594106944926811657656925791063739128091743329342766
0082347445122646817244791345411283289744575514650786
9565899052824665349909387151111696795153643282615119
8278986688971093101695910417845024882873523192384486
2422636297349757684392822631120200047132187201701783
7109757566855371393824758533811898056805524561758925
3290124104460613862625297201449551884592486107615753
2019523792106931822465517728586422660477869383984421
5191391884947712040124803222614617113859479756568591
1745747244418275564366755031861737105312840615915488
2124971937173021267439200820411175498546836398413567

7460883816733867417100470330451223726996329675335035
5683742559327885052848439709953415176590329384028506
9219964260910899068429037128452934549490793498403703
9501594365634463139951298245813335313896483039546853
7774238675823879959507312791666339121357622930082381
3749950880424143677063411867945790202225251977023599
5448842923046328754923496722087300742261853262394415
8468650826153186560577857699729253351807440463828973
0436121312487416649557732058303198526492284003382122
9619829400035721889609222760768921737321375612817140
1278947680860184617347613358304779955557011084658991
8465264716029062432682309721379199995026002007677928
4281280102189046550368644940685163746947400979670522
8717466533465347668328529930583917529192568422946133
3033502992661474903553099705929443017566334383432230
4415434370347646640492739502658109126488241517806384
9558473212925984863914337804053563760280606861278168
8892152483564543619165845905112065370194484792425462
0558790155833335432558659101839153275563432530479137
4070246658885855173264155785108271162140919115020187
6161758170251311700794414086348631483190552961554113
1867774760309553999860807938437497622730370371697493
1622920021830001353391812999182960402359294816224037
5734996478981565262268246922466182266003346563155444
0691909194461335922947647617530984014456964949854178
7417212313607795570046231575740701647612827340963897
9769774019876160194157601529109109253122761838347867
9518524193706079791655907057515180512954283101853591
7318636529202913084205780739267574115631345641060048
1485906597772775589233973047601761186109466683938317
0513676765980608645465327538441719332482103500346143
8700642574999821742521821218920424698345969460171454
1080967173547847902896490057093695658507360279962669
6168464311182371981949954017555507937248780196933765
0451347412370232785817170527287540676808677865733191
9180654146507023469304107602604380764983984187762433
5388381358183139633229783045192735420001244347701439

1410202805835137615248683465400922414855558907372029
2194609496783093804725225427153971564461393207175620
5105747947382556303404499840905518931225258151706446
9135994549410789766052839379950219126026120477205145
8836877277420393934927466174231696612724488814202863
9142022781241935332974216945094679545206739569773828
0628009153419557209296208702178162359573153985804940
5991309645978436746168803327624713133218716363946718
0936674734403357555262197725484420249996339317486166
8116858002416101935718158763913937159153107634250433
8406774911006239888597954611355539755839653053924251
1338515197295715072567194931591445438867480941270092
9742577221170197678077755411531464441588881354640461
0034367546339543813657379417552022983704633781240452
7631159587874291541420416206624891261623850077703928
6348472623334350644174622655488896432896084716921233
1084463333505337147173330331901721153077481815975318
7403206520654666303834024724044364192958515662077201
9573193519488591629815330055105279952540010923346567
9859706454051429106571050430262847937593593880541200
8468120716572595896827794362993111922904628749893381
5161612418075418388765972717000319388966533656527359
6570472983705556510026802792385167950336520665309178
4354680053222118117845683243680044326659025462859459
9103854757965209123438950325105958302845145019453098
9232771984892807878455467496436275646169662618364866
6203671557849813839868252876195638573608520419332025
7641086585308207834609373542674417458791798165097760
6748034235879437881661119989595666794464838215477158
4452235751309630861325983230445664681920972502934490
3578658988884052055288786406593898270941098662152737
1617524492691264722285626744317906706605132503314572
0678344046379514260173334959206264618127328737794001
3015357157366237612685283211039112016619481155877955
0403940865122360419497579357189797408115563734672074
2742415773774044410912384855584519673648350488868309
9313898234421248549562023391981900603548985180440671

3580331408724132658155808563355292356505562434062217
3863587100591091666902011060085192106206152172988987
0836133794588258419892972813593784638640814620973215
7488764585454529056485693476254092899106691922560246
4254529100149820094514753886905850782157499016423832
5826612333030842360171331330197402643659281051697460
0611297413499543611415077890155231616362758211607345
2193855111272300896003399371087363400847445858281169
2917284014715929271271973822535539639876459247383693
2611280374299401387580817617506936047380882591607654
9966028549415839794291304417893741349812851299433175
6758244076734176951024244331173208054198122503155476
4825787016508657667023097852712134326096082828084722
2735516127980217943248638989060292519154480885294492
4913238575321489641284456505833651534398394353706322
3690868184748916340829302685671978579660416012152288
3409864495030061363798475278879019534177434844467274
8004363482761808233991610087405112235165967744517830
8192167027512514354946996687266873708278491919916820
2590371147765982606846429716828871519648270565793794
5663048499534258282710020287457514253134878869888080
1239182527547486293592025797500281821975100994629178
3785110346671609157650768942405764348823245410625517
1455768523557408154552326601776743209479015645205466
8753256510525147976320111226602675248332139550899312
6832634930136851924284030942602387905323204787679384
8815781799120758839989511824201625492439937529250292
6833808912965124723281499026982302378886144353189879
9215072001787269476593216105185240507686364891235175
1671937073774216243588562940623594770431200526606256
9762509684217812114888298800266160440592223293316241
7612290874337902228780456170135772375061952160342686
2806290537864968871393385712562416964079324475831369
8859182729992757829492957513048250436660285323714020
5496447338073824577555825709270075915358213622487873
9519806447536509228338732197378945098948812224316665
0106987396166729899209644205968175696192318398661917

9340847425786815461594145893864060229613295003812003
8389767450208633855782667988106569036999081567827785
1637829345993619433669806529792215221536662888399402
6803862183878413895499792007228937116950775706172400
2344872898683808889469325821863378234356912074028956
8718856687096061273862193498732632240960659606991760
2005453603816589660214387177128755309837099020713308
4712303917975574483810050683280951189329272191231654
9409066402145683598744632162655757397928373087028606
1293977236838581491993925841574254914633515482041412
8505256116414384738621579485025909591694016701922227
1520515924463847867368440341019760225485058596203745
2021034019586721608171270192646070460795992871312107
4803511882506823353044969812655209567080884541941022
5351991313683529115972228197796519175109141257490667
5271987992843990372741064588671698118350910120568349
8176773100954846991006642170375101296402795266807926
2013464908265708373127308877034985381688301804159107
3508780278978144452516540770812748847379655033182989
3602610515170900920110222071001669479948606498841609
9255778210332925422202382431616937945524450771166127
8195720289949059237178763071379162077328051900435950
6302783720524286071686319756831994495359654617582379
3319549221835571406382172621701189906243630164683498
7994306732489748402990062675636866393449866870295746
3295592763582274173904767366468327098725400825658274
0793730134212501578722469359336023202788021334475471
1692547244278823478857202047848109668249573259465069
3818393289944085629329548452346954732470720957681550
0031362868183058735976655244562923337098003920244653
9708198088097516775900082945233933825387379751663668
4848199061917189230530293275282052288738997779857757
4644330667384284668342381977782239415152243638987424
9517010656678530267666437085462406145075109823826508
2329219231169465936055216243701274300239492460008464
4191371345373908817105432971996259217859583689002275
3546934419927056354499446464846356921147395454234209

3035095619125962942760332314028381564195812399216843
5705926115521864367293908114049884299540135030458261
6685615119192470424806777874883871318987189673826192
4739749089221696564899815767170428901744966620596800
8681990919566483987127996000606600983366508501317267
0506678073810536253324043615610980110847675549487749
4236585163719465279328497990577018451049091701533568
6136324438948796590343675903495600216642655815244439
2827872271735945948307893872024847342032229005213360
6846053129294097497598893201349050155466399788069991
7008737158891759567768947270618115030196472891325767
8481619091938845977305288817391379634191013912282881
8689576669158175066401906642575911857638875482934362
9921171912710549773853730155778381018844418606578305
9241045272431966922764394688190230293660368939291435
2790067834549205228896117886051875408310491808917759
2609625711828832708643634672781816274725571485025357
5093560194453370570429727931651832436970736387485609
2728216957075593521798282931763040398854389505725017
9465534119664384061183281712258058093138653669016323
4155042355394880397710071250704105678774162025859908
4007176820989414486746229922762892025508580816217315
0838975538840534942791905673448766048309707910866592
2529310175974785382557147194645296290878451956517409
5947894396337685688784133633407353907274376637220528
0224459160534573754066183716958052421718003218602285
8375322597888350188042317887568940231975197437444613
3525974578974005546624424324975934404953762682364015
0573472695398011010025658251311957538915849382125129
6799677253621276460763910672269184411105916667182317
4812066194772280535025793861898710732931143196219558
3590075732545492654405344858762379970486963982129062
3046526023715456974639821850402406160647212247386285
3691454227582815893032578671920038152313070301231450
0162038355855970846362887285686618295880381514125979
2427122280065807217537599560975302816813243190882675
8112139786458977991567077123341300614050720737874775

4227133477723187913877806041160283389280732294629861
0943899480426877630904138282008249327643784456946668
5596913509728092965602968378424831906376648975894022
9747652337370707590295732296764107444779028542205710
8331864160628348328403937671338148094153081003836346
2098674092314162577259260164241310768383853609677439
3896453881219871847087835760284657585006626430131835
6377598343942325631956738892178647425115391464830610
5617618522661484986211735299300394196267978351154324
7197921009990235995010185045226336213662954175879041
1552911630059509298870937205111995320951919761191111
7856568563145823742527336348744262178943734255594438
8272109955258344048078153363012318172504761120985862
1391195038127685768422270602288085792280227899270135
4502689828981286708469480858690773687310488241352092
5337799281517828072247304329560570662234561899656929
4079904301031800558838515495600371015362883393296308
3886118375725134429296237436256902862399081808966747
8407215416482153646698525111835609376953883824779268
2054035562293103398234617217749928761143107112618698
1767165101021328174843226860492899962137442648917874
7080052177899145979368325769082544704995573654607383
3295445503760545561693846299352695559825481436952274
5135159635012844381657623878219028344778419434849167
5433220898865725107216380125745592050062613832353331
0017463563269678829979522312213359229559877714784256
2152819660095824803979607418806864814622218467350238
4649962094829002372167471513017616183486648569009580
4452712924136107744855016454016588210094695318516709
4965320283685563439427552586762309399026264688025210
0234839881081031395915672216752103640311682769802044
7084682751021600765291285961812328923919989837615465
4015288476402389564008009111707771686632584715188865
2134181009630978924681427776744424909848194620720991
1860617837882725606027748920250775654960922415372184
8981913999493043561686936215964291771095256950970699
0063231006108564855544823176816949189203353825938397

9779552017806002615045384660223480584280668080054097
2724248870988991814030172108374085197168446555068686
6825957621317618534741443776409811689746202037121131
8615031805348163709928051005793939581839605381572799
0531735646204725646756465733752324960442866627542283
3411947710115861482251329057472336964545935077863028
1397035026933558672025420653201911364568427852227113
0499308474015508320553450220711151829250378245841515
9542385729092055909315552709371573043650713919706627
0720836605065359257807538799662427829626027190863678
5842034261794272927838720742270392586478996988852017
2854373296389141798564954987123131742107811174584783
4714810113022006691881171390633223774639144278713501
3391013614654758235473132163877978594229259092873266
0398061705145105189352613846356490075823486329152025
8655103110341381404910157351617886075764396461883410
1756544414874774396958712593182068392921811680883559
7127226537119526746391354728890901025277774299066816
8005319870623247556963164794199431877289989589073717
4479138221069168303144302108832718736937223637159824
8511277737658543952206649876302769812346134536976210
4739754844398066496855150018242872413962930020784239
0407701717530874006211038869812751613311076710422656
0953420656279648030919591712222568605086606974919587
1952851172018630145722311125595805803005959045178016
2091560033927158135619396152450582940942383067522310
4876027568331931395370615940569008254671634076885187
3806202839376494141695224789764482741592439389073858
7033683110647482959165636319407597879773893685682536
5656798119405558294689075620974863970515862806160495
5380199298790107269885264068694896100332327284203406
1884542128979432188689707273827362778501372701149638
7088466407860794265555255548566482536726238855301957
0089909441814119681927822521474367302375772647906302
1362700929435193537520244824650445028177473530313055
8784042890529565361669557574603044684002013025834728
5787860347964296622856338390938504059599633820520131

3459775949549591528249136758265577370863098534310316
5699186477493523558769856167640256943679496508265951
6493618563906776134086344949610923085281595759473244
6929978437875069577965234066270348934399220039331420
2215964047530798772485471928990319033113606753874099
2626587614582952454629627447825306071370861009325656
1091062901593703234578465109425415658486336759717141
0368246906821364595022593823586898034205214584241562
1097712594198860517418460981052180233091349159305532
3640211801353993827390760960188127085700221614994202
3522853011978515148254092669518565676534104044454854
8817766062915333423935588304097708038262943189443850
7702390408475017104093236868309790449784932389215598
5002143587007713428715847006930247167663122853028391
1202961584833228762187312702544027509886977752418671
9725803984567834528672337262681942591376892237327986
9636995719475082805749299080160929638564887577436681
3059933003010651656716864331160038178433180947698249
2426608263925647221085630282122858435912911420360327
2060182523237946293109254102512454170051664917496978
5017658680012863254457378725293267127745162433436012
7340397859302221599617751736148666793967656332219514
9342983767903749308251701618528493383442425041832303
0264100557818831854428936408740320396600889234387100
2340926852388496732284456687365704234315669893811311
7085498055633424110903902940206987883668650096416369
1705281565858356477475504883191164843064598996632703
3980010970627143154871743704811217006209186084162459
6321962758189168759471508236368927617174851631458451
8070543563797072327895745053875844710075645587473724
5671626075875582631624163830175894812372734658328426
4334984211990679033276995061878866730634490328278376
4859091686806539844039317138256966595923657348223568
7609504600206957367369539435373448928789454142994492
4229654199206870717179908202751123228830206374093324
3082840768023659962230745072395482913251001462315228
3856699636464816199306108035011934098551287730815954

5015497909832610070008435163220914097131668390509307
7706782579384891592132099286598075166477627404210228
7069580316910197690665849294116301449041755241528407
9855841920459422247224085795452489966149963129567459
9317874497834135019747600248558309355697881536973136
3214452511408292841812804502491992959845617297886498
3652967173740653502757578465341870784213098057357585
0987089232183386027668096878674458767393704250610529
4559344800033794844118691034384898199272178004569 84
0882561800277402469715569634535370581771324966543170
7954879525776642112068569434074073694165211045301707
7751449502966420150856734185613308793690799085988881
9541774261880314414174869352930128628687697963497164
4124217738001969097486279960894609364253067910417459
3571283190402983113155059303861120619275400347429960
1297698456728568007574786825685265588805504465028247
2340621226723098765095246795551167576075518973671081
8664873391355547303871771482599249820906556362463688
8744281635475973802009270337279723575620585201948873
1175736415208568879983962553950672045765637086678684
9616739928990516639547348064688416321612696232140430
0430349787937658955255912612733494431374931875585152
2035048877154206128323215542501036958420117770605811
3108574067217688447392412151890674297679952843460462
0851042298929559015388617177785962596590245374799644
8057375425903395571736901793975160019987583699094035
3460200600611457081297272864924415558859750242749010
1997527856958345334494325002578020434444086828907507
7439617367055383761578786385387000953573335902594668
1197512373898387266536879554300184150448072052764944
5702579946868034994929416867471047452363136504711526
9827810552059962650022454407342871399194980253833385
0588539364994173363696431899803695321146317461717020
3880708663490647863404224584691355904242451402814259
7209433680394042646957622035197605253746691968646840
5748522732221411263468200732680991283596804337124898
6512484713338659995815570362412843119237138052069855

4630522396286001693260924761875232125700995941645450
1759791304831585226900924405531186581531978459314051
3549679750197159130563640796787427438869743181215963
3202424536950908108540107486745322336694887417447584
5601897763958449021749345971047703154197947217559031
0489551507130337509226428947436615011461711285404898
3628782321775540335581513089008600231119089283171979
4615273396398147557956104816547218228209282412622440
8661731611829531462701196213661995941087935835643209
3296418935628950752183416094956286605476082023394390
3693829441070690737842159371100843550809934951258480
5561426027948881173577823141092156309775563343568928
0906240147043040680967454142850010531291101440719318
1060055619529375994409816126454374436773978928455823
6168065730568681889329055524837773886978833848212690
0338552557729632948503625724161794560688062505674398
3435840688586527984713205682033268015840118612253710
7299459297219831403982495464301364141209379446478463
2967308041240315791671468107172154657975950843790654
6392689441643670202671733321428687279300067525668089
5947246054007739214366237747036693706479892806834363
0666235735491883630674089690534541962549218595948296
3529914425068124219578493976249362699766843201171783
0947897645342821592110544192539567389068025874295234
7024625362720586244529916142578749920154783492516043
4238534924384341030380727377077657014743734358079845
1121498902138772611307493125184849712899174590950039
3219062568077239325454554691767350021142782515915371
3922475215102619575181255589919223775605562518625777
8715204024235643008015440647378686471774548533756851
3303957730505542984102745208488245638018117432441150
8866694172029225138714052593329218903930234849521783
7323533446532626937774732504109205508276270136010762
6880573492834106150143211791258410932812267491152949
6919441405798335403820079492052726207312385833278588
7564779056721104165441661470761288361000624384130531
0501400810107987555773152250424635872420813465170781

9681326647305265266870097539010235484005431910303058
5050732848566621892307361016097987301096045787862725
9697182149777459319121724208552038322309774373362726
0091079170854159065400695404757694645269352738958908
9465660935532169427129426114018933681755821233660928
8093686836108412959313689766825463416180797337993114
3491994506197674140383673959104332508378960976546321
6344296844750471808775387966011594691058469856934631
5467151967310540189434725327351012330555664924462530
8979855259889634444538294144883825709671236053389198
2931364903499133212284219660873657571369432863633838
7496569415447710706138034367393954329945548894604432
8621170422810297576490108461303625809885204445652889
2538656183554374669050757948211069811160943622727817
1946884422390136435558225224330140937115484011366213
8841082479809005429724928690877078639433835697580891
3448554837537677719658959365875154327502949340116362
8628419330604817109392399879190088420348729434372149
4612397017034303370791698316245766050613624590549183
5880524520307312984242588018709607849181635833763214
1776552648466086267449477751311633746098532615677168
2160131478190445605770892030801852154088126888224610
8542068433312797584809219544443889667113144461789314
7412936651281987902593291945652736873448363988933899
8436121168068986579756748851654888633769003537529198
7757693081057355173951443795272703800447049007285730
9252626316730990740006849045997587871320935334814799
8072978030685925274921540325080620629679368029096365
7119655454749832657557609467247229240892061371056217
0097933992793206656709458921208390498460475864801144
5523278136028145344579543873365991854029550601001878
9625823206446714509639808919967566146598237014128736
6468803859403266522240875088605288410671979991408544
8700729302201722026030478638071088617263141531392374
8994778191781040775452555369360545903637816192863942
0022669648039758682263453758128535507962060646362022
7634100156253911994632578678836087025243725262830330

1021044894326226275322073667652910916281998887191616
7866976986172106895009023640592917572185945847630689
2124704365027536328350650043346183189703050828391535
8506052517223442293318962943725776163152226873950058
1359156379950090704572007260968988373875369824262186
3149512139883571673563880630056290325475145199661817
2677820789627279916563774802991022509472440941980146
8690188625285100423366590664307650167100236787351898
0417508647603805560882711984788639116966057125758111
6143322031623986195399608106448911512990383218924496
7115181985798502770446978518416280732953187522170737
5427807836745060608436887769304983023041436468913983
7008269396406069456288164169529166544373908475728196
9614641195715806636881312484878296005192365381669691
4413164378212808037745822319142506653727318660158555
7631000545881914608634151201340660568618305399109284
2220972227727666420070998758231590762943912951563496
7208380989724723042038732783508601474116048285204200
7439596077939676657454574157343814376292956109531158
4820920001996834922274622233203492297879772093259373
4589318530363220021852332922660432632777386992952544
0037460654804476294982724042291465600529045986614905
3053303411842327747134752876324177468600510351968025
9504893354161773876502389319108406652138046674665296
4357719605228927287905823336267171801047870741509786
5327441555228155097901531432569914109132995250276991
6412184798903534173418028880785943704747003746871669
8072913698781051913482317431997181756973247169341124
0421932383237581583407505032512102423272159997622559
5360817163965955459215200626349383277938717450987669
5534287878977466443638551706502650448577147587895166
3905261261876717387540459387759244936972468702219846
8051519182603434146351533751735439834658403665007800
8251337619581112539605958941718804721787463604668507
7559560876646149799759072572546693095018122660597563
3832045120846386446499477556747110488258186245181148
2160241731135115533763941801619860829325848285029721

5641249354354937014722183714909313527465161404229884
0347258735848047100304971036786199690396640031890270
1410218747143973930479667127146967452485821961590443
5885750474711610835582761860115988679925232776700779
1348806270678430582303768443224083728557775850821627
3267775521575385493139188944331137097187976549309906
3043708381210797273160414688741044273294030727774363
7828844239775948723462917328964323648950423030339525
4772328529225182097863294127922776106997648396446154
9803039103687476367007442072013486804209783446567078
0850085124892609127081572378968179538971471665316354
1792741324937455510490677088305382910466098618013354
9273647157882317517022952925574767294380718432352827
8938787305850717216987840870936084891275829673184503
5233009100824500894600016837289698553478156770898606
6370243799181871271374835942442963643209472757271104
1804433460130510702217782472919884095444291559247296
7931476641686467990945637704603699887007961285734670
5087176357992672641907758648297958015096149717986439
3231187059023097451683435712523358744257165025130783
8439678124895410287899686721555835181821976729237267
5088271913259289045705392169962315734135981016260634
3841974111595598712194855707915404099126084315344947
2936182597146663520940304301794361263079707780953877
0794982864536667632635334220689453430306269748572960
1880846564902097499552566734013380282307886298068187
0520541252040119990430428919912401546306489607552348
0019294548752880557065055452354879178955987440250913
0074164191809399729468221035701898118676721549044950
2844625996837678251688727079295334993551398511723804
9116955666124418804935182119542314545793295497329011
5497632797004257252928851676055670695788818916668926
4962782660684281788855135682201059864864426897310404
2064203886202141593133433565060797764837286117478541
0131819538820963263739781867501567020103516138276223
5549907816708176258006326519090723097311326126453948
0612746157639746972903819915758063074587538517334833

4686076088964622701214040165795597908155136431792697
1432781160003950929530530156640538544014468195674168
9143950050601298995325206242564025699973540563356851
1706263129378820965576305578326756161629221703945185
8995939277954633373520501688984643648882073146139928
5601576461906088270052183882944905283501856406504336
4175350139349857051006350044232775532805166325015553
6001685586063216180167882288985927773980708234430121
8764982988219507648793745273249775716467943768259786
5380777090931582686989321856754102218113706628915050
7191692405571751537297985467954771694404608728583402
0071682558785035025806898027943094618467258018625571
7699914366692567766247888256367102390508929757982489
5222709418467443414664411911976248630886752269173795
6084643416367635588308512954865371125074490373228827
1992572715199660002166693895605038279519066233710710
2964526112535482018081623405931612383383278721545090
9054427198032006442253236001249893448436367593719142
3227785159626576845253075848553735830841651524777849
9835560996791452905537128993380417358033223313804348
2601916191028053475986623854151203895606113270055649
6689281316755512979967633665355404709079398886689530
6857810173302660568853689560211807721622589219199243
1173048925224971255312282117588225282650923582291225
2041383700828638795994087533142292025537883192590178
8817597894307727113160489156785086783738812236288755
8552661186573367446022611736332880225562068614958467
2266053779357525558360934098916382963659980073078450
0136994588212051974271662951889363661788724519387988
3914985074636701161462355909180891464878247698323775
9787096345593154570680055282070629464310763848171836
4124284488322241634530641777649280600316789700191373
4414599529081813008273571202445417814606123726714401
8753822527453515155242450079079753687991871156710284
5303187326563112819187358142050423077462972254036963
5233574830656204856108640934273833183346343273512715
9264239390049127973028883783463744236046441965815843

8405319108290323239391500637052537770106594292190766
3931808108787655759690707840731732397323550634413585
5674600292812281944826263869185813672160413954633187
9072561693081540996943760021476848236668959583386445
8433913973419577395445894737996499396501973180187582
1543881858049244015386570718876787890605894343970572
4396806766233077750215424777082377904412690412076066
1717514583290656812401908880652059214459722368760226
1730245554637405620748081399377467009412522215327344
8841706368152443582561186966261363834029286644970066
0379967501793763167639180694376784338621490896235618
2010564061401237788509883566708473114371688891394268
4794853876466509841171954337028921584835072586019765
1524604154356067627464117810379580551105852750942447
2965571294588954450204682256720106209862067718216674
8688559677813336730489413888303965661219189330587140
4778455132876728030143220992705290210617713921277375
9052612468035783233612231672451314301832827898769529
9046550698848503343483983359922741648106833596317970
5040480171635758116961108988752650503945545081890457
8203339888055273361740589066076256788760234489965581
8219507130679827984374701471305670367804002790888262
6096875179843306216168364975773394896181344188241688
6502289967808162797562569227498087402088768810359436
9299203957265613050681288387614411959186240022364452
4800394799942440582531726824651352094895966587269366
3488401509928375984646475342403056155906510544916912
4218607878117680038993076090690483506727951214030034
0459488292084535672716329501200711214683765449414070
6925962143328567874574683801853930454624313698559873
2642070733736209526825330022465956542110218831635555
3221022325835419869926664913531929631878234901491587
0014774992191089465012801726186584241199577478446380
8881792358967236549615822753535499698413029394932056
7962137577698966542096156183935754851061018736631402
6732506199565814384095884354592710427524744855320153
6290028879027363715117097611575104474448575002325858

1485607889851283509556122124413532387816233318165611
9292576209991851687924286242308017058600765585346450
9921222138629193200629166710405344413253099405031484
2016003289992319103284017248036032641173958773736439
3158054756344611667442195905341694665636800499746089
1763261639369972680056711919008111646000296009990629
7666495080108508005158670385819718130552317324630173
5287693034978985330460761267069151980521041879216938
6199911313682841025843874830863102275566524082814123
6888951950624472937522403669031592181864323402691929
3237688715177080767495238989921492457417629918580433
4862960608936311062581001413806306012314943627930687
3326876877147445496118241966037173011732267211548894
1447158076434636447645759070334085887937938855117519
4674235340453941225240645707121446590466567348092426
1538841750263649967640397964039506453603305846710659
1608694936427670638418752507639689615603173119292686
3383238674463351811330830743913035433722790714103079
5282271658841103444348578821809080282082955442812200
3801952660218635952461056660631495761270251582303335
2249470789046610507885186132600708705312521009241881
4033311098034325447218753564339432204046595402692850
4488556461425105265479584721663057299456357882715607
7280214821750447870011247793657070567309892113872193
0592905808649783986319434663257922428340202752079620
1076674604694071705609535133339937607492713011765060
2222407844780821493839639881200547789380056657803599
0431148710163727703521449472844806598021024629628637
4329333750053421098402485860977159460346265070892757
6848032018361905498852289280953768213151504355825172
0372860169596095876425139501382209840496122262422817
3404340280893997262257739330366102986819922133775791
6373560345378075501813255961569355010328329942384997
4751524338100115195012131780501387964656284915433248
9194337183269470192681676759606159187889763652603208
6512658522642452411995901898187884508287694437676631
8492238487992413740076722940680731052803953540236603

5209984205504305921203882755655930480839116597306245
0177252527879079885468508425855517383838338519943442
8891591225364411866964417124240013588796072191061349
4230833029789663443088271107467000523629974326102318
0271422662226187505725439690773814742635221552448324
0080437569669907102947264051780301516189102680092635
8769818418013034526647105519950731606432675504048745
3177281647974984931681635188811325461499640318314012
0849999754505654405665114835838717438107104444681995
7363462868930027137176430696041478322732756789030508
0957691434783086703540161620281849114412320084399928
2131818484332288134255124888668654485270842304284009
8831385549010037940264844762363637536465110550081029
4066099152481479263173087744064207095391999167556393
1676247589083224270729482954432815122952903164750609
8107151709493216616813022200848999073351928484090014
3323698869379175997792387280564485863635604716964436
6020445259704868221514419781591232274757877216398657
5276098408998893737775089374040658456115404534488982
6356794496286422470716132658749959558344008024494005
2564753112758294582455523833993886160216709540903950
9622984369753605794438167782815663517171901560867850
1022971066937877214909889193684543867259730079749381
1034294535811892130291048572049967356221167333663500
7643262740588295448615756961432047896330591725250029
6559546804876536262815497576765802778755902367873481
6250424531570274183539064997237143086239536428333798
5258809365635808787587211416735820002378585791714654
1612112026034272557384221551801587463057767793952936
7891253297770050822625008193741645114168473736572522
6777890874579968237345277327360629946429241699671550
8622928060803167877150190204161663020493507588377690
6618674616296470167705634418308967626618874403297051
7788145243401251222679410412177617218288850815972164
2083847933339829566994034959379072820339780087960497
0137807029401457061832275296034853028371922261005939
5644991241509277875461366823146128946998167258824711

89854476146641359747240400116726464033832999040150263
35271218579913818751838215422530479921538890284616533
79294723637963334793127084664272273704354107685379112
13190343311924506345207673334380916812903926710929877
57177148028222733070858590225913929052589740024375711
02169955326576133351851876638602761997003980523930522
74289337090169102367520745177016964047237538638287655
43190430290357981930446828632045430189142160750516999
66851233644518831394315814046520685035597675284062099
68648400146329880263832549562721325827573448535583000
02225513318596228864977249448196664152819040702879711
09505677755838364707508929280129921465508984652700722
69657168897401324328795719821723119028109909224942100
69115194270447735875202660217787299739380432917832166
34672128872843369790316934859245577217598633216922911
01312996493456569456831267284809584292509355156153588
68203373672201361285171957991790678887948977874155799
50785828040051987951437931024097351375424452291066588
73007865462514188208080730719268983913504925377543744
42026570165148549039037849153357835239195091842294100
07958179462613046216881844121746806220722871046251499
38764917833389258535941543991358005859024298540855722
50448942910311306684106105252152943640589428225619511
50902988534967011852089646433204187932153336684750099
09379474586244050094419795259305808470573044171422800
77856570371279475809345629087704798834697169323551699
60591551290394654649194697695658010447721221152971788
85424206301449359990364704881686963945459873956649566
84468008279740648593976288861542063449595204778764799
60222248140451871122057621282895120964242624397691077
77918759891509169674884969014041781462488218992047211
53978970100410044519163746354849377767240489630561766
08574901906641992085649882441665925913641149797211055
70920048346356219112592053159495207728572853502277177
86911343170950747417740461125977105440663928887571833
93323600024450260387599951742135949797649404000414400
93986809319328642332313807310726052347022269955029755

3364133333637683830769912223911477705585997784287425
6964525973045897989161844009118754738104698043805595
1700629630329433750112437691659207229530151254321394
0544337789162781914062155168208847363453419799988795
1611726102841063233698534566227140898250206912867044
4116902582047965765068060833893544908621143873825659
9464349788032327217582926945169986312673587510954845
5878463140759717201962433708521996779288308204170836
2821886710429402426005844004377358753310704188814221
9209246071491335029636905846644883203194741017346112
8786735179422094145466041853403015518155623214316574
7332666107989803109068170082688732101936459561785851
7345054728589800787287211541725674024419790288432253
1541019214013509123867111032321373145940511561470672
1289593263819675803769072313032161582473040701388589
3346366335976771547070197732495488145171495615889159
7270403164434951218597470414671715097311329473848085
0210707300489521237484215403899818595132249014418572
9193570943752415921554569296311501449384703394893076
2435538342354395078579177058758873286872636137723131
7957631881191749399736458295599559616847144784415189
8543077414559430091627277706400678452622218860633810
6724847269024402642674133907219353005842440622594642
5394836856547845053434905296743058974864956438929352
5069687282557307388653479795697379637394163125122113
5723661242014026468319875234913753259196515806193872
6661939160510493592652713216922096224639699245339494
1681487697594502275693160173729782522593211392279726
4469907870797211292701007289316414132897554051129860
7130045424497219982559230173355939919666258862848902
8016102977414728147217996074304686368394358376209663
7059217800358151699129476731548326243472252980038009
5958755554513635248529233660366613345215784920268506
1519492034529021461785142032423310422848635208968797
4218454003873494172832011762737822647963978467771365
8735111930207072225600375074940781039463389519984544
1663143229731608084404982813543030383363163531454052

9914831642560125106820856569001603029729165846789183
2210586994891004078010769247782572806721865866449357
5923770660199972606595255433273364250389479833660143
1993073084809345161508804807646366675290866716936206
2492873981488799043653338716396911672736970273126537
4284086097348697293255278854199301904168428232139585
7966024873754065439260849531863413469468678923583360
6803394455761856487011325964277558202631925680997158
9448934540735451669323844921499118554933828244577076
6882305254697961282244041599668923715929509392373211
9547894507408067744489003806244345752246115557238942
2683859305152775497654543180834902387291984674869316
2608871792151248292476158935141491415890423510507353
4967969487491863344304793625203651055672156988823952
0349805230153122385212513261664494737046124818609901
4395654637271017556216112211047224792650608818792187
8564564770201918708174098274263885178517823195293419
0481931571564040017826008047464154536425857968822131
4712021950687073703931215333223942964710143388176399
1811507421555422604821990245008205203155158803107676
5688121985750384512044736027969238848943985040776693
9191917803851311790463726457872800566499501595762530
2767342474903557787303206946697620679371095314087874
6609071909005478715022757386156228403119997936014817
4018140726855934642470818651372676127973427764124089
4070241225057591283320448767508382482335490062243196
2572928264805660096775092853257303888341824250441019
4438374908292890770441518151343279012631862709344102
8058333197183938084511247877577905287996142480968537
5809766676370156948434874317475748991463889163350433
8362739885110295590997268995590471511291794555912698
3594293067385743048698989855944326198964253434921711
7176194986881381153736011925283763481221877710943925
9322057370956269816464526459305254130817680476849179
9670945909756270994574641668731299851777131558862076
5543315102630236084922353201840024644269498222009388
5619814174235294211012044888786517620477231007235577

37117569645402677378698782932384884658685482430725 13
22459971819517637820651677017349639072911973231521 10
45083889636900343634564977138841805680298414053230 97
83687878873323574584371677859623193118212996544264 22
74603311656218995807385709140748170907770720601258 25
53725598818255400017096790909741338551791505034624 13
62796294337527980392121612449422857348055409299617 42
21867552670663871540197164959258041982845727233943 58
72738491298062505229908230414417964201863239335975 64
08562647211409871027568423284710544204769273722795 86
93432551623728706130624894831768300595031627353927 22
21555960371912609270563209001688446422399745990762 83
60386145156011467908671952274422534153735630436368 07
65820929448168157562440758354209445041481836940072 47
87199371608074714370480527241227205762001482655673 84
25852761520422575616775663448908355159040347559705 52
78114985130250874121655616058542729230289933165473 54
99079156121786647178134339282499415905014092363201 69
84086805996772364631180032309172314490659601839443 35
73246799472136366714309332268725922769959786634219 84
86047640383312151598246334815753891362137470506267 76
09493915654344496650307157560190525614934341239865 00
86334976877258201426160358764218865753091740518241 74
91784121530322238300418806639385455889178762006878 81
40487669276059762638850841876717239068821513753446 90
74205279687593862965749865441776294251870300911496 13
52844389205145007155110873094664959499070899793052 34
01295734938668817859272442308152159066064996075502 72
37608127238705851213727455288861773544544959385158 95
68775195180268779856482520266240944486188286727054 20
74750435367998458468021181612451191791640838822097 78
86418275681058507677565728648482836037024932871581 98
06043555879980375757476331720000544959849872516688 56
57063033528760680930815901814105937213785607881031 51
29253175041105096097516542537103085517485489928079 27
92165082670247752463749983785047234114872240388787 79
68562165891841573565939687030319350750298138289529 96

8303573043060712075466299805847951077322904191430681
6287029509007188141342145828415611632764589797794318
5244670333572201518300806773009843428145985559436573
8971990326286100716746911509026594642792375562493742
3512174450803121349987410210504026254115763114123064
0337384023024844739361327771431778326487227872000031
3243799115845410732008325471765533577884197388111987
8308116128253343500137910973264580456753562692848345
5102531756976137831443682524778543069370631432550964
0762249427096972762106167981630745864773136210291691
3190193505391736338772095930772880211384952253085233
5642009147582113215081416345593732766381646209964150
4181427926147848561122509697441807399401218649576170
8774298539083941990118885877336373113130171013577790
3347562044395262607677976568538504151780028622026017
3983153578949045444271657055964920522231883544742831
1193469603711941218609396474369683521630084113092122
1376123619315550911877534644560429373792151668962024
2547168037781827463859079682073564093429943342717920
8028875221125433179011414911600479638960331877220471
4551925930589486933504992233576520706393366578610808
5920057759573577060563469345760388491080506695516093
8106943662128758827331613228648314314717672115704619
2356146500377405387217627411136601782358558451731002
9820778999364681776875980577196904429326564149288895
0616174327395453482331666399791748498402747835405359
1200222609439905312070766019667274321466731325059919
6153749191206109264878195377790614253518922346613960
9531960625261784257158699243782660916171746497163472
0477389613148671942948249029198941916758308889233973
1174155541726809475331027377979970981756504505473602
2767862106975404505926143883778151617925379010606402
2916738026962573434304645300421104252766230305520724
7573930679272639371318872288012695855490424866322830
7022774015552803422055731726091592927513287204433777
2363815466022426272279552426404790691285346647439567
0390153666448251186234027804025378088666113535664410

6913769723882365405370572032648513307118001886217776
8059795321806543675321022250428000439940618518128895
3614073372395066311517070005713863153021329368553801
8489869696302851089301202179506470724877503209994836
7568717247002905581456984051446746945071887173763680
2873473556196853175307566120156930570344309876149723
0689528664441564074834588089865256616643797202895868
4422039218194317151275641117761475637140593686400010
3588026389125969238170622763716762874806283816022759
4105114626922888091294330277664959472497384473093376
3274600371084359078599766718005586870287301832296672
9256651195926100594158100365089290626039997891076469
3101952271744645199443616999155564156412151087143820
8088680752297850814802286234135318439205666397115246
0890481318445192314929106328154027922489378228251545
7682716245961176395668864617423953715865744626643996
1554789051637325218257833325356445898929059519260586
5979867134482744782626667898419196273605935202214966
8157043655690416708257527445881757281160956148185722
4369546475050830284430753170779235571329348761178390
8130291059918355226223746867115757059377490937975793
8195247331632266235982695699804734334402616879654751
3042934616242661346074732526957031148814696916429336
9071948154548179082929107206942973187597197310154261
9933564615328361822870151559033107061465304217006688
2533797013234495060714168352686098813122722054090309
4664606618585799991415397814484774156408225890354064
4906463510615433719400401386160350714559736014278623
451486573479621797846757021898995133364438192919053
0085773995045234934957189684612711376889575979332349
5332089538145398467702851241091399996240942861535615
4952015641889962125930051264420968659725289941843503
6681880480752910597233600836548235701919868550926035
0048765737882951629237418327132367686584946400059670
9506778345361003674425949188581955959269025123931107
2595121211563382415896067374800718324687784130780969
3824148291518956042755017542065174420881340145436070

7135560267634995975759600410361609612137736218202235
6398010145592493601568971489793336585499186349730410
3495007905509710373294892197640588699532018966493350
8204310048852305942984868017855565645389715296386871
3982393892788628313053889870441633874853236655056254
3023828613176831474399344615609310765384947584648931
6215158358898933919567329443347903909096450020152545
2974223609334873774857090601864807051699125755933251
8203044120573389116924949793744441817210180048695274
8158248607577122172413829852529767035268850421330346
3703205901112769270842312247374039903446761895700102
5917858966147045611886905543180001357411454538480916
2384560193982145769801540367447309332421416472755521
9087739691741737350641459518460785124018137745458837
6298517906609425179969503658723513291154069411855800
4057561078043579191051543895293017860705688578117217
4213915509532072119708984152215425316479193046160498
1176009994043413190991189215513651226115501813110735
1940674896418609402848692830550221199243438663096612
2998376165898127473066900407133131532581930320281496
7455702892711980502308342949072461054910957995507893
6602634698465662818805549010438789895744093141529641
4337769026050643640983268217633628709882627239743023
0055063851675289226483750950886137219833353460698489
0685568590244467888633643960437818236493160750697952
5366177704480786285210468209326826682897220715910980
0819778019264952538304724636079589391737003693289663
5802205065980285338702996809222867542712913386999402
6633577360863754047202114992733399559638671394141599
5063555038162271317992876143298924595866321022805072
7201766328290281395136246392598794084119774242147849
7488792853481329226175804296953605684964163335883612
4647760471766303398537726717373232324351929797334237
6460670072590569784778225901022471861849551087004140
1552763492243058506497917469994122470166700310104432
7626530993015284206842468595235910530969681058431185
5103760808536810333170953490813488313117223593774387

4146218392650171560903279402818993561269449639671433
2078290473191666678085182551977172880062773545399159
2789034010786288963661157080757926371251575321256434
5879767582279860562178539046344387826022476983164473
0911677313769865439441397481344800381829810375495058
8539835429146322753291226062391782931996213986918817
7111184244196277187899230573504472457753831194338517
9322128576603521216877901140477658976778435696351365
3291514930963803910475450116996758780279997955389558
5500590455332979356370264077033348112055967910966088
0465445819911756967335381794020297742084467140554762
5380016657961957199126200807816682028859158624857236
1559940162554777079141116006764078260807710789473437
2899115676130685073224963159123163419758846276472881
9202367627163751947669533254204908916102491648373349
6591727080014711527101290890296121104047246206562282
0963283626670888972846484919450548524147558133923773
6269212766280901070396032994626272509471411769121429
1335397513015143177467168584029059686222170801110366
6071463020620642207396736740275444511153186803573711
9706126321435523468555443824565325519496223092442226
2761618107635327121848671103874863324167078904688522
3329211115019790098723766740155479167534474858911628
1268686736042229943560768269783317351763941137568187
3118531093914733161347146429574802588661209843333626
4478923277992171893811049025750898332957523113851163
8411810192449913293008778472536273658801679273239119
5667737734602923116714725275438773239540964407417449
3088103356901689944732650629356812407468591689254650
9211091423164339664349653553999052260304711487117501
9510860362143788792840744982527033242516917795343239
3380537534154286334492005727579681918742184272188588
4666266028133915912265087032955629743121006084764623
8240612020974088585109713450244534562696748452174937
9519983660135959959804421055339305799463541215659260
3739545481307090026816164735807530907005794698595121
8576692820433593133365802104393580161079082794266448

7820352801574984777718753666388687146928492233597970
2018592163752637064707239232807117749755236536241706
2631546327005902663040247398045335302040939313049739
7130791718151488632385160351409187151727259632060397
7518189877379429833548962149298830651687972617323342
9518602919791235420914661761858081206578509755405181
2624547853587142349872282450762802185554164393735572
8734131770795331826410695802318126782729262172479047
8673313230260287901476485433580999324437234918849958
5994862583067600012204733634466868003021774428308956
7321206573109092985212685308293535203316260961238719
2704749103169411516483884747974567712343355744298126
8446143275337106037702381158730688628896939413236300
6060504289965200451060374867696136491725117214171045
3972369837657482509286253199176103796050507004745275
1987069243830797208133651074580862533987045295036577
3947943751943255366001421055646414822436061646770791
7165851176561085923563460948549764477962116551131870
0969902914073151483903908991815918578332650277953957
8418251970561524675181074563304570829594428891506667
1592976041280335474515510043994933991135740036810821
4520100371663337695212133753239590644515065233379074
7504285781596952756961817847042381784203159924171121
5728175313825528990831722270803193340184997462466150
6864137178679359480593272851964335736880274143158690
0765208723454663736398318691202096562075413488741155
0435179457052021920866286215704650129595131279374407
2467620419226655674453334447296817148735449387338480
1665428264237833848317565438333617440873218792199714
3097193907561528997991933481684566486989431576014380
2862633533136185723793167236606367549438005252967139
9740350994071219337375857120455594960284445640461306
0336222636216293412245761511654193879168481328096246
9524445695462125087911893539832219637899949870575517
4877188610510452587091200155027181112140083303394599
9772865870452341916673040685570047172861172633588496
8271071745003538903363106665809112216112279535205973

5631542387862792211740027929927660272309100878896448
6719775106448528542367606806783287027160214912208907
3835986791677907984654684765443288633275459268997647
1361182191936371970943091897609589330741950915357899
8159456268174031091186213611238703266328745925123801
7221859237596420397178011973301354548630311562876453
9733301035351993689089171658211844720253940470931783
3060123964167270931216369379193323918425977305276147
9229302123013163652956137623330528454637744966783855
7241630555328610532755207843894044247233087001494007
5648539493897085636662472351155496842637074224198534
0721884331711808624785109999817623225805812020490727
0236751559960385584667283973473259596127104496948996
9280704087235561355018834860982733449421192795115963
8914217013371362540595915840065763710336218594354090
7214950797192642474168788661350962013130319398165644
3184231910367414205125568633280985520770932399557422
0458372892438309481108423300876415366308472416897637
5194193998480863927695317901643727802977688806162490
8419337641036450961260406512736947334321364751668674
5418754235332490452514001261991025504942206089908653
4891218519778520803538297935164736163639485284975628
4971488562703642543761525303485679142181383415467656
3036293594327156888851139645341755011355523422660951
7738178180389386443090830539927386531988392370825144
3497669579512540664055821324953476082446423795952046
7403716910402286506016440118821281688727839234273692
9260620640964091959614590431451723416161791510706177
6717415112970097436263571691798097913107607554440072
7482316585363917076912591900555112850732808167705134
7490741450119502481084276777357730810360845003755565
0268658270894906640961146299690429226983808434968138
9149247988622487167128124089262797006509374129142801
2018819220654215938973633819322591270713038489421629
3191100490714922536282186203561764468544699594307641
9072713387818263384790269051413488524088341597040931
6671764584851653904600109634729323170245268608078649

1800770245426053385920091663315079277873248325901604
4217156687494057915189677115913189275017804451824993
7438743299329143554374680946834026083464252681707351
3602678441171175476803025782843274127129555092671085
7402304746960026445711893018058112189257572500241791
0664730201129469375495333839271076783815855808875670
6132999649915893949904087497782355039210513630164671
6340862269365394034567695186527752685603128680881568
9169916046013679356000288784865017387036118613661682
3370063762490171870354839165300888065752373767990681
5547888893864623380433678814473862636975144463533151
3645033652509877954130939941467601122228501278273455
7551595619844872672888621691139127864441826501071593
4333181605528809809313757602195448423668918140487612
9698357403680117551891330057226994759192287243969471
0724497704047329675133848537289891985144879126933995
6272762863015717827057355238450193665288694250301571
2886490989930558977451480649740071081376020676606100
2833539832072435945672059494512168440253056141611504
7237679687125269315631930981608232979504258981667480
0878152648677364144935695842879538795111120900413882
4350699988820915655540328925022880514169678792992662
6862224670525490667495362501326970031824510114073519
2981527091168287631615254533623132422680452228896149
7091739711353525544012360861881545414708532046722994
6939071488188603326828261722826964785169840975561328
0910904929942058902099758680270118297143811306166501
6560694050941744708413659317294603683231488678378340
1584666526277938110347185652734290112646968995135220
4381388359254084508757429340483048052570263674681999
9711139249943082380948147319257601152853824735720831
4910527160816992228141867532991179552447748792024698
2478357701790581768433766677768902177649062193699589
6546765996942872180109781369213674462209747830040927
1819051376356123254861272145222616805180293256818310
9314139665924531034423688433970673528726638300045419
5146442303262301907189759856124702358650054207598252

4898199075031653803249502601693723058314817314752430
4359424989148791890628026340912272673533448537779853
2768897047616726158528835140603525270885199292171330
7057857638749393745559400967615375217782801162690377
2652898962034412615988106321682532064438164061291711
7212009556747383916722296235557461243901559905448832
2626441625687126870485003449211415757614315487883822
6244938257190720528224356540306686433949527866391978
2619662128890293170809150693354760936306950387796483
8065009708771258420744211499716985561589989747876513
7505785362724536521780662897775073271570349854774716
7890295666395835111199772543088210830083871970300163
6037548232031811034519634199719570801626375425606969
6618343629726907066223061431318636181161133168418495
1612964799463540815516628864531220105617962381014438
4620141325246851026413793411662166660443555433967260
8390029334249856059230477254301604859689878161532425
2348894799274995680405750878596158465639968827705058
2480803752624440992284265581071965313962147422223415
3507700313618665229024242427339752232201197300895968
9104985405447427697563805962622690878847643676551937
5681951996304422809024719659779814112299761130996689
4840654703043061615428405289846055561052774316709454
7976542569994432561515127041177684024726299051846873
9384403174909227786713746504877565400352618233613582
2096915951653100302994702612137983269955154794300452
8250404116178992299479111764121739926937741658202028
3502426115579535771019286950264605435924118006680782
3341749833422352511940395786903578680997957355566463
4818410923535663805321625058733961273016517920915269
6307741603539343614876508656958944166875931028197227
0842130060698903276812481364340882914506935350078426
9002833896928900367663065196212569113708251495264130
7300205723426006143479478418466207633742474019652349
0639302966223377308206402287040880954039448926023755
9302757838186727111955590362643818036944102698956099
7022402685189290570563411576345663453530917836449127

0655146521452745160957092696019819351482504230830933
2402085693823257373246556197838050798236783914896441
3212119032538371930512612143512054346721380249172084
4572406756078389118361442061721960932418878715390653
1193456242314305059597581389680014593272680369903153
1485898178421841408627035413234057140637242334416230
5201146005372433545440858047849152738356053700832984
1944194087857728942894298905564111848901279881742427
1309417325022464998977618499584448243196333877136064
1700505758811206260189035461258593451545618175684097
3147338420149518937581589960120875257562760332950030
1183188095642910867929936491408742632266721386849152
2412990329146293202682373490956625790320642804533851
6755725663359643282983690679715448949144144284457366
1312147165257729283228387225191227818503331845753752
3118138891046873011202533293433033228176744479092066
5632501883887499178312452779568780325185708787710821
3218175422991370299903463408243198220018181430169501
5867564772318455173516019353974118068162554986334692
9742793638368312286209015008476329602715420554092347
2197748755577372771253584379299733675504135390096260
7546017704783200920900004370304772062396931123619969
2306945192122807512806261090339608085511993936257664
5605845474892984566105164377632302047629334883313664
5533457348047357156744499773471782198157392629435661
4853325635257380075373424585696273226443292539121854
8350084718726153761193599211755449468751722095340217
1496733230085430312773430084421703922356580523746997
8119523847444933383738577485114274622522039346757212
3278506610526913279773063462887372622241958467166720
2215168082910005267022364151265227407760046197949668
5044241492903303752615324755653009315314557741560785
4888437204157140600876512807613311400021517609289824
8986294506264798639727812087334479298478545315123293
3405140684725574692848626315035477092571914420142218
5887802572791283311779822123368077931168758654777139
9946239543986001782171404451158779337645825217591991

0881923830051663310282837236134127214072246237953912
9338836418793155329932894879874861538613915230746891
7410066261860777226791348713632214751656850844199178
0694861954601934089370819232141926382775337591945703
2645023630434756871734529583995536709739473113745139
4332819779112222693972545912493837982312660709638222
5967019008381453286290461060658685632097801508542233
4848110590617385229862052817896049500732570427222020
3936136382479583103543259855072621403409859627786017
2168955987503032882817680409468520938864033636523649
4428576533381097953342025875230660994737779174834099
6405620837330431676710875929826666843546700959970485
8953748415115221450224994544152838657802928530176585
6291013881441726693837902070500341910121386791346354
6522874814071533820290191923514672126838275100017394
8051792235759103106294117826715838186378195464884312
2973630207590729496131322642355108491026499847418870
1812740398720306793583123154828787803868672076345498
4951991134450991244247310505227252766832066034853805
6734851263693194665299251629026264658941634139609150
9721872364027550026970108838683249414212571204886964
5658296361609865368598837883902802070607029639962089
2916924201175646292127178414438660944484153071327538
2741805124756047008456141960786049544859255813071615
2717681871096104170286462445106386992799031329802393
8322923078600246111212562537492992069623605549739779
3370905509150615995807462647693070614654733657295388
0108465930773709264393270961733589798755133298517353
3580576198203756071739649512102605682421535394322065
7878065433368166837918392543102962997862558313815084
2902346041464285063318207802667408575042965493539544
9486518527564708814351323195973497899171415169373256
8833893316283389645184887032263989305568945183919124
3082932515654023675385004309455227522986219363499930
7995606896844661874598947488234136640851885321936731
1437589463565702142223037174148120127262829105733185
783922733479526068004131224044469069570034326579109

5617342284655138302877708170928004370327526445576200
9029489870172647182289327617882346799595389668011402
8668705263367060063042612994608494995638275599060264
7776521970253758306411814612875438760985782899634221
0595022534150439826096187609835216523165433169772144
1251770038039021598137974891320292927755438711703391
1632248075246572497296231247650935179435674838114315
2864133302908912377714661246904486455116492679934634
1556211882281756423024051694895444281683141404904380
5788605901073700671829849936504074947027855738627203
2710842602732695690064120155580946913710129842552905
4495764506457560037403149458790821054735591136399067
2780648145919170643387069714773665247784433863025569
8388102589879309501971312840708918719696749394002657
1940572215929586883457866981031818359493810271931161
5251530174090403194517238322459633052678626421000745
7363367972646143529714988846055291907822957213456926
4638347921759405780513036734887954494733446456067966
7691278267990494200362880699002603522166525266488097
2246721212946167822822474271783410535858490938180843
8207696712262215564925244641011600663839118183087308
5635422672150172188913491114434074231672018580154409
6839417218455292470306663317439699203209991372307939
2087063326814950270241836323739355756594835586434275
8527153036475346746011816231218086111379932483545148
2289863062536933279374737264046931267375653401997300
9076142621228650115856894482080371428361204858316174
7503907712876046503361236135224312142049114096204585
8292255435749009027171143100562027796642732820368408
8351421899736766128515417417015505596692954335533849
8868702324902061064458071692286334339185539443465974
1831033154532910259130360646226668797794557349045467
4882327531737599593723227310371044521133115338289304
2477397241957274401165418484315564894048921358055708
5576275584955348891913856437916383424089396022097880
1958750476141645787338434431980873515751667496820037
9153796102973494432109476073270046363343661259071179

2603829657765048983399682005284642342068544946993038
7124964664248581160442000466693398574168555172983698
2926358491044717933844683250433844717587252699366862
3375707985863799511764743787742210295932621738817179
9211256496076654905036475301128460597199864223972784
3391967774038958231917557325994193790085492825980660
7678949854843333553305204429781468642262154639070566
7804793891317765192204993576166388219632235722413875
8048818728755477834305533714162429159181440724910183
3736072586131305858393796369137316050463865378761619
9765683527896039165412211971231637064638435087505880
4657553196720080481063208311821537956138009835355952
6093637000645317080644202888377266908268009424750615
7736530695369994647344426417990880723658569162389963
6517578076237318613662803000677595254569830359350209
3103401066548823876059063096671525803190270180565107
7417965996417788950664060278847170680779275557035102
2237147306795006509607538053426398202615407127213785
6032274328861680241733894597905050321379748466149030
9530174023009549575261795889698360970314291408480458
3842017705933308727898829210653986085497841770226800
1994317231256072796693509378461673808145347108132937
6345219647441631933117869064998248237276162056150244
4394472323379106960839688560326743659447613243668623
9105834352637258702655272723546810973613675379988543
4022478297321958647384707984985141728538675277923065
8409174320605010991022389298189386457216041689492340
2085594048059798887199075389944836245759181795872647
8548243687178428051181657010359994896167564581441177
4359994155741564054198094077706078181787327808839235
1665272981172947045182489488694025397849704040125785
0170852522948003264485539829339541025049341054446143
5613045371236961682202427087546803225777224676453869
0691735846329099659789270857241360685294722841899888
1119769492577756734731492045418824993538607544853832
7349316024944583018400520110059712112248818992601409
0339058430141050559807188444154763356093389295582703

3563839189207244115662413634679375541673890893091868
6080312637892309129166075500989808404308771738687684
9306238533350915041060003830601639488536879210612389
4105743940346062401637185484252177167545163976002550
5022643961152599429430869349869074629783759970161295
0308438036606600589226585293056378866958466784875726
0025329183930718547261012014353181230082628245390756
5263848166284306712414091535323017373577722317054545
3318573303986361162909280796514000762580295868325211
3035625213499854006783290579810026266376780517206247
5401635370252168218735528720401996359618873606934730
6728409608128864989228165452185240832827912818493863
6352722030085982754459989899958351115743687878881270
4855717381485740307803629420485942064433415790169383
9596815335852775087815743971924322779883170605463400
5330969611599543732039412995517719740924937281938691
0424719168074575580541317281683365537965275951040258
2937660069379484763050243686693087498612911515579029
9089147511471436165509778108916891593890432286116309
8089616015436542397071317339876255613839334927890605
7471453816915692648820151026214721832503409165624542
9353117328396837417555069788772460439855626108533737
4028770997288047611491577857651047529089113817806546
9222072171325415946797805559574054495325587792843232
4750482025729610721193054272034454311190184326515998
3295119242549956886629245120615544354851877843376022
8457318552553020385780679964233347394328325507976814
3174935290365355235708336227295402976036224596787022
4679610872900653691581103297724112719687184631712015
3108720228291216785136832868288998410063083059699732
9512401834379280807586687788984960437727540275895229
2936685392267513992823716159644737329827017509083756
8027446669159111149779944667113569108892437919930942
4721308073081984262593144347965790856700825628858836
1144633070690191063606859951853870417106238568043241
1229940699769765217189489934971880450386432175982864
3313402327317350344875527937834864131041994965955257

7066904567184355021562018967279397342682162568608592
2488113166647341429891013875712757048305145943663609
9246610627201124409872399971042075654391506863102013
5759846014673026511990342986506396760006966879582828
9243397825905874856782626926330346837233212406615776
0295156537226106822903836613368341504999895934280932
0186542470360735907656081621959975934382017246180769
5817834722127150399123937330805981643494623136717495
9999463042117638181478301910213344735692656258805710
1644687984556617203758742814909984339304652393122000
3564424865028020022103872381508554360806108595381742
7853246549792311015101812667413846629462674003407290
9243067756491785793427751652950984600098628219865193
5014863141311338234081864181019598887229593585603437
2242367239601514627896565485335331740074198424260136
0667355298407544025731777149540219275487625663663209
7951348389232398473093428272993909549226286852582803
7625710408140414070921533779824712193346482520718387
8537445007238525936056765957622040219451924792912413
0758546485918127845559512533948537732743954653252016
8622505372850013045372400046474447907459782510294447
9047597268994937537469280893311554355051420516123636
8344100734984299470708653487282616826119499544459888
8503596079143671119639139320911200954133512885508992
4933928537947665616415925452758853479068034859304210
1431778577117245113741846243215533722405654121494232
2346734100328640923722757147317038093058466611410528
6653492921570438437193875825489184989446589748921123
9804355925364919086589066917390808867500913233054266
5482077157364025208162483016589873036086598083796157
6736411773627346697615666539213482934564239912928078
5997987881520429221519091416907854973618725169309999
1322700067499723351557657951436647470237487661496440
4926134860829976097836260492782317388949792246852097
7599508049882697239249576598722306469511876779916054
9567269969085152582265295227385885439302173427475574
3918744113766339941285948312343848488127945760136710

0667616597439589632554530670816842945121140912212009
1086669899891500102055692484852372255421310716613919
8282765742981882917518337208417523869676828059102315
1992531280144537722164743682595086088863644346720804
0799574561042901019656088083930982871606160491216360
4586908622289737564557413574307159108936724233166447
7332829682418831492171649497251401194936905670952961
5270432919617564101018514059608395422101125300432032
4477290450956867286869283797899445334732540783200542
8354880451308874136319369558168287460790465669459004
0744288418738123256769967164692679986955886052080637
2983832112386246812028817043481558140629498830626345
9334499888403865260573742230738578664000237741531288
5909455125753539336940869444293940752218284710010776
6580951275670201477540829825943655390077790618030037
1040301910921852932847824116551890922870290124041600
4521493170935775360816342085655233201443845388958342
2068417138823995322706356387242611330217236088753196
9248201179065222750848154653606346843208352519510833
2160684317334365846805913015740878701822959876582283
0020425283556693204501981988171458261171598400011723
2324846223680233784983905719584208339418833025000410
0260037883422114836730547496096779242970449998379947
9044054349710896265896766913002849909603860630462400
9333797909203575516255166640057112187717239030015039
6095405184581699938643044980401039916128659347449558
2760668348248909337338629266989646970531741560892296
6624289143819273723567260303050110341597015039075941
1599156179251165622892441767557720263927108978526059
9471315335700459048301245358602245707716058212332185
2958758220305193728902001773432061942873421475237886
0830070299796531556810301128799258939183387796470067
5202703688772405840664369190902874387633882097145801
0174951013464584028127801131681398978065090074076746
4220963899804533262076514960825977452275842390413450
2684618616814579533717594622683030636661453659920280
3008432528514981788177127257386753550285133836792305

6743243686962027275690495694721422424679884360411922
6316915567882764842223911962740367145989741445431800
6168862933763562397525481610920180628944206508650886
5174388445174402936157089106653051819134408352417385
3908952947331269090022881476173592405472755741008722
1186024807065527347854646708100332528804948728188464
7664513871948464700273983663967869110872249068944525
4499301361359823021009664966265824979074179330260447
9616467896121763047354709410590547677874362769811146
4819594676539533132602160451880568520123818538359935
2509055867304816958939312668888710724516375807869185
2980464437598493901498640886729121561514693505446800
3927753771628002844461708872834613320160278469351417
1037189813592856544404738893533643422529953563067148
6435758226615070847224212139574905881236472608079566
5391821078069759191962729961376825052719016801350182
5936503904314892374222182997294359105047665101984354
9677158363903560509027440945473762000866255189537989
7398695524944209452836893729162255864458818572322005
0973402112420242701338138097506387073687862241346162
6607614186589036757567280494685139294924694749767044
8286278503799394283278791220332971397543843644722779
4195248300533133083265941268165481431836724185190716
4537118394561885376718611463451009876355610396882403
4693227431638685638936692078262878666463162305865623
2080344670332241489658442908620119179775183607898117
8470876262961531940034781546405063456598584539593367
8392047177816196115197815991533483239756111621222104
5289683087138354599880658578013548593749042639560201
7295786811549407988789989527859449531291247582481713
7108859096914070619330361800303323891913216840242371
1785594147938175122615373552928204846201190878355782
4107679589872826483801883630605162587458132380717021
2707031160159931956653211055908684463723011219393528
8299328435695606597198930148418941924696514195047131
0036209138468408714367868832381248718738058221779691
6687267705286949172329692975741293715765031048614982

6499639425425153552278926558176593281228135199049986
2833891769509864987093885286522464162414980091336048
0941616720693342425001725333590241224520696627428380
6079157097461019323432744228427903009219716781979659
7905954912721055386447240760083100058808181907244787
0343657454279475046660211686153282079367042283157677
4109787065652889958092151207900910624889386465174860
3366304880883858583654754195906352901696079598667197
9195154726753999884776218884785107366055679223745515
8009612734637037295470996414489544035707005097959712
4970791498940575042016503073921008375739413281657808
5719802851130424796134514504277763666054870573901649
7996388006749276335699901742142470860427633687015388
9542554486051966156011457074311012678360618976337654
0859558441673969989899171468666484090241900149311117
3462062958230778705794867646385567587210995136468309
9777160945465572016812285393776737409942830395155749
4531692065820371452045827753578337982711615753554754
7595988090288825001513269030621837535588152280080499
4621992631395147590076715044412010286404223234675721
4652225543337464540769554429637336518329440811481655
3112317888568534493625651092338223250887519700640217
8562624050439203931155127424198127866118045752037903
1352263821500210772130502406624183002862477655911114
1308947497641704427632877747366695141529627972984736
2290196362231543631914184179969689628035927750615513
9875578536726635378140081715318318793314798031663073
5418382498547746143475821220350330394913462672508437
3973133134613497187865221548479521132806558974510020
3248729792239256892753749027042598668561490405375284
5046626144264259912295769949845956168866129342741521
6860453695085827710553806842470639686077813289931062
8551128799543943367009919152088861445556457442279253
3094751032786399982860866477824697766936469596820933
0630483252302728616288409185402107585750595233549174
5175635055894316749129117086207384696004878978263910
9056257395994874924959790110901916146590802074849626

3935827926505836476767083830119688555050518615798097
2185298078202754679077735284594885548642095718480957
7355026418379662201056062017672410164759618231771444
1988100961027947776081962562468208545749937593917557
7255504390164427090909940333682021018189188094314468
7251194488397607261064289476377050838168092847287045
3085471610727866631012203668875829064624965329442150
4042608960795596028483768115833106563981887154102229
1856363755468086146768060626175912547513262658963341
6570626511726817054930109406586302669220422989483023
7644323671421946364900320018910295105537319742039399
3038094287078665529147692881385645878149664779808233
1482664021665548467134062401859714121754244097871277
1828778534134383738258099537774555668639030970006149
2840773024700917201995246206245391985859237134507423
9998517182249542889513234350318233448378831929595533
4879010992117189922536529429365336258259539094655296
3581374497293699746511875338537487094217708081745701
7422516040904388461215741244528502023798390369991138
9696734778804970420888639312843891579868614995357206
3769482149209306281251312280804996662625532242838399
1735202566745252299084099463258646834111304208458988
0324142878241486082621057749475330037706021516825542
1685888255205289137193877249869082025393362940473047
5050099704407094693589196990053347830446358119649104
8316381606807432397475188737745048539320801189209217
6200325412859001921285078008780108096121869972156727
8787836037834285050223359104738610037900336819582153
4795312420332192379118697973810932850103678278806081
2745282883103918315704414803727152691119589138273502
0626616788136738930058943498719262754217586783775861
6929115641969549780500502717441482142253177166456489
7625594375826764309491260552928577535565273114929560
8791108215961940175702492044626207794680537769554416
3790238444286270760013358829372290589561353109924487
9237739700126380103906210362971005400902332662052868
5512923889365400766396639839257245082449689892624599

2439438450876557890902818985683433451099619638409116
4376705960548419525350568787230520679162057973980086
9587500465611961545040162976777009654701852843349776
4467949602803722422923120925815523806451507317582639
7848971659561076294495873794568519845906093691349732
2702428238293584834762009727957320397390824404902652
4593739625729544911772960859885910979831916191923715
5743147773056132758650301867128792233399940922106267
9094625850811692827966923817237985512762993523860883
1771468972857155910479691135290085367737548995512471
8689039574466000484795401447802882232082425670184511
4754583859463997760412123218294512072077908176823315
1618455421959874974455890151199270623605189634073539
3487564095315603407069359582576081667695587492098240
4806665275624551923220032514529417553964682719023521
9576379637897594175052158675420192853655966620145045
6338552783571256712051282275196092953909245784188554
0127128286062203184030416252304573349919862033683941
8549191744312814170274683480533123636546408009522798
6799808167861438767022991764204219357703030288924063
3823371188316668923472566531471542619736924881856774
4423683067628580803557766708524939432726759182713058
0282047449868310923884518396569291282691413580228508
7920263815575944588577839176304153824777450273630047
9676856895905742907753282403183887419532847250575823
6998984185573609737934792621075648001276066005684856
2895296270365033407284992457076821660600430851921449
9399461592740186790670249250918808425497538250005736
4115276562788206831683745613611646610594529609876478
7617810657325699441300696040441275748480590815431525
2191475930781041409918043677066395667263443535624561
9356407513565467271679094118794884808680923093832873
8225004285130861556467382313448911976530330590349488
5070904364085511708968159681885355596836776708287592
6959001222681306340790521808314175342883875013165292
1876633335743143710101634779665888361886988349983289
8116400702596550204761199068845930641188013743352809

3795109830522058657551902553313247393551842157260882
3101683181764097241796642821705539235733182359972105
5914675809909601571254536259146573583451080849769670
8071726694371830812496641392852932420581483522627446
9375858569571687034502237757777835981585809545667455
4627297947307975693205085626405035283025567228667754
4482840862099813897930370925326425964038534996170546
3860837230314894810013719990131970126280108496986832
7908059952507493775423599997878374465778371237535757
8526777106381435613678907947372479242618159928019155
4236358409224109269519498675837014008069125945944555
9254670473203928004701116001559541702028795035929201
7703968601134563044492333977435455716545727149517353
4529046221587766932103972565405338682391305809660021
2123248311387049072616349881777525063398803735283044
1378850405352914195734486262214803329495448889085450
7924805426837419406691885551675721662211092089973185
2667678293526613290427617120717343300234525891953735
5747509268743152936342844964077178567399584381489421
0462851810384964342773551924438540110560036409186090
6598300233736845914995840144499012936937258939528898
3481155645581061030794545804756646504893576785275211
1834079945987442056755069846162829748169274340319574
4811212692198913243550319837054664442496096363374848
6558714369342400084268306946647268678216054307605555
5157113010549963694214120145286056715492814503563608
8579354204498831255719599558707858336533055121083928
4998411268479057129722024655013870820524474927234191
9503603039394603476770851534725433807691354302103233
1182709941052543731637179189960813384440367350920911
1063173376587401600018697304252984202448885270331725
1469249747539319823503525226176160943848097053512457
0875131868926737005074427741207097904073463122620529
0006391892909903319643533353377827730338009564355202
3728118934142222131638412466218762656292621316537474
4095230545401691905912102983325487384431499690081791
7666244456257105500140603668001332495809064102834877

1641935647143040955057638038573852209871616795910485
2220807083954438166529535008774606811360273082488566
2613928667728371703032378323346716406418651059916248
1767070634565314676407449265899040024929433820347665
4827303415226579042351013109297567830484813632976397
6322746732729909893927449638572144129084870644807046
6163067183262697301895505226175036704437656063860897
5474809199526394403536046543998237735619024650269312
9589140222105988734088163222562586861744532441158949
3035550997748247415150737341194757333735565276274185
1916424514620104987826294536895088446232117635800351
1841835926759864297128139853841446909691719079951674
0467818935875027443503551180799677762498706284759291
3272616884488849391411287453205724835916236067416163
4463880131235116126332141573507358737890898646033191
0104790495108805327624363438019532053136413435545936
4704198439973273192818302576008576753025832169671464
6834358840039142037072426708260176162533902989867152
6756221981306035295948446464040396885600648176610903
3056079065038768968446731694854343891836893537933416
8987646104050602410935985553266431299975939026367969
6925137693692615910228511721654148892157262235977366
7701463984585521047074763889722549022179557834738536
3008193343789887481568027697599949512125441570271376
4902775150787799491095614895742622739879403322128257
5324578456195527149717737489231107175800448526108753
3971440933423641670007394747605338766380254295863552
9369130347698689906133362459084735380529991695237406
5057657039349015473906556528925808194469628401221392
4551119876038074056609941052311643601925684722053285
1072587588370887878354074617797601531730495005829381
2475052503021700236109573670907840223511622259723813
0144407984791813303210544324431102719791005913738093
4837758636139977194367200655924569938147027619184666
1211804296171866528310695776609243771615983151245936
1728010390163655204661937092525125363139655920117782
1427680194280476586534858158747031199267709791335049

1107205165243246231253831574487512954156035527502636
7961454610934441685357810273138996999546867343518103
4432552595169300233005059257279978159048202235890526
0479203482507504217173455466425312528525947844068421
3337897972455998381452902491341277243424509717951186
4461506281528392865023721292643684081326894231693151
8881895385268180047631030778038992441215187198582722
2544989996778751458791602397076666044133305940017725
1968288276181389025401421411540322218807522149313873
0394389483863488048858725731296054402267540119444344
2712395569023790715007138610641985838887956880548633
5172808444644724813277238595610521301309101510626429
3276316247222859852571026371594629994013226550518804
4657398980037754421846299791933040060517616187986026
5557607689149567962403302092035338230064198575121280
5490876812649998662968202160083935774256923900145096
7655189683001149858039500662463386168022550668707591
2748319753000145540601541198078411349482946568060170
7418219448269420291459193117972944253523513933101121
8688316780023266376919519389893049565596364430499826
3623322221317379586337527290159802676243820849089436
4283124967962162500163529966893044625845804167249071
0904142795447432277640558606449799377481597906129222
0293061983352618004261179665496758054829018106894737
2216120265716263043635302282129337245083943534370967
8543243805831288950527866356571712880369852845487149
5079856246655577931705079028998685971436459307797350
7014015227544766934198239263898929453534319021801693
8758502877868797020461682319735199280766975865127606
4682383916960148671115009603845938820051061526646256
2727086373994715025771810723089620436264155255715212
7904583542959059101205185086051998335829520276445242
5123577351536145132912213357834119671870758566063500
0297664587218996568468098354225556797699786152958316
2065743207372109968440619460852752139972077460283796
1829406778265980996983586608974386510365624255625084
2354354563015121971111673283365507058324791727356514

2769410984986357066618802528434536119774234900016041
0913598065825325104777234873758588466697858029999224
1973665041155119628160047321576507005166289845399637
9194144719612753684963848418407835392194951607607529
8476708413827460440301770757999666767568612536105140
0391716817256780501389783718658379689761501720984806
0227215120876367111635529356196130402092739641852869
3604726513966875563040087535856868313141286860922825
5512422695956799303502490113667709364034990374845874
1989108901894570519857812478440357578671393197085549
8938099970654105791690209598897498738447272613724351
8065616499316389735111970331039397804092809479973433
7726502122497143409823782365196589288627922327799039
4182050685698376711662377215144120227663379491737437
3422831797279401193745390490579714461718360256732214
0552194621862859145438965623402489453798119559649687
0707302186078051319309467852848444202128432499307154
1356442307938627252658520849226948439939485305352727
0682335863948608170577407516973852120210628959417716
0792541306991346081438246328669352312659073430366809
5385956084501673932290965428288540978637787221825907
2743419564661165959340871344812029957960400057641468
4856384192084025832685521237954962489115862600940098
7654135858706192511365371948148608571077083760209723
7465955311440307339494234844861525237226662531720908
1622694000211227591834255298281689971968761014385191
9012898807424280528355553325232571548587561447752011
9468011155094405429657310193621595918758219482207475
6153078334803008549943308739821340270700031128588792
7967396273660920712697115138195377155464106337455584
9196316917552555991892240897978331242735453841796027
8459589864706009528418086646771109464159843150009597
4947022075958749369609348923513155308795226753192288
7519326905699269590112428008437808975802237272111281
4277158092157861903143823282221584311973638679727227
6849581363281470182765236169987038316548057146175201
7797801074900520098622253867134378987984881044503027

1773860682106718234856661032815984091857149084770741
2577372152962362895142819394934492310024755294687688
1422826882633819921070500314822969078127068502360967
9524561563176253784439086838854547176625054868861453
8858940190650918410158852088336961298773167275269197
5623864276095136945838426185218338957051864132632025
2293139134838082128796433881362984553904237312857385
5900623287197915909121817103349208828736572596006751
0345169173034840664773124728536498697032255058028512
1031458139716550054021999125846712475229623304794762
0417483573396512933884625986054669020687274310798920
0936008629962585264955692634224434907588987120754055
7239177888983747400628311035989363975368314316379155
3508445994998151790357191664595797264053563579622852
2323209995625290729565968625666476118217436889365526
5842097358804838235632703846291741042746340327640442
4721791963389233300452352920028757301356303821117289
7133169436372206175085815202908497246369055667628921
8426265178197651953852464303642562032129591608958533
5981540706502452521657088226438896955320033071238601
7719942697998742711966035304852835818441460854913450
6444317130866665673247447943228054733913760627581890
4283656586598954664885617098590233571893647110221915
5448014160810863286526572025037347796586980289697568
7596956655171572997817414912551945083377949744669806
0268643181523422924316776326510155053746777092041564
7646247512496723011428848395397060605607246531422732
1889583835501398502247980616382349453661289940409355
9178092658682360641984947863949273925514675962185564
8434028716399842416564079224320499215193527094274925
5097209837640554799507636962370089615853142082854978
0644065756946107401242830116770917540880488062665750
4874497001006448172817033718571687699052050431269126
7589423546424263219268161260713525593779984687684876
6637463737084830913023031875975519252439917826264027
9966163670943691125300862843502978866714838773557009
5408509510925426723708716285087204991001466660693435

2543968132422775052041208431177836208642591374140370
1893905849130885307671803377759779815450406004508428
1692653954942424173439682579794296332233121321810780
1293219793602750388852631045872578879049933017249371
6992903363545290749651464056091275487528815748748504
6656478815713364324270157712065088264725709115452935
5415106455103509471072017880017924641359721384299487
1045977552798427450697742656414883406830080923054646
2603894832672239604062464659457420252210842881282668
8967527898024346826557896265626637974536891092934689
2092484109702357131627533023890207698673143258427609
4818388124585758734014097068671895618131122294700219
7844824158154450981988915294242890693466056260795656
6439433934279983588235711675751163292556214994709518
3522331245810913158655773591758470369414848815038632
6409866052441608994421423724467318576854052616459608
0359046160459477472320562878910886992342019575526455
5491518628810370985562898184920845362817607417557822
4666387907260467755483451744348189049688056376503250
3535926271374514988624054829562473693334496233090273
9113708105840712372655403944406661654666981279401611
7438031442545836113138700180051064222504742126749539
9707590971527103141655363347732005496354914667194992
5664377111351964712248442133833448299147401916929709
7827330482096570236574416621640413859757199401107550
5481383567509753677020862148022476905759312845796120
5201060652722159599745589563929274161013544777148660
2722228078789190310493304860642348889412653650919804
6804726679290797077022905020257671384436845815634829
0022266587224538924207586781264807425984310372314483
5073934916328568708798935308983035538438395007970780
1508993166772121535578504768244806681206008160925249
0055065606882005394397529404297779977739095481218233
8828159978628439344893791861555638458479425929894905
4384503736976473332256852424021601959044724016399444
9258915452220947381972857356081416294420120829692342
1367519056921053373592377816006656687636414636657904

3573782443665166851049351957765790791933549005424537
4836258264105168437864450295918358983964405685936888
2636863710502768783853127777056659956030015723582371
4715472190567142601433687966468436491439499957272215
0580428992181009895079934494421419902144689255392867
6917510398246758246373438052397350830280687187344606
4187629932031217040704830461291642631982086285078695
1729018997001434289682624417807819162644171950445075
7666856324245674581755185913604359995857459882503553
8466741328204525370108050902351148404819530588806416
0408235523512941281404845448810370856942626000271639
2301008603721424248429126495719569873219052427563394
9016224645462593474567030056728375084672498252798367
3495617708983651654782788942618610518327692085036036
2178003391524933714844455014157911625050689091071382
7580224465050986098688327677971183257939187162167678
8356224198867538683931575658977686520163945282738867
4406517875660068949821735574807444325577590692736378
4818051335709627018971520589097081198620052276793499
1330405828456720358566472410588945530875223646183843
9640259601125485287660886628483076360128700666722802
6700036149423026125416550458296169931638437058296750
7053229269109161196749361273729163120586368847905250
9527351543806222193273002895992407939074438573669093
5294925868944010734221780243707715281612425319088227
1732715433823740147454362184333256802959940771301498
3323704510969965453202745335070217703707061138191051
6358803074747708198082653195105524034346189080408772
8558871026191299109229595088225181920585188744198634
8625188245665450780329536348102634843303083724181136
5562783019391016183517403223845097914687521623877144
2392232324557636410797470012553983247122601905304886
6664933322848632905380868351709643540440256866791168
8444342169402517726916720054236587952458648729193194
1963983705910834659556545737455427472252563872049196
4846804561216346755578001839114580507202910409177461
9788296504861235552872697552000423820381832076425536

40963208933924544967598152309215189473050197853510001
52995735305428112836484236594743595960959569801620 27
530239594199533446282082649792360794218868041106 0241
5874150857519458061568880834301854125378195459697414
2367787418706672158427522319352770701887762803032337
4028626604207305052378520354210425772442559140427008
7490764352482693868107163766930730372723741717542458
5224773575827029599664985410231151013874323704799159
1787902994485501682558654515388125485742541460429120
1232285561488953271771724012262799441082686913997298
7438255681581112623873262610426424919146730313932407
8996737861432904142084411467415351674268973370321906
9028747706020088422190321251655289117173856747773113
6615373439175196270795492161709378005434045787378968
9579406584026092226706663974706474619148511446524438
0741520552128686202006717232636847179672315491533594
9245342892887485931766426993620967340749774505307056
8431410133263287775921305762314700874373845076260305
8757749787324207140665136179949569456010819284319373
6732841798935189595423519702289347297697710497535864
9956731850469509873966280139531524733606745957465367
3422509659501098766962373414060683935034819838271821
1868440601761576560254701139537357267834526945596079
1770944971727234694733436780387223775747916868955520
4136215380142825486737769528370384142793400440013056
8978355622798607136070596604468533433322470819960517
2761522010806671065237950190709749374182163352993865
1891700778898834763609236188052690600640828079713434
9789427595928872026107851555411239737304609759231788
3468807253835105908002186604490289791189672593675521
3964729068579735073675718216974036670698961345787450
6109712035014795653751641661531216473254416789277507
2459436281923825623362981037565328913282392322272506
7941709469131856699622676300847493329492377248202516
8055066631619903457800502962160950971278313497549748
4970807501469286177972053920877355735426344654046917
5078783629783037122212634499537604587068254665673292

7753972286036781924732607587537503639605557551524704
4790468927900741744080991621535352379551593168250391
7084115733894721483377058789369541534827316071703023
2024929094546655312055250126322531741427372940893582
3231304140359670910492571838352019535775111030301893
7387366729568834759002804669015028041569220627941078
9768280962696610121374881195563119677798046812493064
7837405316274760628345868471645943824367753427608269
5576532344276053399712987080520898985614420659347221
5753513573135315096432568632699760118167688723309047
8473878261088300271876502608262925233194476909940416
7002640065555971369918146697911356778065745105729647
1196382643806989602338811353072149858536870862858717
8926879629780160894163936301209416355230277734299630
1526343577512504135198341187362058334547311853807457
2684337206905220856255010506100937942856140474184559
2117933927270859725115288005694028053614114492402092
4620879487418362248544537016734793590201500699089471
1195009547716996064515693409809576087230611679859305
4474249458559276375426550790985078262274524142805641
9619579470161814101885939670292884088175071326949126
4514792458713883472209570125453762871154613584471013
1132320149549094464014760003023763285717139536547149
0013555869633069258112640479200531728092117912870096
7881389373295949068769162309178222864353334059339679
1602428932748444663155945748561132045178306464916622
4181324629576750918590298833323065514502362940434740
5492556117642216093884711734189574071998503527366986
9338669851702573938066023027910628085352549353166194
5852938854013476198182979019270269975539762709721332
0775214288831363827940377954810436396846216952494822
9844322968969208533555308531740953971002744873252835
2757362479458012780445503610606455857803573626252556
3606477349056863832460058826457299672867064706881971
8804899591820953876986724126105812313371883281538730
5324063517160488373186348319448785524534021310596054
3269787362789902736235815268667728648413763217540668

9989734882611860180029360022362615884959038938183834
7815021647310891383695373808683164369908798085930128
3735287622060053622758728767946579168057635814324092
5305502388654829492572512760977104308414241327149223
0145550249153801165157010725991966088910334458778020
1842019868725579834858927941157916548984180796559816
5292440028600089283308995984612515413473641247553705
6580724960733728968639565510344975858300171880139293
4081593465774074916873140199038284277122623332446058
8756739838593500769513118556316845738386555122929408
0306842203625672459181138606350480155226167063564964
2867323459656693799243587293291166884983936420697970
3901915931945597036125926270637083717136079722292448
3897365994926322185943095293445517054009459274870328
4351993881402670859152894963595076363807323470534623
0932441509575691850480891957173919165310002415714293
5668690970675538485026108040743406457426342832522110
2071034503745383407217192728609307970908786402740375
6034196203260951802333194466047043934800540635869102
9418314381980766263692292015196267454788905487300853
3422088159740328925356782478045723448555663884299365
1785938154287147347054077625040798071086832571272096
5952470280931298490597903061967508059944421798850698
3161096380431857573493208970279214433939134282900983
8902927600998103497167534005535026657548513582069817
1894317365218737272703866524342059269683995858771658
0753629304917458210275330126702362227330521370927475
7549275403224866536323928428878807181194344775443943
1574633737742190514462638401483845223060132636502788
4514717047905831805834894085694942499441515548386334
2377204069960193358031375344976028444995154109011381
5606641323294310535403566334982500900534136214959747
5298028239846197283670062105846139778158274676579826
0178472965764639589418774249633169588422839119159056
5640228193496801758163840139429208142088204546902994
6376520599881978317544801271199655622131732442716080
2193166446071845067024516046120117976382723921134833

9438798962905840179686360943255300650628889732392513
6163270239075239598265348939946806580594876864627514
1094064993115341987321729914312591009777188686945571
2444493582861381259764637551134279845737620234356256
8983122530420590214907099967416032154670753881650296
5863991553151342905513331653248308850347019549055674
4841010321876589956758394558238286883108141862835473
1947552527471157540254348046617748038786859837871569
4913427853082947288735412043190232090595295432861066
1326976076266926611352114662527698412773408524191382
6858280955075837578751829441953816699647593380581097
4419704088768832525737438561825911089750196431479372
5720708094058960553970982858004455630785998611082978
4519803998823094162509868026351628228075608165070483
4896441683618365946329769197332664504433270265296440
7332608359487129720563680362999226922055550219361313
0943929256168982589380953115434813289518491654287254
6286351978103023730034903799179307688612204532651318
1013816898791956784667866144331058014381259912791415
8876671702906759990712229281727452785443191763775186
4885446905521418294754607553734560608556346420396176
6709575287454949012046603519646368735772929742823475
0549678654459273606276089892469782418907136691030092
6678191303055919511695693178319754079624033842110466
4465240458186863932609646353033471129213154369571442
2067237270190321612831636606835353591402798852609531
4744197670576401090753060472138657067665499726561399
5625908185085303045559284076141246522180996543530716
3185074886478933137159804019105801024255417135661896
1206011069761203387121769536277481470240462879594479
6568929166656151629117736946184946177683166359452851
1716414008796109655867194211638165457893559437474165
9601910402650699653760891088490540908807662422362445
2531328521785687211710750728725802437027495835664624
5635139772059647787347672137096977874372222228544415
0516258147590016001034987342164287379415209172808743
8528706874529967585063462361565683806568468586659128

8392993987491928045097599357621915303453396402416281
6375645733798596901282927418796276250380640305799822
3939350958921995278510291646380478329362091927804077
1504187300689178573817825379312653226956429848057505
7043385936993433934560449623259372433043576671477116
6426205667161937773757701820493615978206551775960445
5742714015958506242086143202210279470032864409749853
1194933972205256017243806783980806301980330387140108
2373702021780239991965948474208100416246313999893872
6781296983713965334369780639466764286002415282487378
5639412927935383607707608300750085468836684968344738
1380001948099946392793972548092617557123230722879914
7294996278293181211799953119139336829142170800391710
8392039962532465712426711480762187323888663027326326
0510227485587685824829962737856303750205262948831696
0894901291631372634885804981924875455352488732623943
9367504501654768934020682145856556617105077511943738
0589413425696041313947581919706682302632423409024450
5409588341087688988058360019058008561994910332388450
1331396041945468828359061480627917027425505698363038
6819040807698607465088442367761705462260845439722219
4035542026520331044552690070468827745821456966366744
6999428841734711480702347951794307783615275740011675
4238104978226516996793270179099132535969256413102120
3176759602636795510869259891936051529316116399937906
0522162182625734473633420263075057264252255255006962
5037213380532448446537714971705777121738555021403641
0911783897214179763547317648609973237020876571662328
6273640666665252244484423704743126027019405293253920
3881245611678392260714780011957184750455611615250384
4822082691866745295001945479554742670311953388463367
5047534119243051728905583063960642730093178990339713
4393158405916100858639382213820227170819247577821001
5039163848666017090813909135953340619014626904240956
8052624010705647776618407365199659831201591486191247
9104532820849003769623579045204928147481448465817268
7429160111256780098117726912220907023785148661124371

4459198466857630947375120942433520044650532973912516

7598491780867455498329658519534616200108125422426766
7244659198500373724670009453141832851380224433867264
2501595675921141474962451849120472064675609405933598
8791797905900136748853864506869620656149908349682257
8568113455648395727288903768564734858702787470924437
8417015400337456982748232469221776707380599850755861
6687137871868309680967655837021661114077220785221064
0268269888730728960516843783520367402250033012942208
7997280733552033208498251879476366174290552837591285
6416497413728193651143325413215966544797508896637844
0583413844824557338593122367508775691745807770666376
1431383703360433458674650157199994935709263149831353
9476010997460000053765814494573326745861063304932150
2973839393544273709938641506048173133810760582530943
9419878575323657233230479592495750580234655485760174
0783004076489467682586409510388043743992695534150692
6095520996684963621960975419902566892730018303988411
5526354875841019186258565195894991754367013775262962
1235562608907598144724510634531684374203926963412512
2549425532446603922385414921802547488287657536406695
6656854499049481491528050382535285564672004425228943
1463574597056054132111347761834591435006804097516578
9396711785485165229206778357107525513953145282312224
6077432648578021471069671258249054953252002980118743
2980385609820847498781051624257385362174946834560794
4013868042725973721779959761348492683692590494544728
5453485554767285509262696633351543953191992620902229
0117916503540532053541557323962865877885010012105094
4836436935545656529870251042733731127036864327018539
3046229863901849849779085270466381007410606419934676
8348792149018237926231535776760045253983094883499531
2973156738110357671380950602689881032606604431764355
0299533074351217835363742263073089411890983341196386
0551522024968443939425305314272823845197724460166040
3995835472791372579948769056309963892047062937605056
5218205421345799773554893538368033652820760812514894
3624690017425014836948910324978639419114600540230621

6185569908610246731548646098802027810734737377704918
8343981529606960617171547709155804539147213794488630
7027510625910079537399951394861744903022978921815233
9558821015764964020945919409001606116587411111980734
3351033819020401202992932403144329877164721941875892
5676345923219909229015572393372888607136940617761645
0934567152280499827475092783026359931981639916855626
0449567993519483203844703965009899853801126586133337
9477552130328891619787933732768447413283162153821350
5022329824795198558425306440627327567040485421579537
1733438301365792271697658152542272531902691721583563
4790655014339322211845457743373186545271855941021062
2948825204347308783812222983168723577837337344515599
4823292339269578929447998242009493926642707394323274
7132009716034757074410728503079630390757270283805020
9591617550700050166277924293181245052386723996858519
3177959038840467556513325575821494315351795949879017
5939954018699586162066971538933325756001809207795785
0127158525013243702679838153216831510588027374558939
8225165079655680674055293358041645869285231652892506
2441034507254751746696957664759912065568479831778838
6244416469541919134116638313595094269840789705549465
8072945983183865462177521063114551043363353661775730
5037406342920891277502236209418309382037942049430183
2564881514621747314953122724966519403326669465481748
2534172522912474970511461616005428188054035419894723
4572559079014198518298148145990271405143266071026229
6349194812593421453378987245830976048618886695851303
9072917339207722568960983652091279311482179747506446
5597613403875759224022396034735708491833781121498925
8295323354443785318333610174372171270755616191438053
2726602449486684750343807525183922459239571783023471
4190223503503807941127277487289504002308047407224556
0012476207315992125045788619133864990339126851472433
0910608559436605624848676475330103363659857748963154
6413465281763741261581100781736701249544796541160122
4900939460349933099940239274943813504008029448991687

9393667610867550354692386570894147039461647784558930
7011895036016968430078158866532913562055689169725157
8127543955416215195210530013133622187193814571974435
4678463679883187413333055129221001574730478222747354
3623380608200983366453685586892563315746920243151683
1234067297731417050798536830021146553627357812063386
9732468790648595875816722867648874376546504490483805
6518022973211179540565437944615042021563406645608062
1749083449296578888295379303504726086216823201549849
3360658850369596066660761363412546677142657669669382
5333353829686677801855476375874137561497408516836293
8644445437282528212759657334188069780403863756243213
5935333827691436403187220519444882699899867804379050
3172651953332428847616800651629802990750232478834434
2406567828812886076963740649602563071565676750520370
5308391361662839216418450982844674305122844423983346
4130153027392175042011926614572750828139037673363576
3925446052971601765375773898946913977067171842123029
6681287204971364532552899308640123305232260540816113
8788925319487861723276315274802047699701084142223997
9291101028956919632942803327708353525389035143185882
8631937429265422212927628526004225453979981005955366
7839998923929180915038777361400928153081940786066474
8423822595763528517849241474448436843425206682156596
6919769728805736670362156355124994436122668933058039
4804546277509788735672716123758257138391394108724319
5519516053376515555892535518269797147560668931735310
5319152122983591730230267583927416493142243938974431
0087449119622244807371585249475527582841371687338856
8451397193571743513766510489523902901706309271226468
4066927834839988628595942396793645641735587184019901
8757254634606762070626278962322034805371836351794691
0877638519911078379366902264789614265281989949894409
5836469265144316565821247179207892033514059366782849
4017798965297981505475435803177585622558690610123023
4010936129553535763584329974630792884081670233667889
7012814595884742819942874986614377118597010460786118

3122285674946913190044825643028262007248787525394397
9012522035179906701088644441733329427038504777629594
8438149909989126253488800224701126853866059437662227
0365082672212322235157255037650385409525310175758687
3383531199693602416600396509280727438071375470484844
5688684891968721231098199098730747953205414010395973
6196967523052442164056190052674275993979137268686278
5354305517912575047157660184923718223934849391969295
3944998838019657266250365175749403311969597941172125
7623713831651147962605579178440278188122553834089639
8270497890725743044303341179108109059950541712207737
3797477503036812582030995844119486799985794011171393
3242395262703619927063772417033978111333238271568277
3420147905472616543401932267144218019610533705026333
7427319045518794871348524986266822221111131891445576
2210422898347390349981259778110809013186425670889103
6742980304913636514213249820339892382623311457754007
6316382512686668485031584111114115588642679072134604
0922170507459824720702455243140352011956531392458331
0091425363495878979074393713659709552725556666007024
2028391007459462466313074544675113059937775120411928
0556497291512349150553278102298643060840537644098917
4443107699871727600381516344286065218306921009179927
9417093193629424774580681733533597017598932860314853
2015687697915652124189670040309591727757081603018694
5971407998244363332875431922140735996952626830385659
5845208956554977810132745394340864386526954140660420
5250151308378086645729974925690376170930028161622503
1870030327371448605125164072390070088238239068114739
9580435559035124122182329827436677789038538586239181
4781588353258137893191305164723901590515007329258274
6204289143926675349521549429240927856129414258693729
5146893142705444227000932416109334447817626188849164
4169256081359077579447386206015924040120633749989542
5291163409524485314231824745868679213510269662875170
5210646078470497466615451881415103516734981578308880
5062902524727849132883585857196877031633009755390490

4884566448974688882484225042274107690691547824158619
8518421957909459139269455393497074170826012991361372
9331990899612447611270277043889271701723488617631963
6850246720826698760848197526515117846839743308317260
4878540303329427864436091148976287974130203367539268
9318594580161832917944010095832498058750645836641276
9529285982657703330062345826549555323166532305637373
5121952849214896392942381059559822709275997303299473
7505687449872812934702606624477615834666170491626975
7179758724292911418797507487821715334199745268055732
2560031417046342203189757820773023738624697850416509
7975844527164585220435513975928752950895465228066269
6943449901488020041811864203977422040427026695544603
2990925495943520252796488734580054345846849452975353
9158379291130570376177366337579523977108739337954733
2118548790619268542240083953610368778991522104521065
0200380051808347709316151054412972685089966422824644
8977642323194767560243809769463100168877605725679893
6928008650248744676082454595750132838100012297473056
5399913766112760678583451295803038405025304166311739
8222113792207474393966300340704960764328821987339877
3338028597936098215354655910072431709715706097105968
6886906647906795150810115197051363575163611207596373
8637573858499983786453057903044394301295041097437837
2747321582107022267675703966141860877443969096247776
1429826812572551820738210629142917928982577970023307
4929888582353993193563942526806194864206708332451046
8170670405542134187651641921977618868029589218724367
3912979296702170026047079540759988069653829470682629
4750079920917805450108721318367103021403412399398867
4101404729131764442090801109198035244321133365358285
2718284326236725037002411625948225597448808317606737
0154856289118665446550456305213779045057328185120197
9665409430270049744632546121224141128329366437940321
9844792766561092717076355940122055359024467307073784
0681089160484022631613165378822613062347649322815522
9195223409251839601714955706253393019190674597189900

7765435853945373057085876733937752255618875908626655
7260714813602684304809463378108948705332533469315286
5228184850150399380033663878973389341128843457735233
2199976257543876194782906084104923170869792726685650
1771784535760144406871716869009528068034318933563042
7097277870650886333372970310015932022324797004101814
9467646133696884479094536114901744629897493230037580
2531917695505241625006552842611437307622654208268221
3459467537033662184218165664434775720963001368851514
5967947203601213939946326114541444682752648661586756
6816023239217047451257013486590616430860058855687920
8478336062463041958464174708303366065330053420620428
3686318888324266816035175214007464026900758761089479
4635150844959617000501827708966824263274755293913446
4826875600962762224250729627218837874695585396808698
7312689237481269813501287025948529870938537227129950
5563015937166285821886591605207403805726033045131972
1679291477186756305293557276882390292266219730580487
3114011753081338921701186180117337251436635308756573
4208941704807219335988747973642686198794102854212952
9410436548061666466560953506812679860837723426852206
1587747745045440859724123572813629395024877213229181
4474603528240905004010177366269864170218167035181897
4670512042796054366627945217494148564864034233759590
5490601360969309168410629163679468923218912091274019
5170618038721228430708796077311372544130605417298995
0548378827727046663864191137798869740632776799960810
3275565628709067701614858118516715255720673109259432
6602485559388718412430422561677461510838972883424225
8914850508472905176061899777583007065652520847408208
8421733739107673981148807733322016588911410051585422
3884063672865672508971288503845294031629188371445378
7866130540009170501111547185436355831033207211981334
8586311534236019202937183043526169084401955004814506
5893768715238471241858207056441353048742051561451208
6668096886553647307014517115558399645835678003490949
0393274451441177991637963103382544506126729662982076

8928347383482756376171383960793925089382875193889083
7424828395334225664013005888776657917835977404067075
0177108719391145784682542500121104011567813812956625
7255109176460365967780579961308628200115012516792484
9476048498842037333939543468668590235457523095407381
5304116759191964033950082322121220319485821924434755
2933753017393518181466920530067683549832608976267360
3011784585548752703073220035324122391029670664816719
5592545322134782497402500270274805998168376214118183
3876083487925809813815166160414642080752020537454958
0130513552975387831795560660953452750289995284305259
9586314977990312599592685238675997576441360225760470
6511987193235262649108130193591599676247754200546833
2913608093323184530910326642695702736368861586841986
3558986966216126369423462698706529516460365908897763
0943695328921497180259712631079198634234336833798542
7815924375361059523136767058842514726866925962255623
3388254449153389514807803531600962362723626293210938
3812134429259616897760712961096532858126385635284069
1870726909508719908487588059768061543438498330786223
7329950538595446528725800917384821639524761866919428
4409202032528586359734273652108417792406533769948709
1904006536284002657991027808588169428112419867803212
6766119082120687694390256898318550295073581362833258
8132934978756119965706477032335460135933015737186998
5275952788401552313044666467007044570167374730294477
8425837971339579810234192743011416633563104720020623
0346672004347363362091860740637938741083779837265910
2262266281668368174614508881058679402091696237070267
2270785567024669662152359232489065565411423216530012
3066831581370950117516497474177077104786711442312703
0825964970280652309726522595350260937603204641810401
2825702971529099663017972874967160781873864346043656
0312600913080199055913449698243059810448121422323919
8832330517487617761060380242248693360782698348799053
1871201936565718811379882854700036618033764616418080
0562110656657135445738035572162702066987066596116302

6692813351285972342273474043550450301843661570597586
0259189729717629378207585152143663005844137543530152
8736386393755920249401991229614783120533902040215249
5716237517713920740481216320695616878374406751427661
1819357040926225442812572446747935669902240001616279
3569997737936222932889951096671881472547244744233245
0832816113588506261778134752263774166893067961890698
1738542620711683452020866172255402151315203014261536
3529241762488724019384328470314533568553211634603241
1910969004980661636537048300440101181291865610897469
8069557691913585155593833791589067981878573696873349
1665317029348327448262349678936713440726772268408403
9078504473370916901619483417492844768576605582389499
7626657072609591728102611237088220424048967417981761
0597121652434189876973254051835763906896752464304945
9814039960198336818628217058607337201469935697285000
2318515413576999413002897984546387206092599165675004
2574557721385582160662381881843260855908648842305772
9024583754833271946592218906082255771930310244284508
8023804118245874545954059911879389866524346777606716
2411186100101040907349130306713696907321594348481297
4545321465061611701587079237826776752243663566391919
6042731264289021440187347592854707425670344996917675
2335984781388655657058999833186012346150364759381711
5860385697047892499359044412797604188982091303483302
1495306782619690302406082509918409496241117147501936
6825471966844473398515303878500511069790200649735354
5593857570788333676889246110793462714441980272903059
6670946542694668000365572482505385372657003464552984
3754856057664365484689591987032555090859093742098048
6241309926743237286356117618097163687365885258754292
8998684070667409131052333116913901759061177908405550
4104097301267508771676860043264047173173087489445795
3682806516682884188097638768775167722540150700393369
3798837135823136755015852875240337553986870947839756
1579463085259914621207238560952292201024194226436550
0964372816212365929564921203419310855804805792520705

6097073311003610872573365512443639741726877515322065
4225639339142498791922029924304015135326183042516398
2175998799403271770630652960196594660331660919322521
3978517420275604532822558009110705570160851977806571
0143163012118809125580649860309480491466761077207455
0263450695615815328307044196446264440197895304167380
8332923846545153725133316840331525638965894711008798
8118295323646800461339453767014912177042821942828505
0662218846305088040978357011532665491625524952638509
6275796749475771603492347359628017615562674390623303
4700445381194095286975509385806797026611456848448463
0899149431515248817802441697409989060390843944185025
3047357246937305616185379340688294601425402211423373
1397240857449458623776193225518556891103646376840705
1600360614064357081184779777040599564120010460141300
0890788089377595295745047655039053183599146854554075
7025419445358817282302171840873401560659477066982172
9663865913212771651925166629124216002608240455641818
2824207155593041412847921238148595144839965671505764
6027136104073454888170722718008329681401203966223752
4091762593050119645743753899286461890218741075016941
5517307487965557943412213401861945711114145143476791
1050873858437954322814623925103215251780610220029850
6117151152292468440623367092708922645324107817862349
3002036486152993070136846970506636958762130387191849
6736262861287731353077213166220888069011784522319949
3653227244317747904408052101242682827577605768520721
1032353363551733222846585385579072592997658497868869
0119345292746422752555850487824174159627743927269508
6300373919723243280964233209858067469745511572367038
7955932445758453122600239564518634241370010986150265
1950965012763338281967365976435594000089837050721922
6837562534245796421520814153814035283423274574528218
8653998454027744122254522942265414501024628838852570
6463919904127385863747237719701816615015593065525804
0244909298244953501327780953256426342626343279899516
8669229691978769462307638213428791853401663826586138

9810556650674010634208285394781800204536422696790163
4799177968142697019624314183701793323920820696391145
6865343575178937024664269595259600654326091604270903
2773341248793762208978545694318474194345106203974665
5514720561706101946122268668159690483839429043099283
1677354144429372219257639217779234223773397148488181
9101699820182786211318128453632639843862156550967765
3198854178318558674158239004305345117073737286728018
8233545785306959967788067417964300979384132540544812
3808319030586540851532278542313542837423537587216882
3632205507065270649525135696367521210463206418432239
3563779520998965454720016968351074307701939884197870
7094769498748642008554707457267060217127409566926654
3143833776902401314278999775677872417957135722306063
2305238563514763132557264467597733776986284175094033
9381216921426735638646235434020669430635075139402744
2889765823700307564930106573451869821269475788504119
1961361046093431644074153374395772435885256502313780
9475431957730545187852072098638611622730433453295875
4748551273383279722191850350478718978842492871068961
8210899417158677612083801516188865145400128585971646
3013644965160551493822792834449083767432819892114291
0431061110868692165527920798201649329374581342150765
5118784892767382548793511783235161120855178410837621
1752815007851854778186076794286414843233233191097266
8500329341289140458582044315576107787857625774238384
8993157237395983811060781246778635855689965273688452
4094252870459643592076057934668432414532559635624874
8821179120818397034784641754929142248619881698358368
1912923190240717129570007687463345058546053618974136
5291148748788266701276631750412100720315781088924376
6403677309117553090409281711961362105392341387334225
9376787685262571351224341348244924378633188527682753
7431049035445524214357136931385640290400065699362536
8177991878558718447307705825297435057486641270846544
2603847274453318352984893683898897808289018623074484
0470844908528494030394342955463852740854759015268816

0003747725228117811611574237139819507406741753431841
4635054434243835745192486115808800396066834394019089
8737819244040021983228452076515479211369255074787416
5836602422185462194643077515218613558151988754046355
1761409590543738570052501358093904859672162021606984
4161569078771684068127414376614091118139631085599151
6614810795445153249599213668254996327123735625684147
4541431231206048819565493502823579799437677757183566
5900690204179933690917282410490623132712442494160142
8516096280779063603833584141992709154227684258177499
2219930572580325951237712012994424840701227968679444
7341806792582657934958914767888378915440932631656700
6088947310932561056198803086054865208774564545247485
3531540501116812428474957904372159415539436702907125
7786536897542289496401161850244371314084346303543580
6477212711435043136254843866092436155003196510855005
0907158337041502556889102458400418193352025212449845
7167679255682218023266396231570964589079686994941373
4326211814954738978562488217201055827898453333313654
8489450888785675190404459592653195215811924489811352
9022134550383565982719369493176565781867600470077316
9181645503419476677438018159033309930256659566680601
0250926530115150160622468616389719309021432141704002
1914577268584078652097221680980634034097430869072988
2230975195198310029861267048908177882768775796117833
0223290184600190716087476842315224050774360779330699
7142082661884079404692580632742472750415657425467147
2330996260395728653859055288800591122257746897621928
2389040655521371974945223212768338451551457684661127
7668820788347588586006488657357631652755699941191083
4661445618670681274946339286427366151155891224086859
7078927007503394219875836535973430041069559226761035
5330599310591763127935116297140796065012532936194013
6650630794157005021118804358149453379132085873254843
0463659635754453470236895764075477044155714931768679
3022777531198821848373019117862893069401115893599512
7482991205306812955192982493827214779554101294437734

5175642511716526216166999185835646874649357924097724
5513328023629366757099076978525142266411911935854345
0137409607937600508655283422806572647802204206130795
1699639533247616622891546684694096914515252563415426
9767270965428923669684031925350949067384509020297243
1809123127018255195138346002937882176207767661470179
1056987095327706600308416181059206587486560051483952
1848249925432548561325085851974364170725538031964069
2807156322132217334545187661557552639031833939656842
0029460701119129003740322343976701062591895494110816
6177756219216231211370903139719951093000601030071533
3345290159788935879859588478418005253796008798347110
9265787548954190753861066220189959789540593437873947
0638423676775021833692887286605278345445220240580571
1362960075974665197771111263096946950342384651053143
3667095110760686290587829078822087133464200364390366
3329809885008513378149631661885771080663125580443242
0769864751221658236162083468148414083152527539605062
7066696526301902593848744034126606357473771924725215
1652293940669552453422481299918326565944237591205598
8200472864204206707422287275907409713982157237963214
5499167296730802864864484068322190336849026989929710
2009204118657870251785918357505527033476887756385854
6273960750997614740057219289818296672620120311458260
8158553826922251013825611025429443067962436400899781
0054400678021342767076425499259367101022846748662259
4117529405166755581862694175059399711537675399665098
3301615736970922700557309695955649272385182575787553
4178887526397885559646244749232640748216285423380363
3594937489952680601675414219788350902373105768750328
1918259052125331849318306100702202678580527851563024
3241955539385657113306225224623414797114479325789937
3626522022284579923034601177104157440941246819729500
2911413986761550352599819480735239154528581118222981
8156494479275635533223506290496237602188857729408189
3514505639433389353397765545634086655528265813600081
6463334525449484505955566705163107708872505206626022

5605736292144516747211388601691629300542051242064155
5527240194558735950975262553372909389990752880852423
4255268789623261274535575780817153091024596353553948
1476277196916419842611182758862589058043661548756206
8163743408004760016441984380293734146828677717616041
4285412606425641580937439615912924271363136731776124
4589933859177730611332988466552457482834925231194587
0699891238106008382022702659856257633391903367294000
1264290699668384238179436127469866593189619045322223
2936031663296014943687182809327430491523893225625519
1114730075314812880720642684222698113618552015447980
8760107287609450269249454061487952597780986360069669
1013778141234900206869198389264153767225517687495152
0401488303120411302022780645964589480587137799676887
3493918703798214391672064590869651897225691698509990
3102030966573633018772157910787814264457256174115841
8587769963563529157661151716117556191214637721380113
6522636271278503521453330658424042719346570805618056
4286269988319327273724685080806372534586774314356471
3432991361084587114567017680602415639874524038583363
7939835633934944595030714427694213921659335151610028
0682600186217108027589909163454624602269858671745423
1097729730601950455750212178667292686336651258071050
5001747213369207269807192779063301291903342918220130
1171302068612463736153617639480556112267903595998345
8207368030321560508366401669154640260872605066519849
0757496212963311920460864247010599662752040679908521
1118873939526222171718394567435843662502594075677874
5520688317702101822639289293331994618955621433939875
3777418233490776385599354008719353215578106343492646
9216680197306958779347172225448079118111196392639276
4800123517725719274788305783971136690606455254331919
0322289193609549718432491009045406627292385027405633
8385485901882681494358436458439802626111161717658428
1609795765250678761179511632399263517003261953415092
9750055340488664096583519165979491935086348821926159
8144843692437545565112204194805526823075332476974088

4727787435452359272108804052625835368689199319844609
8971608446742636735916800552847863406269127172173171
7560616771714634755616198078843903113584777164260510
4747457663614383208549936721974573997978665227750353
9806189140888838590932137437527203362302578779561047
2928563860851309157784649600087363339203104897781691
9904548372115769326147221016937339567708656913761110
8691532783540556894860507108229542480918055808095852
8406673528781480386538021466467571438965475808604345
1295539355130958693211086299311106058399394249657601
0657495240264494636552444240730599036528490896664804
0467945517605689027631717191876872772574890336567177
8563823216530569212911505032641281573270750113835519
7893094089107488034261090882741413711940912309437167
8696136360724776571046234861504060685470464577187891
6603821401434750973053691031108440796955045623775381
1982755215952136501877563397073543958012047196601951
2882510545033173051621634109051819226052554631226435
5322592957475728820016262708082360424445903581361990
4596044916754037553727206181988999514777161493276079
7999354053231793037435275849954268171872137473000259
3356153619211112926163891621846956695620335649705965
0933237168855187842033304180750306650556062517416050
5262331664091925223825887095521890288129575052171165
5979171308253404608433079774654768816669196814476897
3284839179217767764271032151745244795073588076328905
4167316031811926102417003817756596118256541525180967
9876026163430172783703279617329250813470476548576560
5901072767352138545982768929873329758329944685365659
9192720272312841968966632593594772667223500113719502
6467308449262860959852620522409521822359200314706982
6977926672250423655929192492054350354442390940880576
2010504653092697731349410857279976380113049279739865
5841989887658332015933439610468750796352017847298731
7304408427266965840609618054654656319302505014959888
4019250855963188092332801473038791291957958251012920
4377653347410891807580724712724027616662968626222223

1660704487522921471507146159607733512382721669154552
7291307887613670400334477105207700594289972711773659
1924299132120809706489631255884391194426424834555020
7274615226720992564456528346798994906603417413684761
5577313440734698003798042141226713203724653210732173
5737604919320627556467665490391302868997801527791202
7247510532924595527398642066245295718008680915733555
3970196512931004832314704135049429359651182657240498
2044313975670314705370985061314615599154596790803820
6332711270539764389461306833524669156764448058479053
1856264957839354546836297097508864072557823669299065
0812706432076742539043688571381094074585565967418113
4810267297801287659770581626728457561593281266064575
3286983567541694373351718691544942980395328095625329
6717647427841921710549638534142832198621486189526791
4830400454230243724462494276958818851304787948051509
2402218472726874326028962048598568318074375214862909
9213391399218069538074436347110623021020239908016642
8312197903931088989028681277449139877816012369634935
3837904833738861649739886424085640388600121735603712
5663028241932844139371526035502553650068469137995125
3301570881693176198059576609611328145362486381437470
8137609973392793407138710218356044379465595763210859
7263805611317386277996186628281065800063606056696516
0500275463200064283833990047068610621589780135918708
0238837576895579111713270218719138612446092285094662
1788009123566467142528458131688323343861702634545310
6357326861714173268252109927195832490783232192898048
5122982303379378592697226931958950333541260677636845
1920279901129435132900852589604906134817684612448185
8634526732494412395033702428955285667455763776548313
0391544907241744699033348963010326960851264053688782
2212146215219423825090781889402643536751568910448856
8329212836025815748009984583205487653084082425613508
9359722960318588388580383581856644121538670876760124
8310104630844744321880144790933674746884478564148174
4590912453981032300885920637101563587565164309596112

7396401586176978131408795073714288317760436689819626
4134750069719405651950455051677423979401968998625278
9008523744580732886706974177396357254506098542345678
9852042507322860709420263948418286692566061665444072
0457756839393231226540781247831666880180301842250452
0053551868684850558453085253849754261205794305353308
0747750826496088529445572785003441395589379335051184
0302252987061629150596422599960058564895723365317986
9136696994427866578912525684626041479811086568557173
9072133078933108525903333113718587728703492627902715
6732916866271667490819931825831668328285001575707801
6119316922193151475493775515980465409283999109493742
0103717085608605881785449005704104136040435137642468
9981526806092554011234653295043491805374773561666671
0469298309567892031648206393173922124840551203478063
2111316813373224063216455415588237846091942738088502
8383123622665497443005580998982995742584323537642986
3146563505528356047709075747328263677043963097923465
6297949344566413960851464371303932136774212470904452
7215406879215426306425972602920189946552981151426126
0490076386714173023572727683904155972345066669386646
0588292012471144178831782234821533891876058363276181
8194332276955531125819048475174629056200134689964407
1619823205434718461102051131555093022682510749199014
9608178562605108859036584745150376384915134000329516
3991062192405572830081035217619796168322139816924083
5763955621165712611219290508716325525855864961663825
4193591482181876195923292056995506376458182685575221
1527870118029943354674153627620774978540541333031363
3542368241010846476374906388527984149007646469764854
0094796358954975461448137636970591635699836811987525
0547930693532075707667801484477014247162419081668224
9007420711186488154772891718653596776539579933503342
7282146054169649600984706979585592643042870363664713
0713147823306115764199132224206460998988307626858360
5552740990478467610760424178421506285175573529996478
6255295428367429870664579433758010140740211618614484

3297657442634285287047785563083096314352787830419450
1970294657577773281674685808745393160393725331589928
0579434631408735860861778826334927746151184911655130
6818467136773488233410851364039479392088768863363394
6138235834479408156961091429387734713893423773619109
6460564244474779082076049660271356168954106444832136
5980829389097296189121183429149061638963861069375208
9534688398334446718982124347807238740745769755450743
6846747135024858818399665568196344528811941833172636
8250506118649003941255205745712036035578025141904352
6718372192138482990580322469584243231589844325103965
4435350535432292167470407786146848597625574461535118
8003143056995492784716745449726976128393325183819722
2328360707522781292813010656941262948730634268837338
1817421706086475482763942423914027532180429519034116
3517046980742335155605785756245099925320178749963664
0473477038985587306507603870997731843128109897898820
8543559550943253902371895216820233442455725753078792
6339855090164559423733966252233516487505895569421729
7244895998825089232112034795894154654603037878617591
5716613988693268737496847305496532937821475648105793
8082853005324470805065692942234001095934829461453907
8890661626402150130735330033192074563726377077099939
9922886212243248802062634850888530360107234368901360
6427581425283987859491799796112196379757651924521867
0960880921371119775000878159304307293448839309575741
5924137528597779729189345385050803831986774590025186
5791723708085741642971538078840607130686803619824197
1577476389507253468404569192759531937223702229015580
0656076047385473599044779967487499697694271376686955
3319512533776409858709668386326392616494560868414037
4568420719405950701743035469182150900466493998551741
3893851975731215682616228622318810967297476060130283
3119371611408747270676255856777511995666748615196491
2970193318084994109618139296492789360902125354433273
7506426062429941203273625582441749834509473094534366
1590728416319368307571979806823153573715557181612215

6787936425013887117023275555779302266785803199930810
8305763076523320507400139390958079016377176292592837
6487479017727412567819055556218050487674699114083997
7919376542320623374717324703369763357925891515260315
6140333212728491944184371506965520875424505989567879
61303311646283996346460422090106105779458151 9...[1]

[1] Calculated in 319.520 seconds with piX by Coriolis Technologies on an Apple iMac with a 3.06 GHz Intel Core 2 Duo and 12 GB of 1067 MHz DDR3 RAM on June 14, 2011 at 10:53 am Eastern Daylight Time

Postlude

Lest the reader despair at this incomplete calculation of pi — and I'm sorry to say, but the mathematical construction of pi will always be incomplete — ponder this: without radius 'r' there would be no circle (James Joyce seemed to have realized this when he wrote *Finnegans Wake,* and a legitimate question would be: what is the radius of *Finnegans Wake*?). Ergo, to find the complete and perfect pi, do this:

Locate in one of your junk drawers a pair of compasses (if you don't have a pair of compasses, contact me. I found *two* yesterday in one of my junk drawers and I'm happy to sell you one. Just google me. I'm online), a pencil, and a ruler. Draw a straight line of any length with the ruler, then measure off any desired length on that line with the ruler, marking the line with two pencil tick marks. Next, push the sharp end of the compass into the line at one of the tick marks, and move the other compass leg so it sits at the intersection of the second tick mark and the line. Here is where the magic happens, and pi is suddenly perfect: draw the circle.

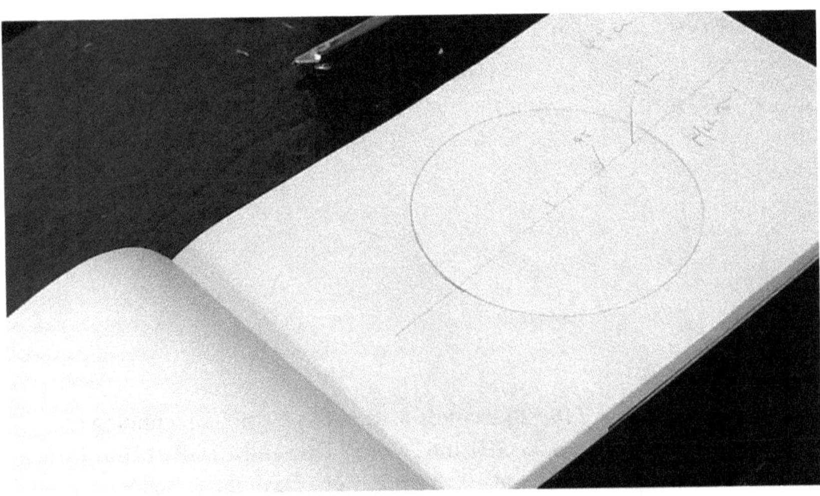

Pi in ESRI North font, making patterns more visible....

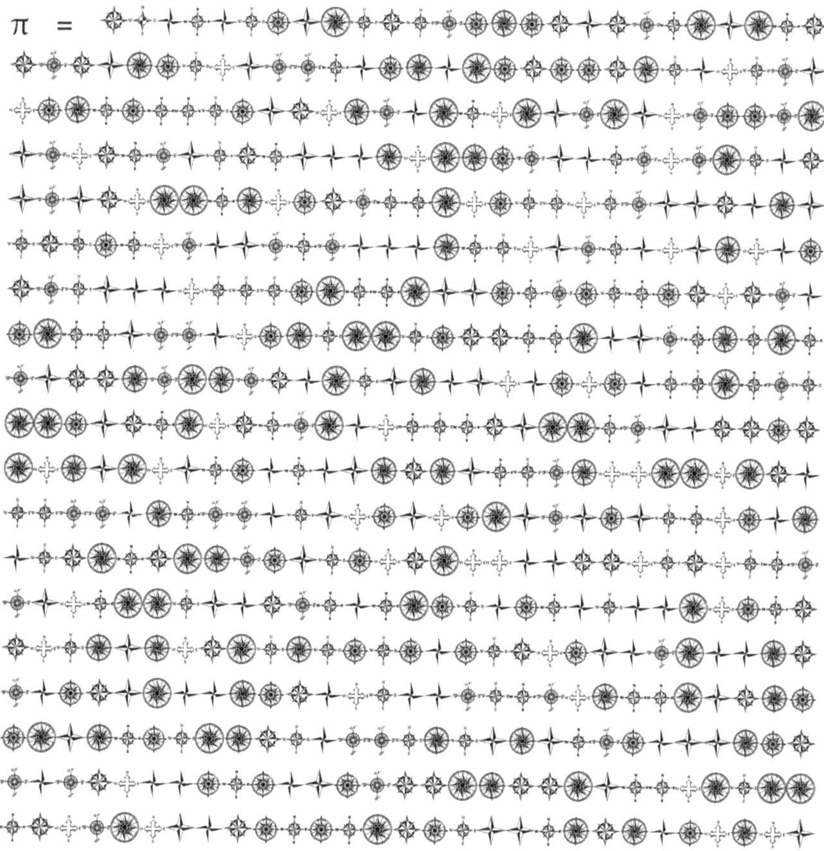

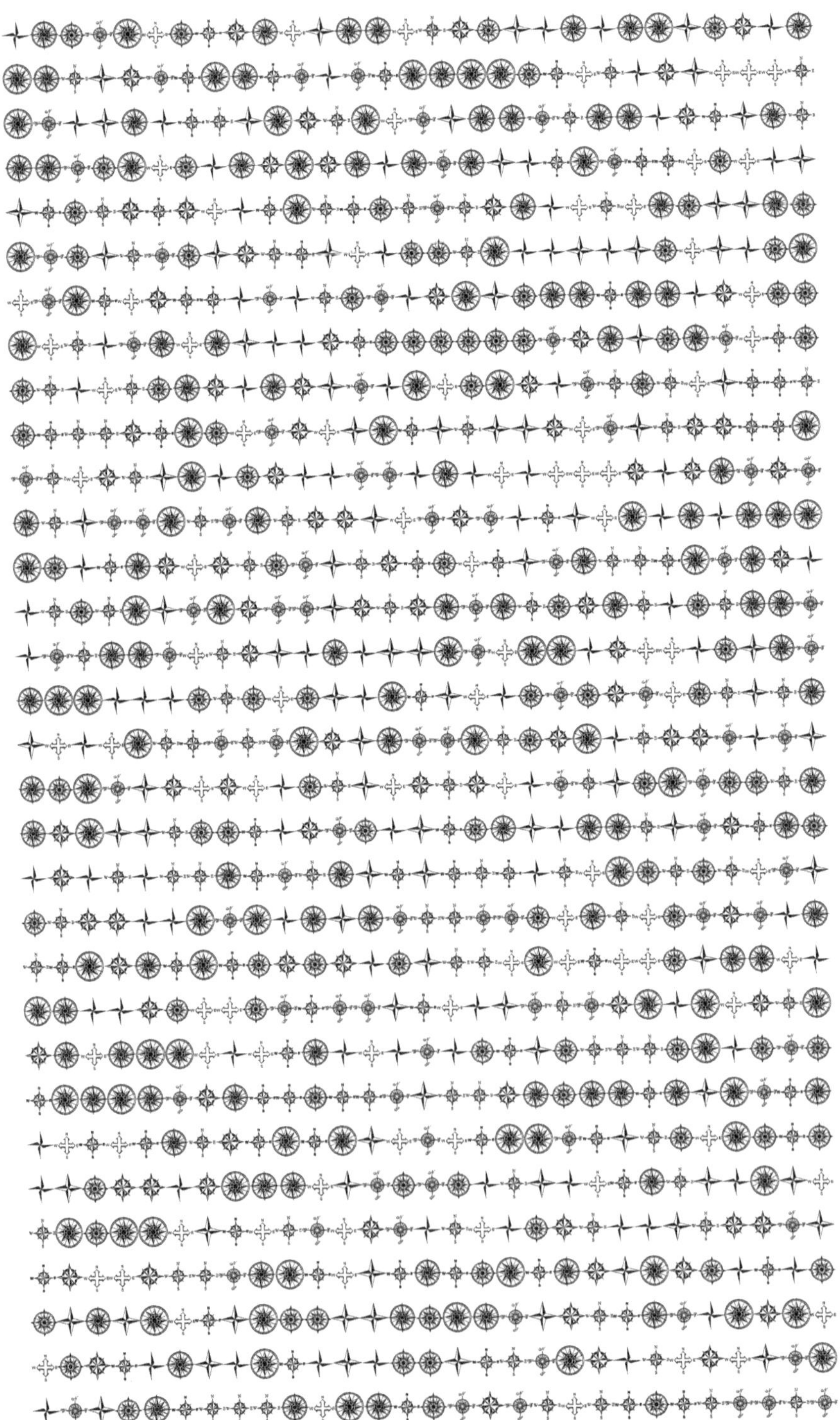

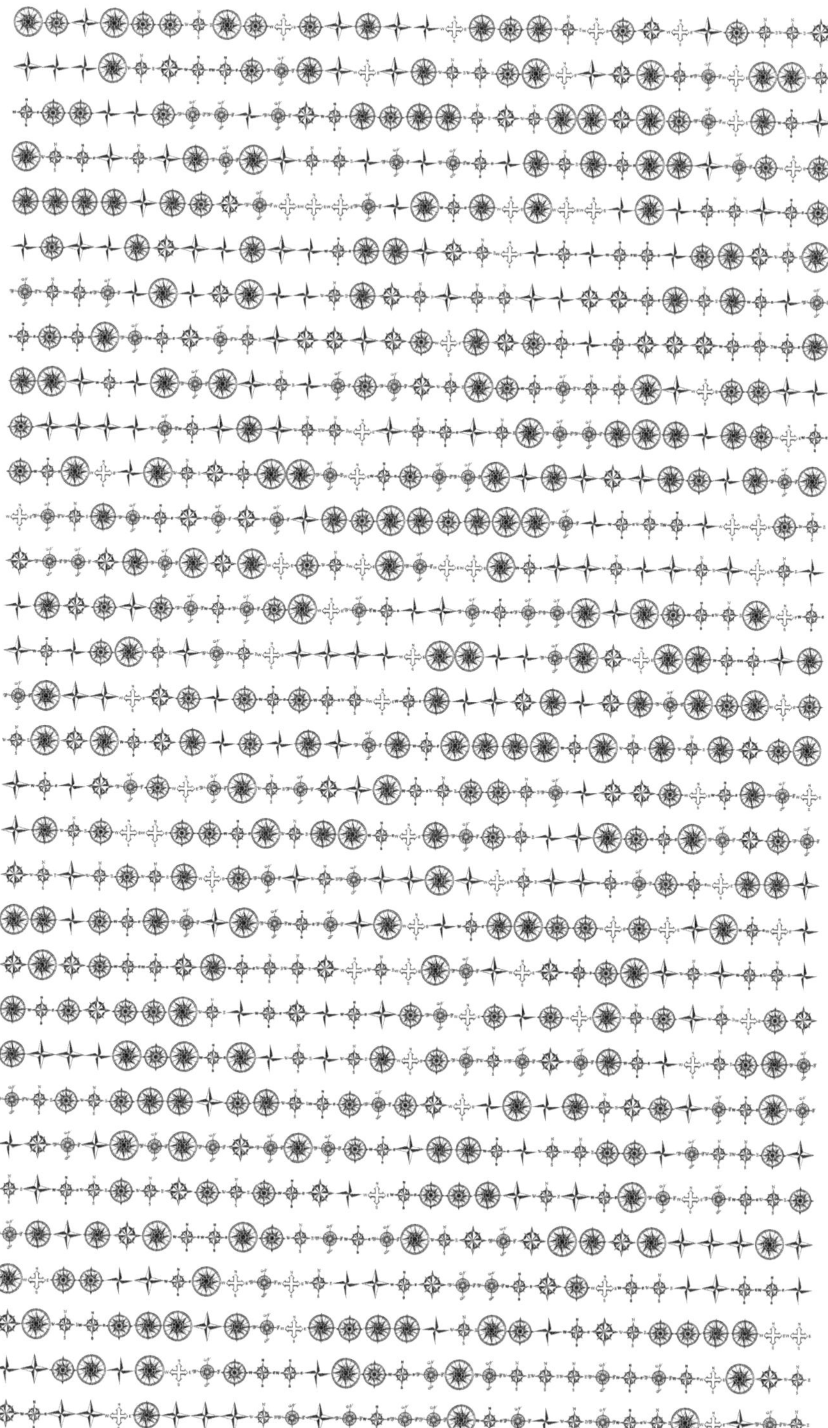

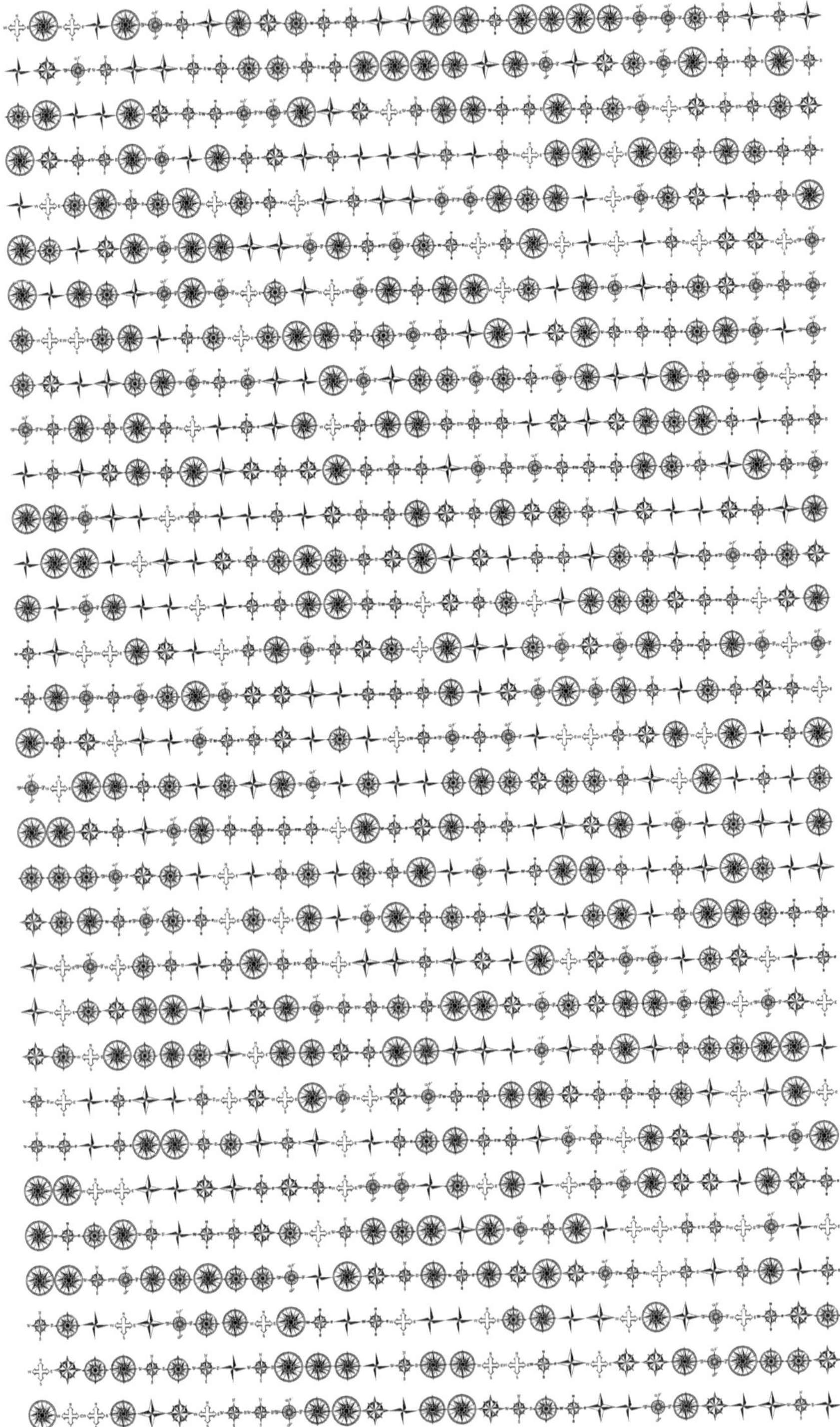

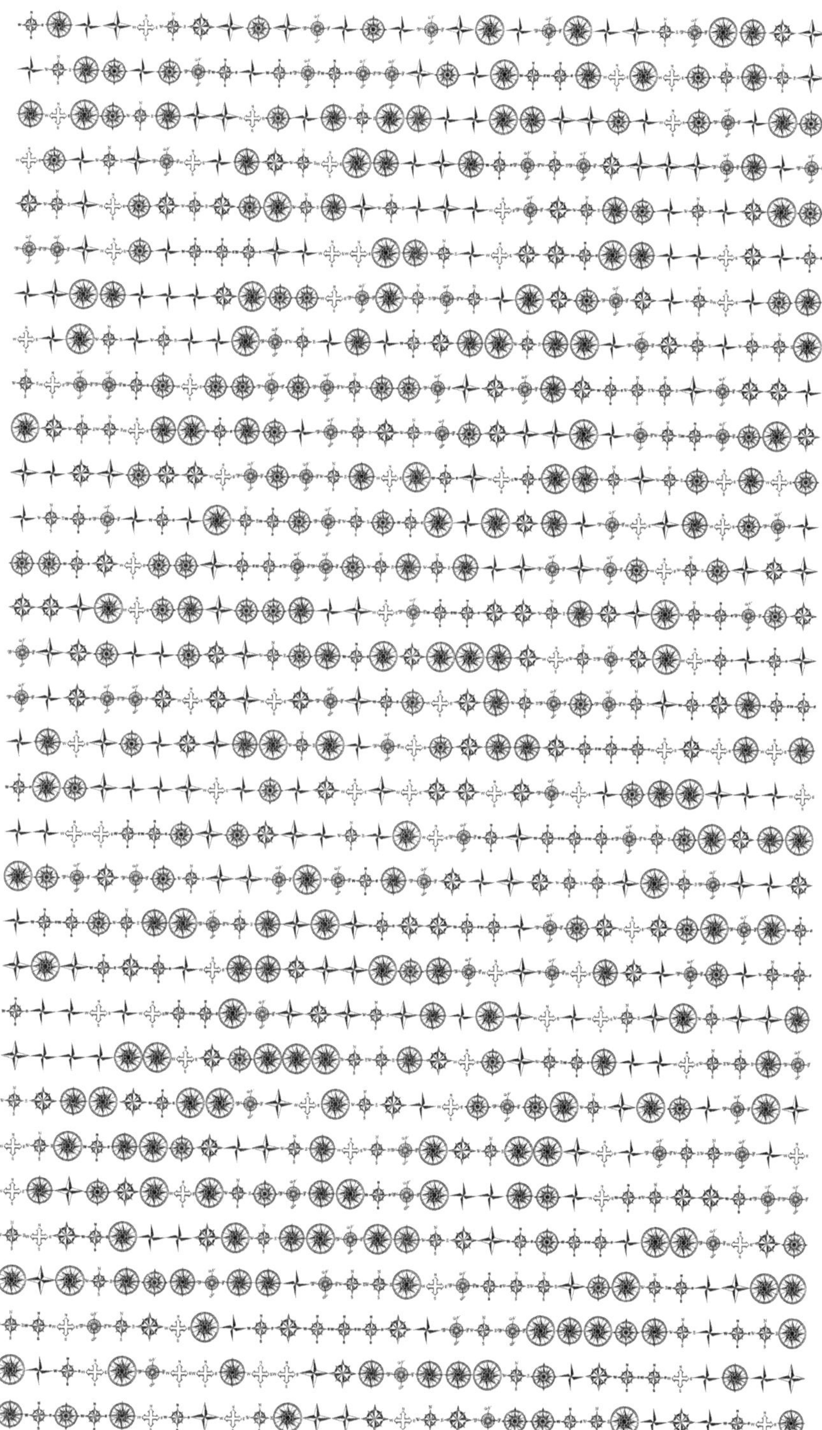

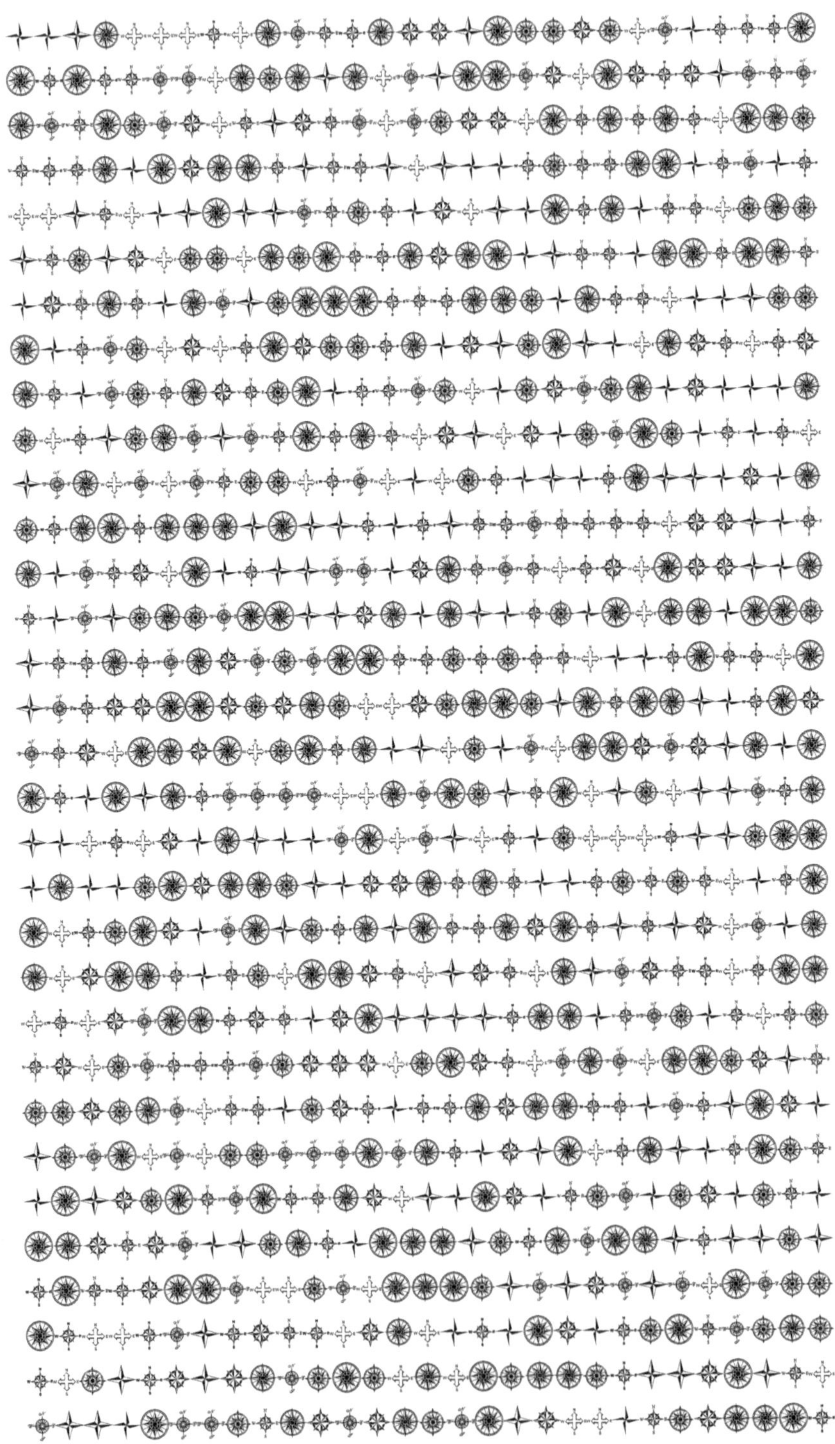

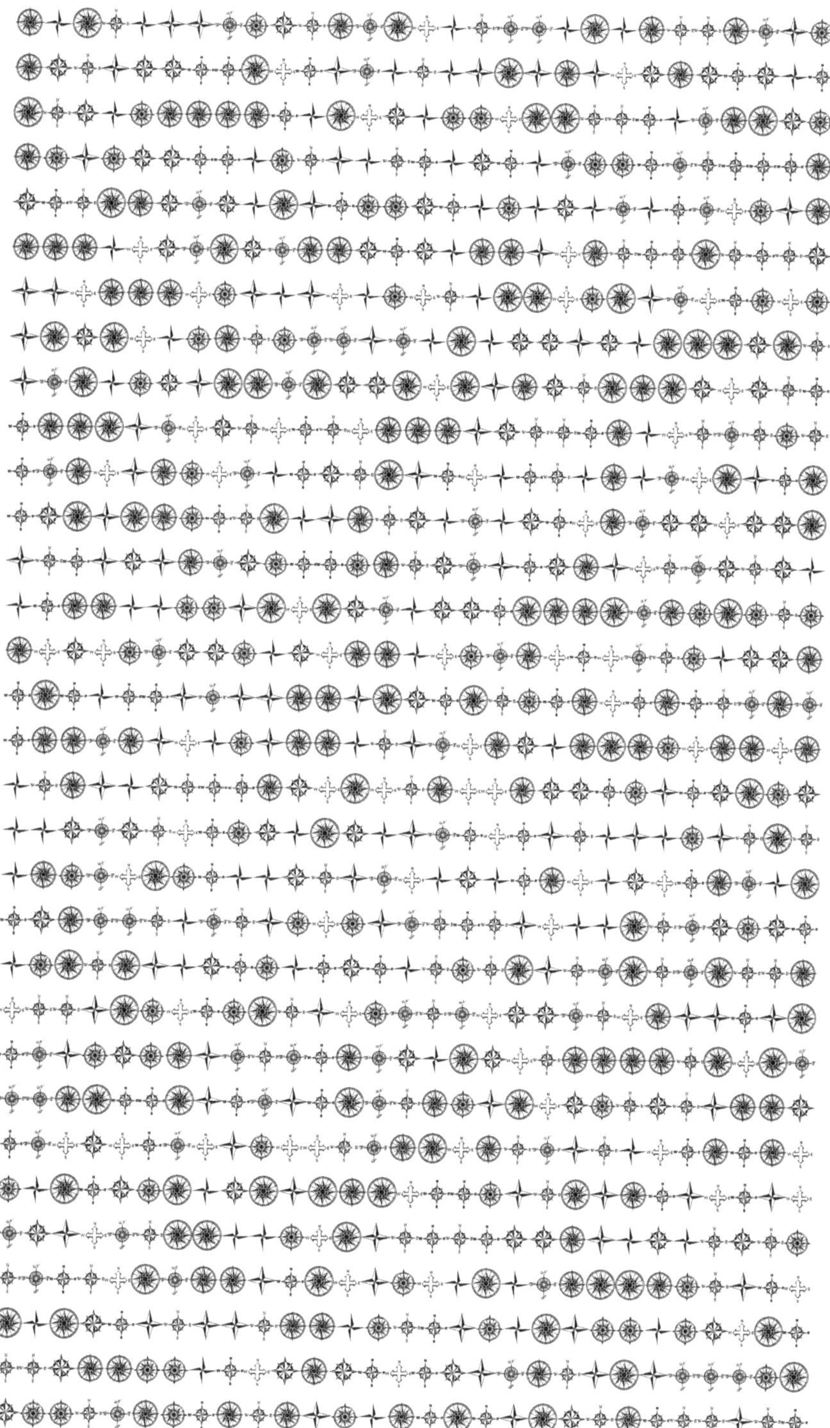

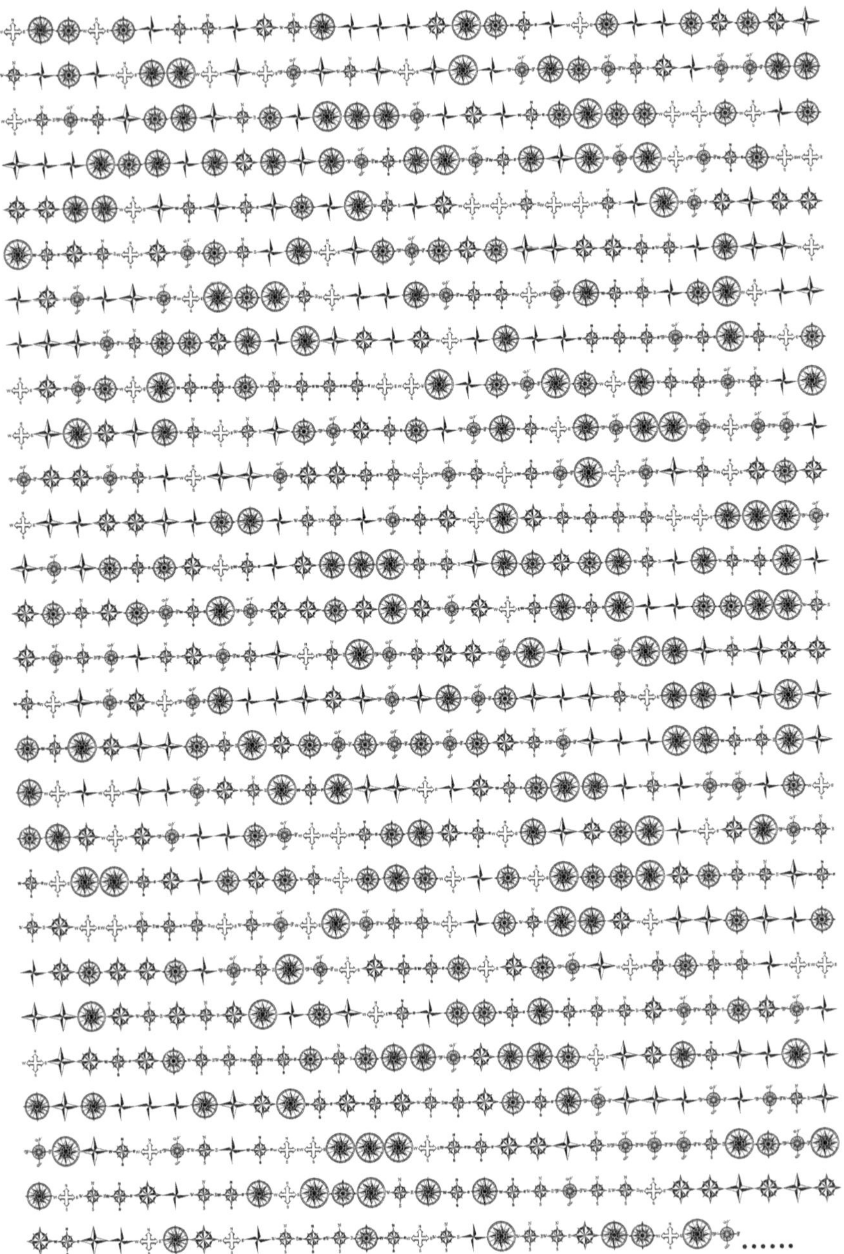

www.ingramcontent.com/pod-product-compliance
Lightning Source LLC
Chambersburg PA
CBHW020521270326
41927CB00006B/403

* 9 7 8 1 9 4 1 8 9 2 2 2 0 *